AF809857

Bio-based Materials for Food Packaging

Shakeel Ahmed
Editor

Bio-based Materials for Food Packaging

Green and Sustainable Advanced Packaging Materials

Springer

Editor
Shakeel Ahmed
Department of Chemistry
Government Degree College Mendhar
Mendhar, Jammu and Kashmir
India

ISBN 978-981-13-1908-2 ISBN 978-981-13-1909-9 (eBook)
https://doi.org/10.1007/978-981-13-1909-9

Library of Congress Control Number: 2018959440

This Springer imprint is published by the registered company Springer Nature Singapore Pte Ltd.
The registered company address is: 152 Beach Road, #21-01/04 Gateway East, Singapore 189721, Singapore

Objective

The objective of this book, *Bio-based Materials for Food Packaging*, is to provide an overview of new developments on biobased materials and their applications in food packaging. This book includes chapters from experts in their respective research domain. This book comprises of 13 chapters which discuss and give review of fundamental knowledge to in-depth knowledge of biobased materials. Overall, this book is planned to be a reference book for researchers, students, research scholars, academicians, and scientists who are searching for biobased materials for their applications in food packaging.

Preface

Plastic waste produced in the world reached above 323 million metric tons; of this waste, approximately 16% contains packaging materials and plastic containers. To overcome this environmental issue, biobased materials have been widely studied in recent years. Replacing the synthetic petroleum-based packaging materials with biobased packaging materials and containers might not give only a competitive advantage due to more sustainable and green image but also due to improved technical properties. Biopolymers are most widely used materials for this purpose. Various issues need to be explored such as tensile strength, flexibility, barrier properties, increase in shelf life of packaged food, and safety aspects of packaging materials of these biobased materials. The interest in biobased materials for packaging increases day by day by both brand owner and consumers. Consumers are interested in healthier and functional food in the quest for longer healthier life. To gain marketing edge, brand owners are interested to lead the market by providing these biodegradable, sustainable, and compostable materials and to fulfill the demand of users.

The objective of this book is to provide an overview of new developments on biobased materials and their applications in food packaging. Every effort has been made to include detailed account of biobased materials in this book.

This book includes chapters from experts in their research domain. This book comprises of 13 chapters which discuss and give review of fundamental knowledge to in-depth knowledge of biobased materials.

Overall, this book is planned to be a reference book for researchers, students, research scholars, academicians, and scientists who are searching for biobased materials for their applications in food packaging. I am very thankful to authors of all chapters for their enthusiastic efforts in making of this book. In the end, all thanks to Springer Nature for publishing this book.

Mendhar, India — Shakeel Ahmed

Contents

About the Editor

Shakeel Ahmed PhD, is working as an Assistant Professor in the Department of Chemistry, Government Degree College Mendhar, Jammu and Kashmir, India. He worked as a National Postdoctoral Fellow at the Indian Institute of Technology-Delhi (IIT-Delhi), New Delhi, and as a Research Fellow at Bio/Polymers Research Laboratory, Department of Chemistry, Jamia Millia Islamia, New Delhi. He obtained his PhD in the area of biopolymers and bionanocomposites. He has published several research publications on green nanomaterials and biopolymers for various applications, including biomedical, packaging, sensors, and water treatment. He is a member of the American Chemical Society (ACS), USA; associate member of the Royal Society of Chemistry (RSC), UK; member of the International Association of Advanced Materials (IAAM), Sweden; and life member of the Asian Polymer Association and Society of Materials Chemistry (India). He is an active reviewer and member of the editorial boards of several reputed journals. He has published numerous books with leading international publishers.

Bio-based Materials: Past to Future

Tanvir Arfin and Kamini Sonawane

Abstract The research related to biomaterial is just in its early step, but yet promising and enhanced chemical branch and methods are continuously being innovated along with the different publication in its fields in the upcoming years. The nanomaterial possessing low-dimensional and 3D structure competes with the challenges with the other biomaterials and enhances its application in different areas. In the present chapter, various methods for producing the biomaterial and constructing its derivatives needed for the food packaging are forwarded. In this context, the future aspect and the limitation to overcome the situation are briefly explained. It is believed that an extensive study related to the biomaterials will still explore in the upcoming years.

Keywords Biopackaging · Renewable based materials · Starch · Cellulose · Pectin

1 Introduction

Food packaging approach is evolved mainly for food products to inculcate their quality, inform or provide knowledge to the consumer related to the features available in the products. The primary function enclosed for packaging includes restraint, shield and communication.

The various food-borne disease or illness provides awareness regarding the safety of food by the government in the different corners of the world. The bacterial pathogen is the agent contributing towards the concerns awareness, and a varied number of food producing organism are the critical source to be noted which results in the safeguard of the diseases or their treatment.

T. Arfin (✉) · K. Sonawane
Environmental Materials Division, CSIR-NEERI, Nehru Marg, Nagpur, Maharashtra, India

S. Ahmed (ed.), *Bio-based Materials for Food Packaging*,
https://doi.org/10.1007/978-981-13-1909-9_1

2 Active Biopackaging

The most innovative aspect of the active biopackaging is the ability to control the delivery of the active agents from the packaging materials. Hence it is found that a combination of active agents directly to the packaging materials would undoubtedly prove to be beneficial and it could be reliable as well. But it is quite essential to note that only the needful quantity of active agent should be used as shown in Fig. 1.

The two steps are mainly followed to generate antioxidant or antimicrobial biopackaging utilised in food preservation are shown in Fig. 2.

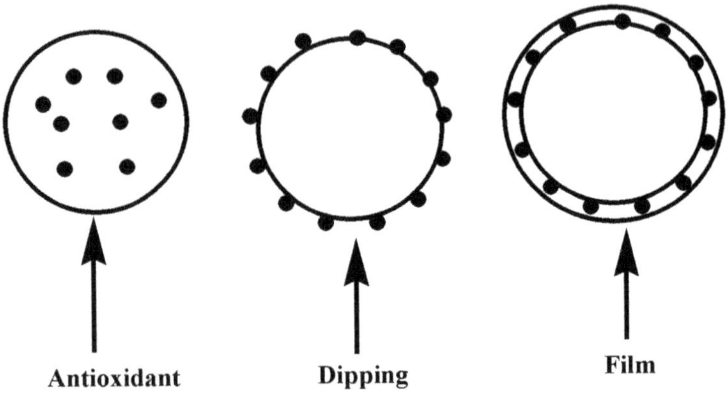

Fig. 1 Various assimilation methods of additives in food products

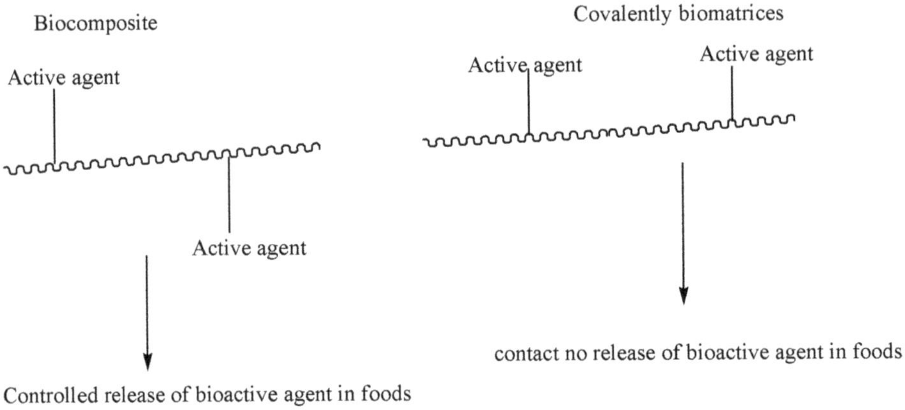

Fig. 2 The various processes to produce biopackaging

3 Smart/Intelligent Packaging

Smart/intelligent packaging are categorised by the engineered functionality of the package. Such categories are responsive, ergonomic, active and informative as shown in Fig. 3.

Ergonomics is the method where the respective task is performed with least physical efforts and discomfort but yields maximum efficiency. The responsive,

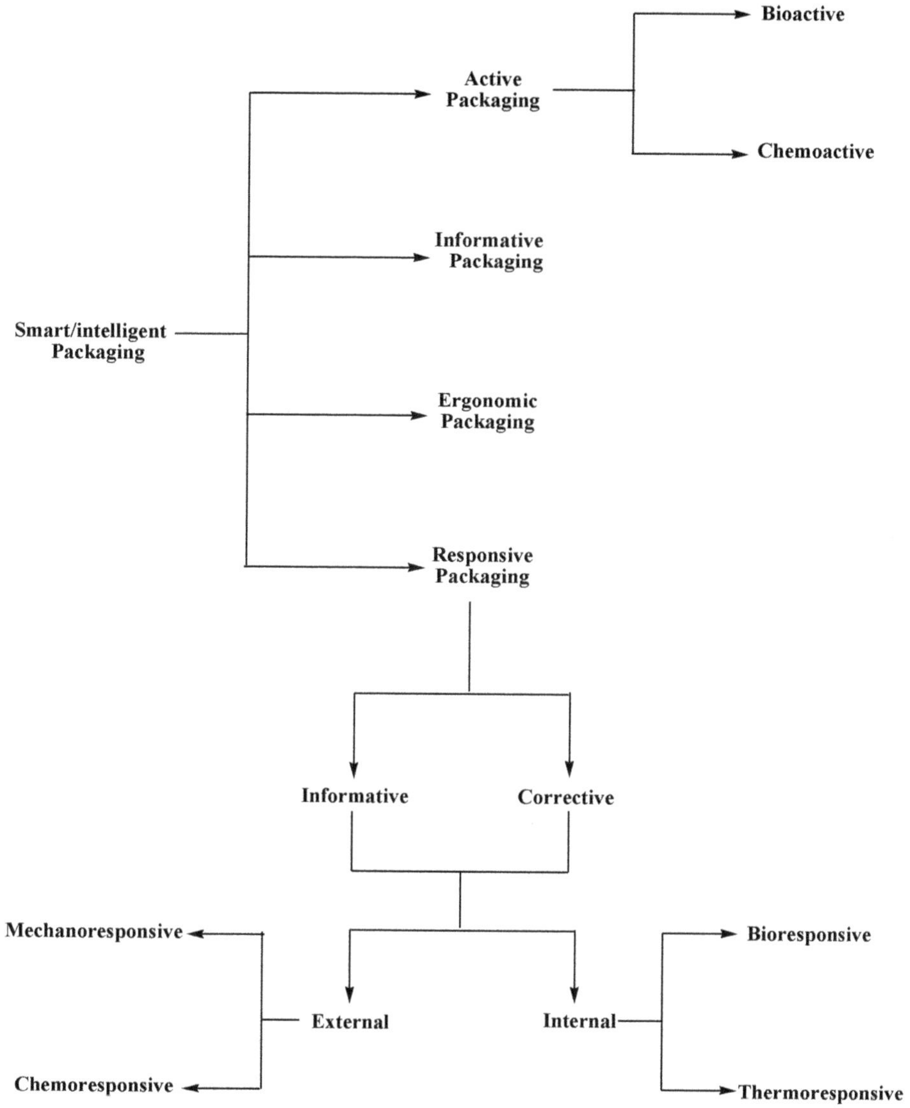

Fig. 3 Classification of smart packaging

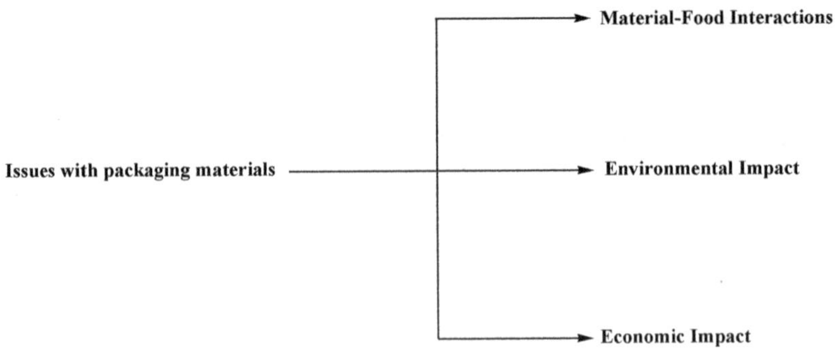

Fig. 4 Issues with packaging materials

informative and active packaging enhances the protection as well as communication of products.

4 Issues with Packaging Materials

As such different practices related to packaging are put under effective use but yet the various problem are needed to be overcome. It includes the types of material used and the interaction of the material with food primarily plastic that are reused again and again. The disposal method of plastic and the low recycling policies have led to environmental damage. It is on the way to increasing rapidly, so it needs a solution to overcome the situation. The problems faced by packaging materials are shown in Fig. 4.

5 Renewable-Based Materials

The renewable based materials are eco-friendly, and it is considered as the effective substance to be applied for different application in the varied areas of science. The biomaterials have changed its motive from interaction to the body of the host through enhanced processes. They are originated from material employed in studies of research accumulating the regular and convenient mechanical features. Such materials are generated in dense and porous type as powders, coatings or granules (Arfin and Fatima 2014). Due to the characteristics of renewability, and degradability the starch, Cellulose, PLA, gelatin etc. are recognised as innovative one (Athar and Arfin 2017).

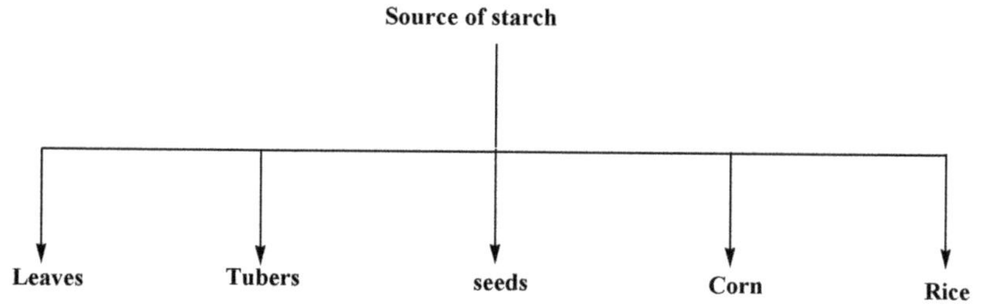

Fig. 5 Source of starch

6 Starch

Starch is the abundant polysaccharide after cellulose and composed of glucose molecules joined together by glycosidic bonds as exhibited in Fig. 5. It is available as granules (Mogarkar and Arfin 2017; Arfin and Tarannum 2017) and consists of two macromolecules, namely, amylopectin and amylose, as shown in Fig. 6, and the overall structure of starch is shown in Fig. 7.

6.1 Physiochemical Properties

The starch has three structural arrangements, namely, α-glucose units, linear and branched form. The properties of such agreement are shown in Fig. 8.

6.2 Characteristics of Starch Granules

The feature of crystallinity is mainly because of the presence of branched and linear structural arrangement with the amylose and amylopectin. The structure of different sources is given in Fig. 9.

6.3 Processing of Starch

The processing starch is done in three steps in the form of a flow chart given below in Fig. 10.

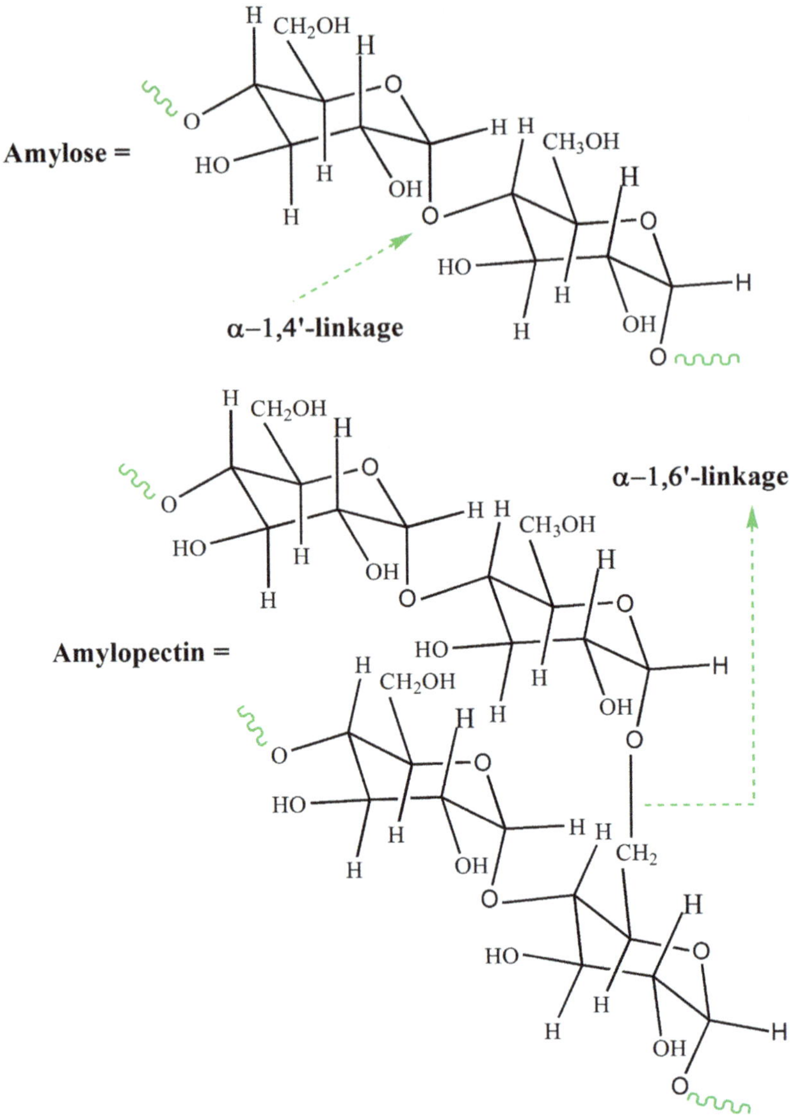

Fig. 6 Structure of amylopectin and amylose

6.4 Resistant Starch

The resistant starch is of five different types depending on nature and behaviour. The class is shown in Fig. 11.

Fig. 7 Structure of starch

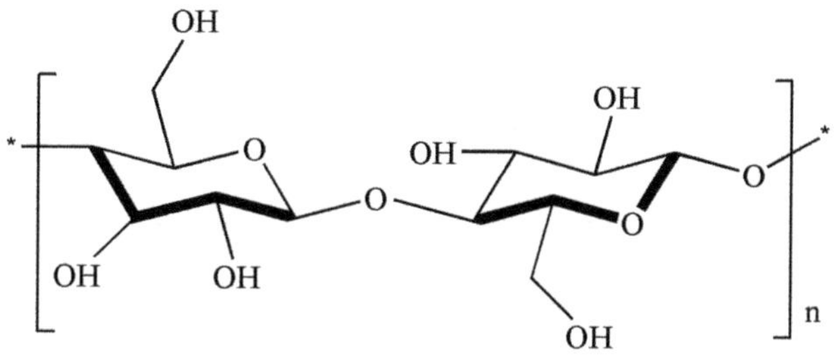

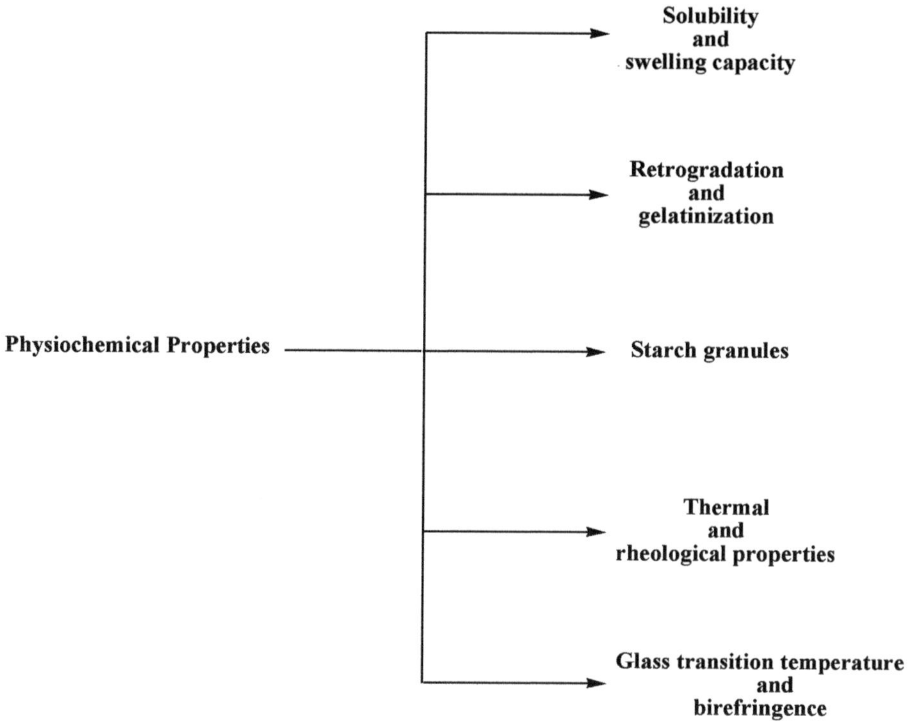

Fig. 8 Nature of various physiochemical properties

Fig. 9 Sources of starch granules

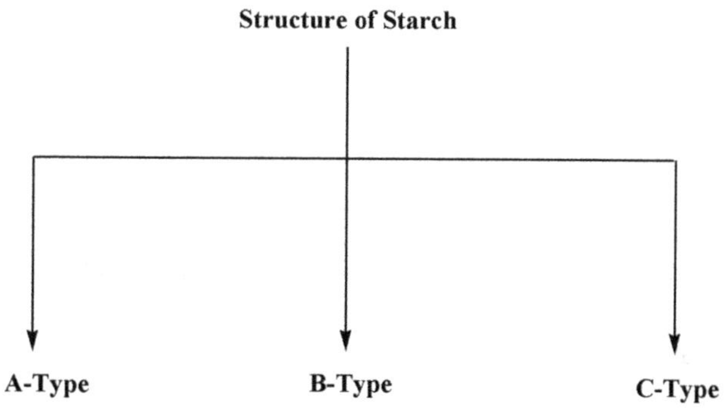

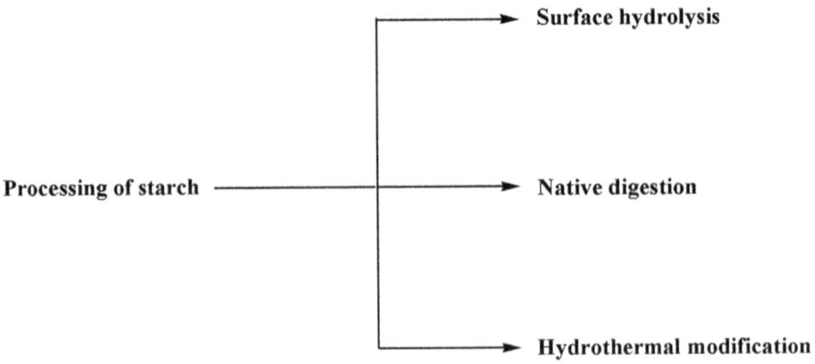

Fig. 10 Various types of process

Fig. 11 Various forms of RS

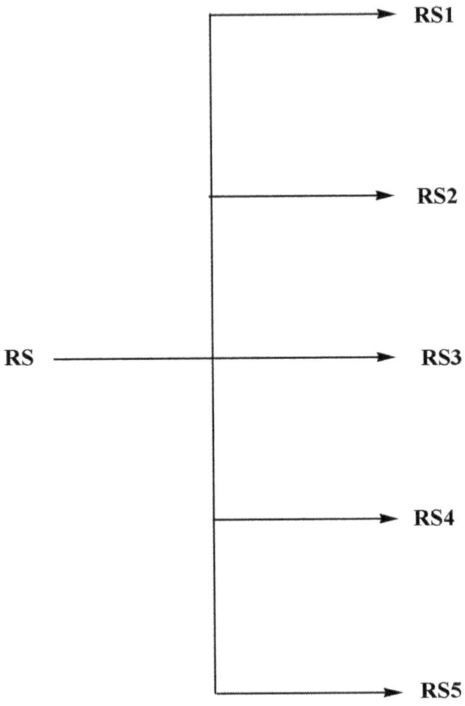

6.5 Ionic Liquid

The desired requirement for same chemically modified starch for ionic liquid has opened a way to enhance its application. The reaction of the ionic liquid is demonstrated and represented in Fig. 12.

Shen et al. (2010) had developed antimicrobial biodegradable films from the starch of sweet potato by integrating potassium sorbate greater than 15% to the effect of anti-*Escherichia coli*.

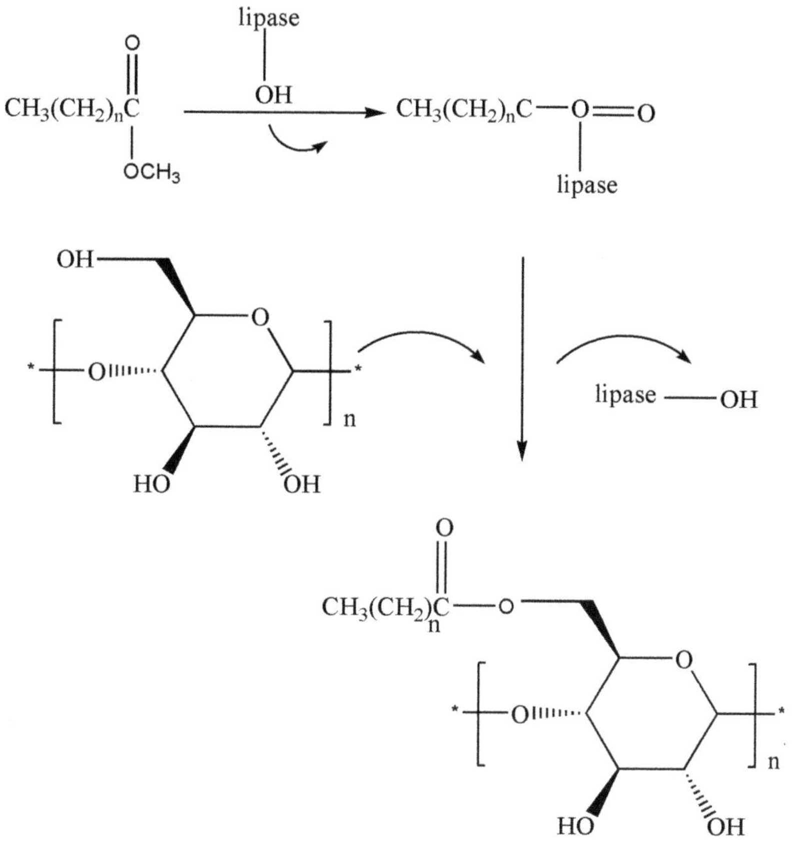

Fig. 12 Synthesis process of ionic liquid interaction

7 Cellulose

Cellulose is the peculiar and inherent polymer found in abundance on the surface of the earth relating the structure including β-1,4-glucosidic linkage in the middle of the repeating units of anhydroglucose. It is certified in the form of cellobiose polymer as shown in Fig. 13.

The mechanical resistance of the cellulose clashes with each other by the length of the fibre. The crystalline structure of it is mainly due to helix formation preserved by connection of hydrogen. The cellulose is typically thermostable possessing a degradation temperature more than that of 230 °C. The porous structure is because of the organisation of the fibre of cellulose.

The cellulose-based films have the ability to be hygroscopic, and its features are dependent on the relative humidity. The hygroscopic charm also effects on the temperature and the quality of substituents.

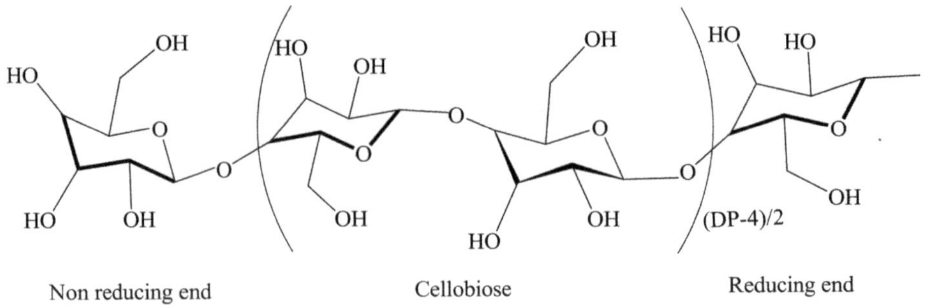

Fig. 13 Diagram of cellulose

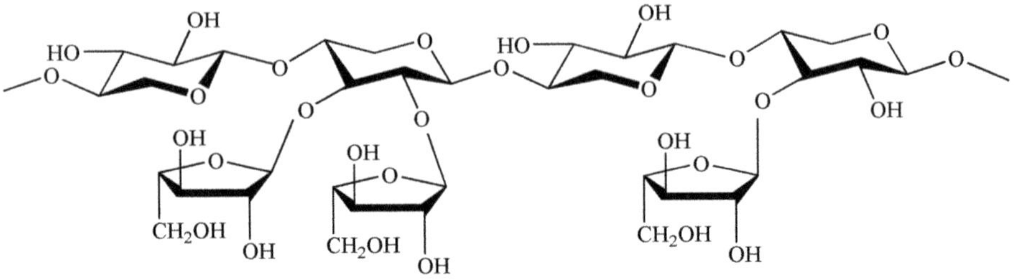

Fig. 14 Structure of hemicellulose

Films developed from the derivatives of cellulose do not have innate inhibitory or the antioxidant features. Most of the studies on the investigation are based on enrichment of biocomposite systems where the cellulose-based matrix is connected with the antioxidant.

Gemilli et al. (2010) invigilated film of cellulose acetate possessing different morphological properties for controlling the release rates of naturally occurring antioxidants such as L-tyrosine and L-ascorbic acid. But it was observed that greatest antioxidant activity for the release of test solution was achieved by a film of porous L-tyrosine.

7.1 *Hemicellulose*

Hemicellulose serves as the active alternative for petrochemical and is the topic of interest nowadays. It has attained different advantages when used from agricultural crop residues or by low-value wastewater. It is primarily used for polysaccharides occurring in the plant tissue along with cellulose, resulting in the isolation either through extraction with water or aqueous alkali. The structure of hemicellulose cannot be easily stated as some are linear and others are highly branched as given in Fig. 14. The branched chain of the structure is mostly short.

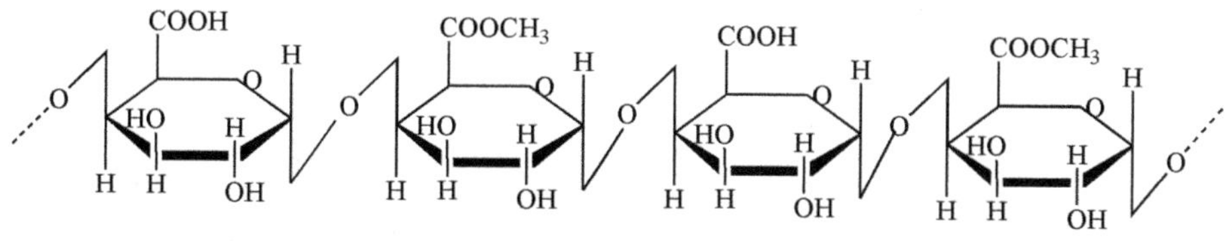

Fig. 15 Structure of pectin

The investigation related to permeability rectifies that oxygen permeability of hemicellulose film was compared with the value of the other biopolymer films, namely, amylose and amylopectin (Hansen and Plackett 2008).

8 Pectins

Pectin is acidic and is the water-soluble hygroscopic polysaccharide occurring in different natural byproducts. In most of the higher plants, it is present in the intercellular region of the primary cell wall. The primary sources of pectin include lime peel or lemon.

The essential structural feature of pectin includes the availability of linear chain regions made up of α-1,4-linked D-galactopyranosyluronic acid units given in Fig. 15.

Pavlat et al. (1999) suggested about the condition that aqueous solution of pectin can be casted in the transparent film in the state of low strength and resistance towards the water. But yet it is also confirmed that recasted film, when immersed in the aqueous solution of multivalent cations, tends to be insoluble in water. Since it is dependent on ions, it becomes stronger in the tension. When the tensile strength of pectin is compared to calcium, copper or zinc, it is found to have increased to a higher level and even higher than few wrapping materials as well.

9 Carrageenans

It is the material formed from the complex mixture made from various polysaccharides. The chemical structure of carrageenans includes α-1,4-linked D-galactopyranose given below in Fig. 16.

Choi et al. (2005) reported about the bioactive biopackaging done by incorporation of potassium sorbate in the K-carrageenan film. They examined the diffusivity investigated about the behaviour of temperature and pH on the diffusivity of such agent.

Fig. 16 Structure of carrageenans

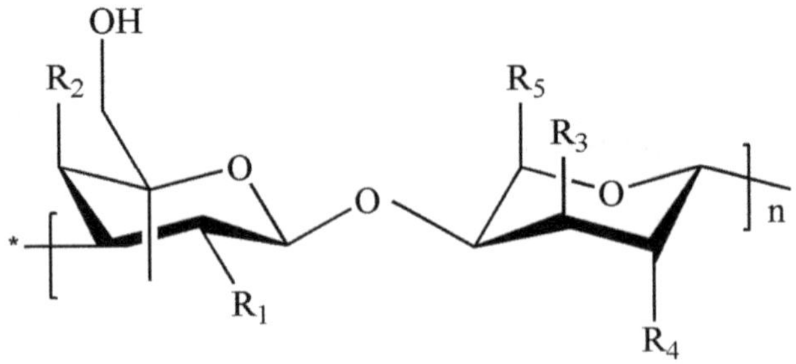

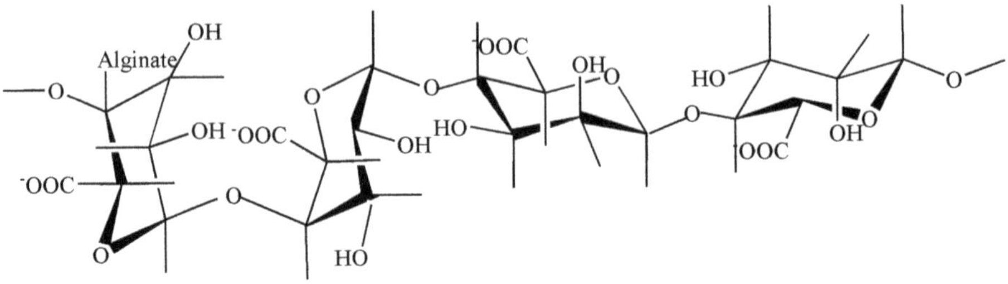

Fig. 17 Structure of alginate

10 Alginate

Alginate is obtained from seaweed. It has practical features to form film enabling them to be favourable for application in terms of food packaging. When we go through the structure, it is the linearly unbranched polymer with β-(1->4)-linked D-mannuronic acid and α-(1->4)-linked L-guluronic acid residue exhibited in Fig. 17.

Marcos et al. (2007) have shown the working of bio-based alginate materials along with the antimicrobial action. They have investigated the capability of enterocin, liberated from *Enterococcus faecium* and posing to control the growth of *L. monocytogenes* in the cooked ham. Bacteriocin was involved in the alginate-based biopackaging. The most influensive processing at 6 °C storage could be done by vacuum packaging along with the alginate films comprised of 2000 AU/cm^2 of enterocins.

11 Glycogen

It is the reserved polysaccharide of animals. The structure is the same as that of amylopectin, but it contains a large number of branches as shown in Fig. 18.

Till date, very less report has been presented about the use of glycogen in the form of base matrix for the material of biopackaging. Izawa et al. (2009) have worked on

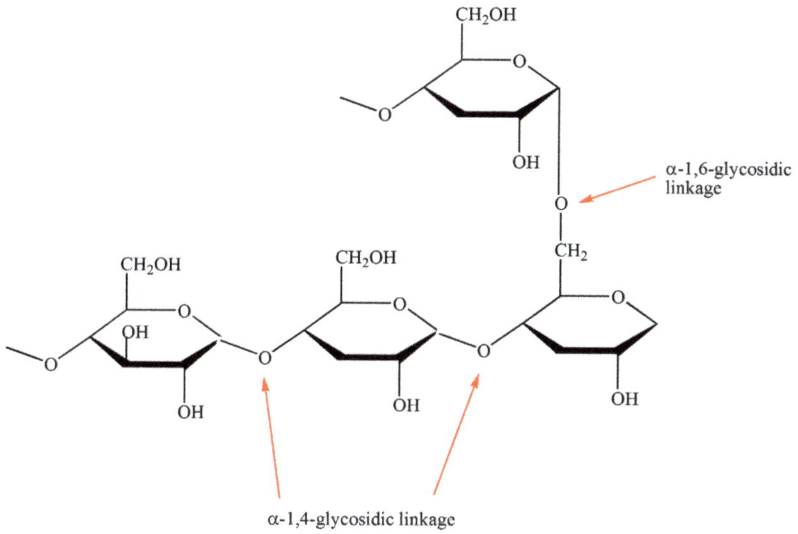

Fig. 18 Structure of glycogen

the making of glycogen-based polysaccharide gel by employing the phosphorylase-catalysed chain glycogen abridgment. The resulting solution obtained as a result twins as hydrogel form attributed to the development of linking zones dependent on the double helical structure of abridgment chain of amylose within the glycogen molecules.

12 Chitin

It is a widespread fact that chitin is the biopolymer holding the second position in nature where the cellulose is the first one. It was discovered firstly by Prof. H. Braconnot, a French scientist, in the year 1811 within the source named mushroom. Later Odier investigated the presence of chitin is insects cuticle in the year 1823. He gave the term "chitin" that originated from a Greek word which means tunic, envelope.

It shows various features the same as that of chitosan such as biocompatibility, biodegradability, non-toxic nature and so on (Arfin 2017).

12.1 Chitosan

C. Rouge in 1859 suggested that modified form of chitin is primarily prepared by the treatment of chitin with the boiled and concentrated solution of KOH within the

Fig. 19 Process of chitosan

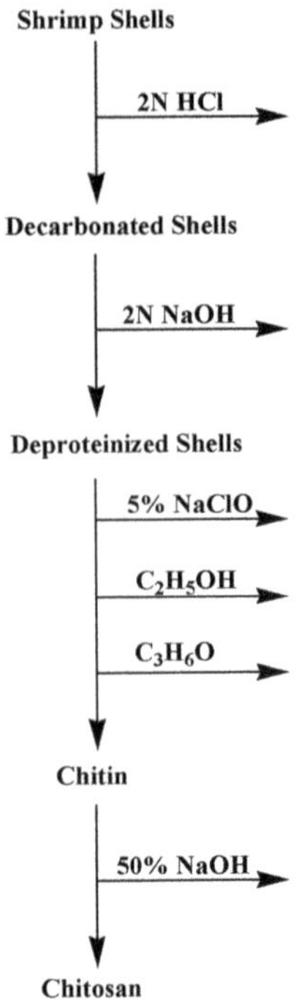

water. During the experimentation, he observed that chitin was operated with the help of chemical and temperature dependent treatment. In 1894, Hoppe-Seiler renamed the chitin as chitosan with a pronunciation of "kite-O-san". By the end of 1920, chitosan was mainly studied as a subject by the research personnels gathering interest as a significant compound in the various fields and ways to extract materials. Rammelberg, in around 1930, derived chitosan from the source of chitin-containing substances, namely, the crabs, shrimp cells, etc., and the overview is shown in Fig. 19 (Arfin and Mohammad 2016).

12.2 *Stability*

The property of stable nature indicates that reaction is performed in the absence of radical since their consequences were almost same as that of another scientist as shown in Fig. 20.

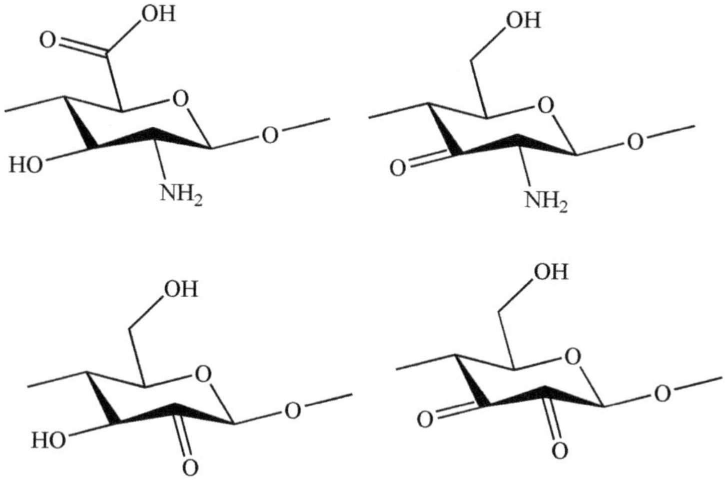

Fig. 20 Chitosan under oxidative condition

Fig. 21 Structure of carboxymethyl chitosan

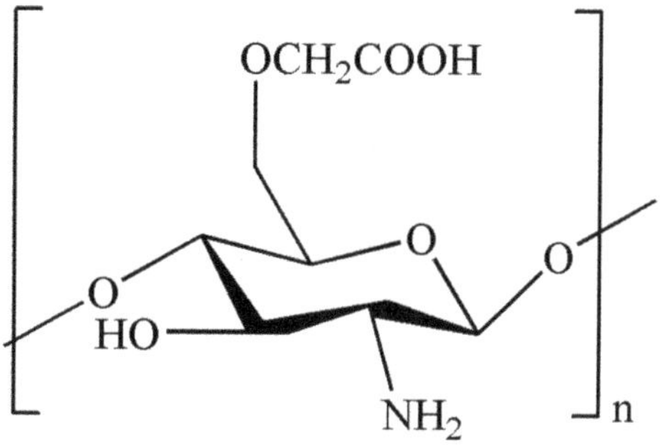

12.3 Carboxymethyl Chitosan (CMC)

Carboxymethyl chitosan (CMC) is the modified chitosan produced by the method of esterification. The products formed by esterification are then used in cosmetics and healthcare application as shown in Fig. 21.

12.4 Various Forms

The different forms of chitosan are applied reflecting that the polymer is versatile. Multiple types of chitosan and its subdivision along with its application are represented in Fig. 22.

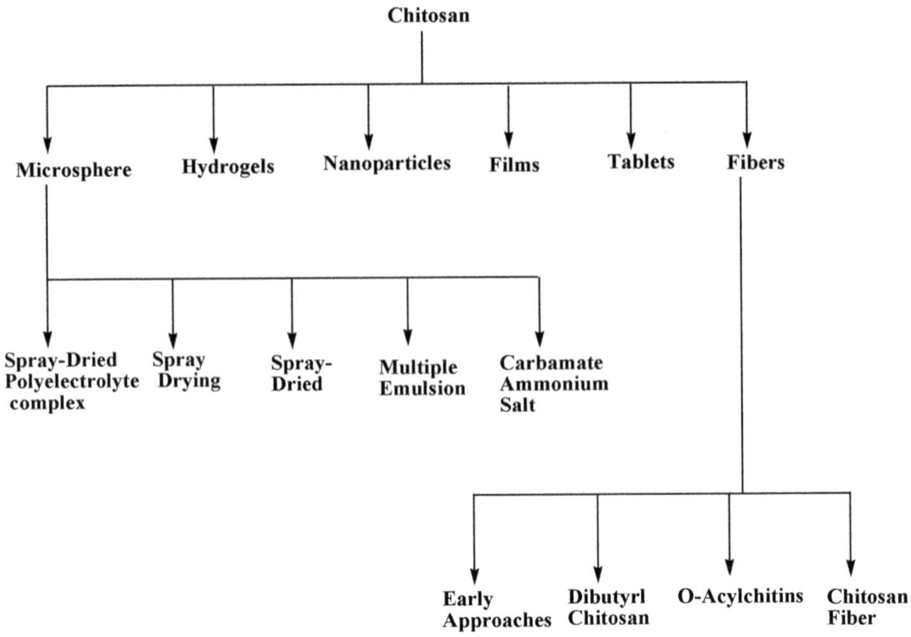

Fig. 22 Various forms of chitosan

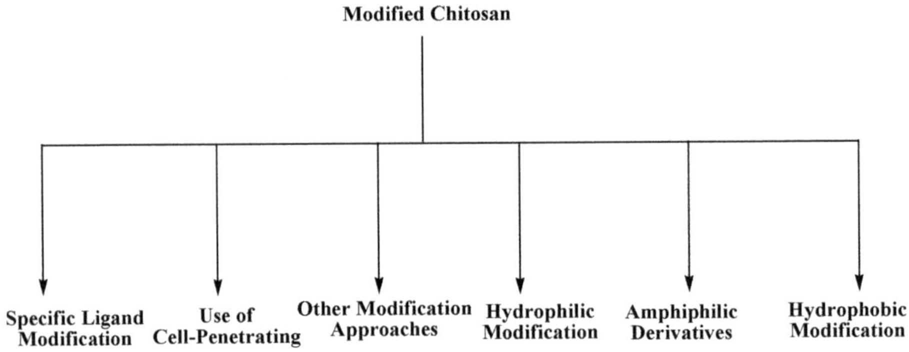

Fig. 23 Complete modification of chitosan

12.5 Modified Chitosan

To attain the suitable chitosan and to meet the challenges, different approaches were carried out with the hydrophobic and hydrophilic modifications. The complete scheme of the modified chitosan is represented in Fig. 23.

12.6 Michael Reaction

The Michael reaction of chitosan with the different reagents passing through the water was performed. The reaction is shown in Fig. 24. It can serve as a better step in modifying water-soluble derivatives.

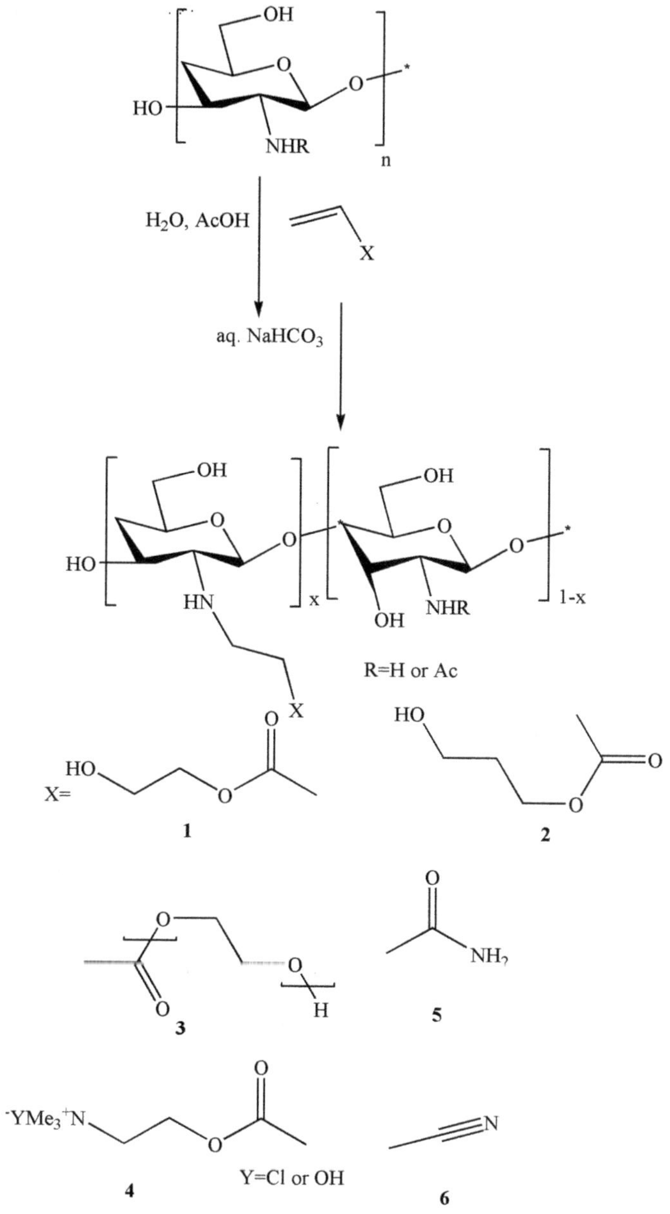

Fig. 24 Michael reaction of chitosan

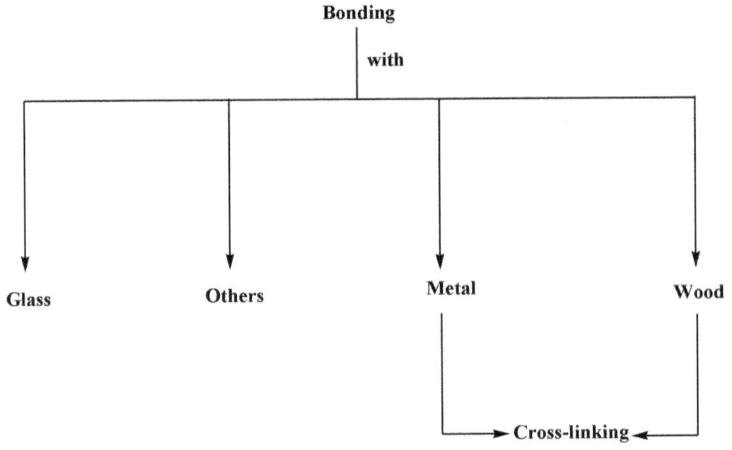

Fig. 25 Binding purpose

12.7 Bonding Purposes

For studying the binding features, the state of cross-linking of various polysaccharides with the molecules of chitosan was put under calculation. The nature of binding elements and their means are represented in Fig. 25.

The antibacterial features show the presence of protonated amino groups revealing that chitosan dissolved in solution has antioxidative characteristics (Yen et al. 2008).

13 Bacterial Cellulose

As we know cellulose is an essential component of the cell wall in plants, yet there are certain bacteria that produce cellulose such as bacteria producing acetic acid known as *Acetobacter xylinum*. The chemical structures of plant and bacterial cellulose are the same, whereas they differ in physical and chemical features as shown in Fig. 26.

Nruyen et al. (2008) synthesised the cellulose-film produced from bacteria composed of Nisin and the *L. monocytogenes* in the food materials. *Gluconacetobacter xylinus* K3 was responsible for producing bacterial cellulose. The nisin was inserted in the matrix of cellulose. The film then generated enabled in the reduction of *L. monocytogenes* from the frankfurters of 2 log CFU.g^{-1} after the storage process of 14 days.

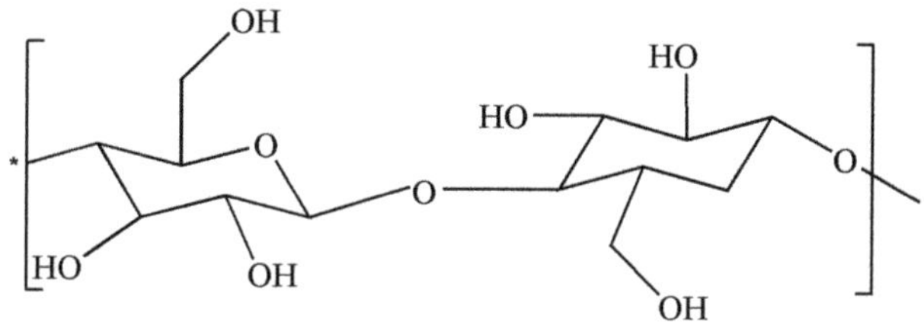

Fig. 26 Structure of bacterial cellulose

14 Bacterial Alginates

The alginates are explained similarly as that of cellulose through the strains of bacteria, namely, *Pseudomonas aeruginosa*. Its applications are the same as that of the plant-derived polymers.

15 Others Exopolysaccharides

Other researcher reported about the bacterial polysaccharides to be a complex polymer posing for the exclusive benefits in the environment. Since they are present in abundance, there are specific polysaccharides representing the particular features that permit the growth of industrial utility.

16 Hydrogels

Hydrogels are hydrophilic polymers made from high adsorbent containing around 90% to 99% of water as shown in Fig. 27. They are comprised of various macromolecular building blocks with the cross-linked polymers and colloidal gatherings on other substances as well. It experiences the macroscopic shrinking and light enabling it to be a unique one. The bio-based hydrogels determine by the xanthan gum, or the other material is used for the supportable and ecological packaging films.

17 Protein

Owing to the various features such as low cost, enhanced film forming ability, biodegradable aspect, the protein-based films have gain lots of interest in the field of research. Such films are made from gelatin, soy protein, etc. In comparison to the other source of the film, the soy protein is considered to be the best material for bio-based packaging due to its efficient ability to make the film. The structure of a protein is shown in Fig. 28.

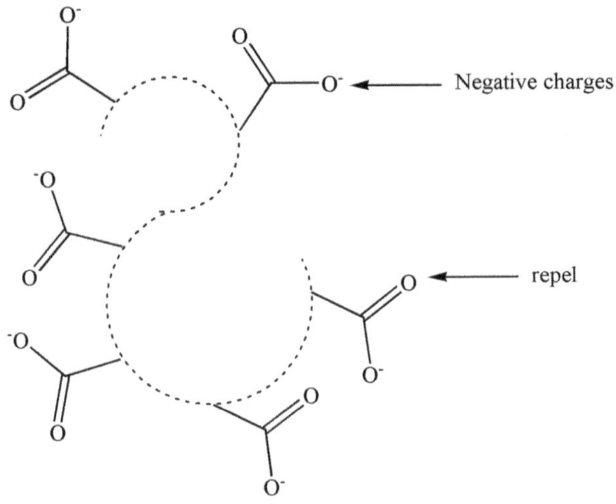

Fig. 27 Structure of hydrogel

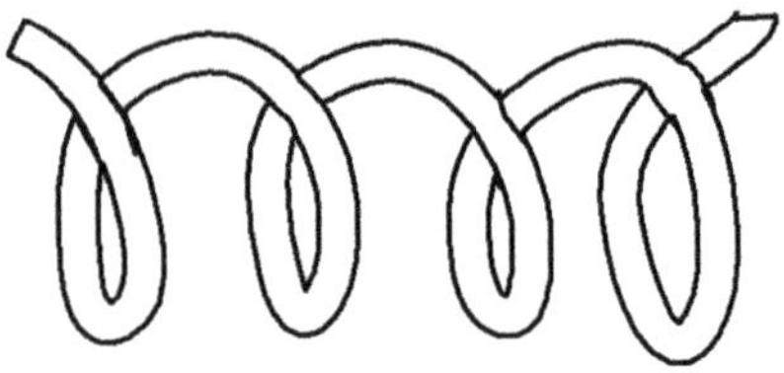

Fig. 28 Structure of protein

The soy protein-based material is not capable enough to meet the requirement of the cost-effective film as it has mechanical and barrier features the same as that of the plastics.

17.1 Type of Protein

Proteins have various types, and their function is shown in Table 1.

18 Gelatin

Gelatin is a biodegradable polymer primarily derived from the fish skin. The studies are carried out on the fish and its skin due to the spiritual, fitness, and social issue, but yet it is found that it is the significant waste product available quickly for the various purposes. The structure of gelatin is shown in Fig. 29.

The gelatin faces the lack of barrier features required for the food packaging materials.

Table 1 Types of protein and their function

Types of protein	Function
Motility	Cells and their organelle movement
Structural	Provides protection and strength
Enzymes	Enhances biochemical reactions
Transport	Transfer molecules from one place to another

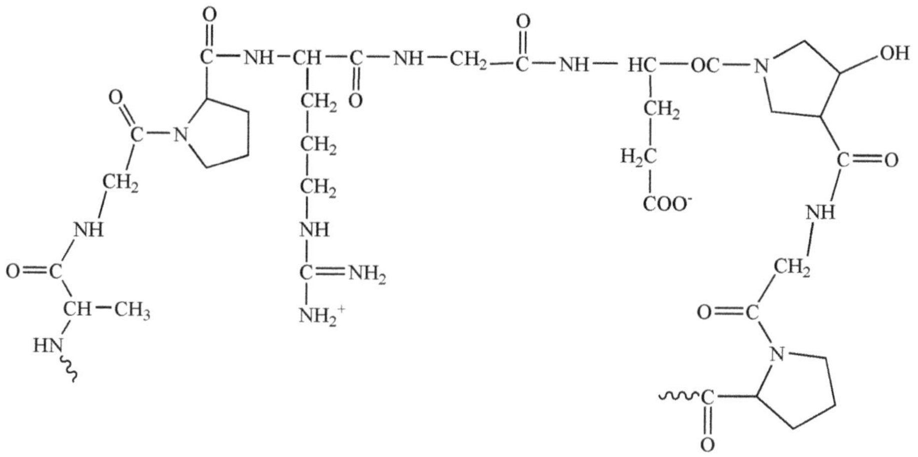

Fig. 29 Structure of gelatin

19 Galactomannans

Galactomannans are a type of heterogeneous polysaccharides. It is comprising up of a β-(1-4)-D-mannan spine with a single α-(1-6) substitution of D-galactose as shown in Fig. 30. These polysaccharides are produced from the endosperm of dicotyledonous seeds of various plants in the family of *Leguminosae*. The chief merits for using galactomannan are its capability of forming a viscous solution at the relatively low concentration, and it has a very slight change in pH. Hence, as a result, difficult prospect regarding the features and the use of edible films from galactomannan is presented in the literature (Cerqueira et al. 2011).

20 Nanotechnology

The study of nanotechnology and the use of material within the range of 1–10 nm are included in the research of food nanotechnology. It has emerged the interest and provided the new approach to the world for the food industry. It imposes the facts related to the features of material being varied at a nanometre scale in comparison to

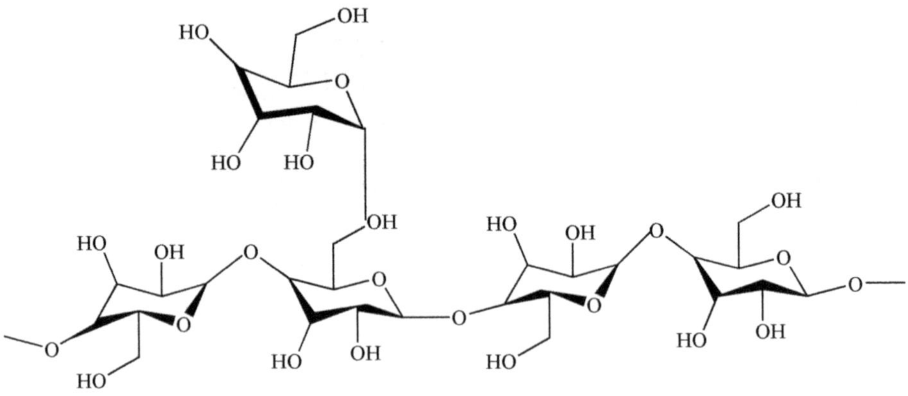

Fig. 30 Structure of galactomannan

the macro level. In the recent years, it is observed that a novel type of bio-nanocomposites has proved to be the best alternative for enhancing the mechanical as well as barrier features.

20.1 Bio-based Nanocomposites

The biopolymers are recognised as a capable alternative for the plastic packaging materials because of the attention gathered in the field of sustainable growth. Such biopolymers are animal and microbial products, plant-based materials and polymers synthesised from the monomer. The schematic representation of such biopolymers is shown in Fig. 31.

21 Gelatin-Starch OSA

The assessment of propolis was done by scientist's team of da Silva and co-worker (2013). The production properties and their storage and stability were studied by them. The purpose of the study is to prepare the powdered form of propolis which should be stable and alcohol and water dispersed. OSA (octenyl succinic anhydride) and gum Arabic are required for the spray drying of propolis. By this, around 15–24 μm size of propolis was prepared in powdered form with the antioxidant property. This modification and preparation of propolis increase its opportunities in the industries. Characterisation was done by applying the techniques such as scanning electron microscopy and moisture analyser, whereas the size was measured by using the instrument of laser light diffraction.

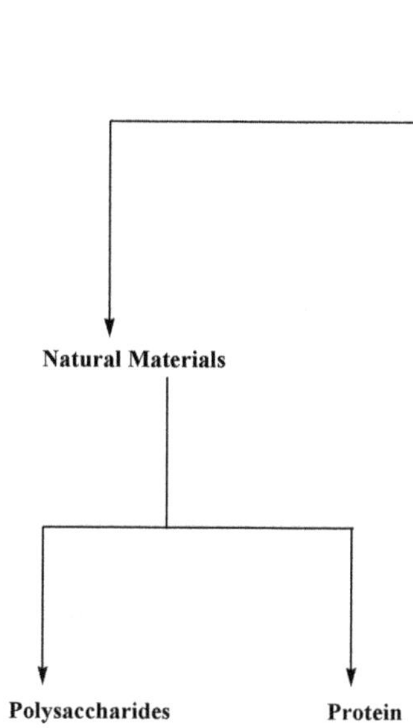

Fig. 31 Overview of biopolymers

22 Starch Clay

Avella and co-worker (2005) discovered that the starch-clay nanocomposite could be used in the food packaging purpose. The application of the nanoparticle in the different starch-based clays was used to produce the starch-based film for packaging. The method of such materials for preparing films increases the features of the packaging films and also the tensile strength and modulus. The central aspect of this product is that it is a novel biodegradable material specially used for food packaging purpose. The characterisation was done by NMR technique, generally used for nanocomposites or nanomaterials.

23 PLA-Starch

Muller and co-worker (2017) proposed the use of PLA-starch for making biodegradable food packaging plastics. The replacement of pollution-creating plastics was done by using the plastic made from polylactic acid and starch. Starch is the water-sensitive material with less mechanical resistance. This starch is the best substitute for making plastics. The properties of PLA and starch match each other. By the combination of both PLA and starch, a new product is formed with more efficiency and without creating any issue related to economic or pollution problem. PLA is hydrophobic, whereas starch is hydrophilic enabling the processing of material to be difficult. Still, the product obtained from PLA and starch is used in the food packaging industries. Instead of making plastic only from starch or PLA, blending the combination of both increases the properties and efficiency of film. By varying the ratios of starch and PLA, the blend can be made. It serves as the best alternative for reducing pollution caused due to plastics. The obtained plastic will exhibit the properties of antioxidant and antimicrobial. The plastic or film has excellent flexibility and mechanical properties.

24 PLA

Sheth and co-worker in 1995 (1997) suggested the use of PLA (polylactic acid) and PEG (polyethylene glycol) for making the plastic. By varying the concentrations of both materials, the composite was prepared. Depending on the levels, the mixture becomes partially miscible and miscible. If PLA is more than 50% of a mixture, then PEG crystallinity increases, and if PEG is more than 30%, then the weight loss occurs which is mainly due to PEG getting dissolved. The tensile strength also decreases if the concentration of PEG is high.

It is a thermoplastic polymer. PLA can be made from the natural source including sugar stock, rice and corn. By fermentation of these plant sources, PLA can be prepared. It can also be developed by chemical conversion of the references to dextrose and then to lactic acid using polycondensation. The suitable liquids for PLA in which it is soluble are acetonitrile, chloroform, benzene, ethyl acetate, acetone, etc. They are not soluble in ethanol, methanol and water. PLA is approved when it comes into contact with the biological fluids that are entirely biocompatible, biodegradable, environmentally friendly and renewable. Most of the food packaging industries require PLA for making biodegradable plastics. It possesses hydrophobic nature. Thus before using PLA, the drying of these pallets should be done at about 60 °C–100 °C.

This process is modified physically (Lim et al. 2008). It has application in the tissue engineering field for wound healing (Armentano et al. 2013). It is resistant to elasticity, which is the reason for making rigid and brittle plastic.

25 Chitosan and Gelatin

Rui and co-worker (2017) have shown the preparation of composite from chitosan and gelatin. The composite development from chitosan and gelatin requires the process of an emulsion of oil in water. Again, for the synthesis and storage, freezing and lyophilisation were carried out. Linear polysaccharide chitosan and a product of collagen hydrolysis, i.e. gelatin, bind together by the hydrogen bonding. The composite is made to increase the properties of chitosan and gelatin. The composite materials are used to prepare nanoparticles, film, micro-carriers, etc. The analysis of the composite can be done by using SEM, IR and X-ray.

It is used for the animal cell culture on a large scale (Taravel and Domard 1993). Due to the use of plastics and their derivatives, the environment becomes polluted, and this is happening in most of the food packaging industries (Zhou et al. 2013). Thus there is a need to create such an option which should be eco-friendly, economical, biocompatible, active to oxygen transfer and suitable to use. The characterisation of chitosan-gelatin composite was done by using a technique such as UV spectrophotometer and SEM.

26 PHA (Polyhydroxy Alkaloids)

Molyneus (1993) gave their review of the isolation, characterisation and application of polyhydroxy alkaloids. It is the class of phytochemicals. It is the newly discovered class, and therefore the analysis and detection are developed entirely. Polyhydroxy are the compounds which have the high inhibitory property of glycosidase. The activities depend on the stereochemistry and the hydroxyl groups. For the processing of glycoproteins in organisms, PHA is needed so that the biological mechanisms are regulated. It is proved that glycoproteins are indirectly the enhancer of the immune response. The source of PHA is a ground plant material. It can be extracted by using ethanol or methanol or water by just varying the proportion of solvent. As a few compounds are isolated each time, hence chemical determination is limited. Mass spectroscopy, UV spectroscopy, X-ray crystallography and NMR are the characterised techniques for PHA.

27 PET (Polyethylene Terephthalate)

The scientist Siracusa and co-workers have shown the review on the use of PET as a biodegradable polymer (Siracusa et al. 2008). They have demonstrated the complete study in the use of plastics. It is not possible to replace the plastic, but to a certain extent, it is possible. The consumption of plastics is more than about 200 million, and hence it is an important issue. The substitute for the petrochemical based plastics is the critical point.

The chemist Du Pont developed the polymer polyethylene terephthalate when he was trying to make textile fibres. It is a thermoplastic polymer. The main components of monomers required for the synthesis of PET are ethylene glycol and terephthalic acid. The esterification process converts the two monomers alcohol and an organic acid into ester and water. As PET is a polymer, hence it can also be called as "polyester". It is resistant to moisture so it can be used in the packaging. It is widely used in the textile industries. It is used for preparing fibre for clothes. It does not have colour and is transparent in nature and semicrystalline. For increasing its strength and properties, it can be used by combining with the other materials like carbon nanotubes and glass. It is straightforward to accept PET as it is not costly when compared to others and a non-breakable substance like glass, and the best thing is that recycling is possible for PET. It is a lightweight material but very strong. As it is light weighted, hence less amount of compound is required, and transportation also becomes comfortable with less consumption of fuel. Since it has the property that does not affect the food and water when it comes into contact with it, so it is used in the food packaging industries.

28 TPS-Chitosan

Jantanaskulwong and co-worker (2016) have introduced the blending of TPS with chitosan and rubber for making of plastic by using cassava starch, glycerol and chitosan. The thermoplastic plastic can be prepared by using glycerol, cassava starch and chitosan. Chitosan was used to improve the properties of the plastic. For making the plastic, a plasticiser is required, and in this case, glycerol was used. Due to the NH_2 group, the reaction of chitosan with any other team becomes easy and fast. In this study, the observation during the blend making was done. The improvement or decrease in the properties was observed. Mechanical strength, tensile strength, mechanism of reaction and morphology, etc. were found. Finally, it was clear from the observation that mechanical features and tensile strength tend to increase with the low weight to form a small weight composite. The characterisation of the TPS-chitosan composite was done by SEM, rheological measurement and Fourier transform infrared spectroscopy.

29 TPS, Chitin and Chitosan

Lopez and co-worker (2014) used chitin, chitosan and thermoplastic corn starch for making the biodegradable film by the process of thermo-compression. The modifications were done in the film by the addition of chitosan. Increase in the crystallinity was obtained. The obtained film was of very efficient quality, i.e. with a smooth surface and similar nature. There were no cracks or pore in the film. The properties of UV absorption capacity and their colour were increased. The film was biodegradable, antibacterial and ecological with excellent appearance and enough thickness. DSC and SEM were carried out for the characterisation of the film. The materials chitin and chitosan increase the elastic modulus and strength. The applications on the industrial level were enhanced by the use of such material.

30 Chitin and Starch

Chang and co-worker (2010) prepared the new nanocomposite by using chitin and starch. He developed the bio-nanocomposite. Chitin nanoparticles were used because they possess low crystallinity. Starch was used from glycerol-plasticised potato starch. Casting and then evaporation were the processes for making starch-chitin nanoparticles. The characterisation techniques such as X-ray diffraction, electron microscopy and mechanical, thermal analysis were used. Properties including tensile strength, water vapour baring and glass transition temperature were increased. This was the novel product made from starch and chitin. As they are biodegradable, they are useful in different fields like pharmaceutical, industrial, biomedical and agricultural. For food packaging industries, they are mostly used.

31 PCL

Lo and co-worker (2010) studied the morphology and properties of the composites made from polycaprolactone-starch and polycaprolactone-pine composite. It was biodegradable and non-toxic and thus used for food packaging. For their testing, the universal machine was used. This device was used to investigate the morphology of the composite so that the conclusion can be made by the observation. Silane was used as a coupling agent in the making of the composite.

This agent is helpful in the identification of a material and their properties and also for the PCL matrix and filter interfacial adhesion. Characterisation was done by SEM technique.

32 Cellulose Fibre

Wan and co-worker (2009) used the biodegradable reinforcement in the form of bacterial cellulose nanofibers. He discussed all the phenomenon occurring during the formation of nanocomposite. The method used was the solution impregnation. The comparison was made between the nanocomposite of cellulose and starch and the unreinforced starch. Sorption diffusion process was used for the kinetic analysis, and then the different parameters were also determined. The study about tensile strength and microbe attack was performed. The use of bacterial cellulose in making of nanocomposites increases the power and ability to resist the invasion of microbes. It had broad application in the protection of the environment.

33 PCL Fibre

Azimi and co-worker (2014) studied the polycaprolactone fibre. The PCL fibres are made in different forms, and according to their content, they were used in the various areas of the biomedical field. The PLC was used because it was the biodegradable aliphatic polyester and the fibre prepared from this was environmentally friendly. It is compatible and had good processibility which makes fabrication easy. The scientist shows their application in the medical and biological fields. They had explained about the different methods for making of PCL fibres and also their applications. Tissue engineering and drug delivery are the best fit process for PCL fibres.

34 Starch Clay

Park and co-worker (2002) introduced the preparation of hybrid composite of starch clay. The process used for the development was melt intercalation. Natural starch and the clays organically modified were only used for hybrid preparation. The storage modulus of this hybrid was higher than any other composites. Characterisation was done by XRD and TEM techniques.

35 TPS, PLA and Clay

Ayana and co-worker (2014) prepared the composite from TPS, PLA and dispersed nano clay. Potato starch was in situ gelatinised. The thermo-mechanical properties and tensile strength were increased due to the use of PLA and clay. The synthesis of novel material was done on the environmental level. It can be used as a packaging material also. SEM, XRD and TEM were used for the composite characterisation. The use of starch is the best part from the economic point of view as it is available in large amount naturally.

36 PHB-HV Maize

Reis and co-worker (2008) studied the characterisation of PHB-HV (polyhydroxybutyrate-hydroxyvalerate) maize starch blend film. The use of maize makes the blend more economical as compared to the original PHB-HV blend. The mix they prepared includes the variation in the concentration of starch. The process of casting was used for the preparation. Morphological properties and their molecular structure were identified. The characterisation techniques such as FTIR, SEM and optical microscopy were used. As the concentration of starch varies, the properties vary including the parameters like Young's modulus and tensile strength.

37 Gelatin-Starch OSA

Li and co-worker (2013) introduced the use of gelatin-starch OSA for the study of stability and properties of oil in water emulsion. The corn starch was modified by using octenyl succinic anhydride, and then it was used with the gelatin. The size of particles and rheological behaviour were studied. This starch OSA absorbs the interface of oil and water and acts as a stabiliser. FTIR, substitution degree measurement and optical microscopy were used for the characterisation of gelatin-starch OSA.

38 Glass

For a generation of power and thermoelectric refrigeration, many new materials were developed and still developing. Nolas and co-worker developed the new material for the energy conversion with the help of photon glass electron crystal; this material was called skutterudites (Nolas et al. 1999). These materials have the best thermoelectric properties. It can also be used as low-temperature semiconductors as it has the specific features required for semiconductors. Glass is the transparent material and used in the full range. In early days glass was just used for decoration purpose, but nowadays glass has very much importance in the technical fields. Glass is used in making of instruments, electrical transmission, domestic appliance, bioactive glasses, biomedical devices, laboratories, etc. Glass is the non-crystalline solids, disordered and amorphous material. The use of glass in different fields is due to the isotopic property. Glass can be prepared from any material like organic or inorganic which can cool faster without crystallisation.

39 Plastics

Lorcks and co-worker in 1997 proposed the properties and applications of plastics (Lörcks 1998). Biodegradable plastics were the invention as a substitute for the pollution creating plastics. They were biodegradable and can be disposed or degraded in the environmental friendly manner. Plastics are the polymers made up of monomers through chemicals reactions. Condensation, polymerisation and polyaddition (chain and step reaction) are the necessary processes for the making of plastic.

- In polyaddition, the chemical reactions occur which form polymers from monomers. For this process, energy and catalyst are required. Monomers combine, but there is no movement of hydrogen bond, and also byproduct separation is not done; this is the case in a chain reaction of polyaddition.
- In the case of polyaddition step reaction, there is a movement of hydrogen bonds. There are no double-bond activities taking place.
- And lastly, there is polycondensation in which actual plastic formation takes place. It is also a step reaction.

According to the properties and structure, they are classified into different classes. They may be thermoplastics, thermosets, polymer blends, elastomers, semiconductor thermoplastics, etc.

40 Conclusion

Packaging is the most crucial and significant activity for the preservation and marketing of the food products. Since last few decades, petroleum-based food packaging materials are put into practice due to their non-biodegradable ability posing severe harmful results as well. Therefore, to overcome such problem, continuous efforts are undertaken to synthesise biodegradable material generated from the renewable sources. Biomaterial had served the prospective to change the food packaging sector to a large extent. It was observed that bioactive packaging material releasing bioactive compound was highly efficient in comparison to the active system without liberating any agents. The biopolymer-based biodegradable food packaging materials can serve to be an efficient alternative for the non-biodegradable plastics. Hence, it is expected that advancement in this field will undoubtedly fulfil their goal and provide excellent results.

Acknowledgements Authors acknowledge the Knowledge Resource Centre, CSIR-NEERI (CSIR-NEERI/KRC/2017/DEC/EMD/1) for their support.

References

Arfin T (2017) Chitosan and its derivatives: overview of commercial applications in diverse fields. In: Ahmed S, Ikram S (eds) Chitosan: derivatives, composites and applications. Wiley-Scrivener, New York, pp 115–150

Arfin T, Fatima S (2014) Anticipating behavior of advanced materials in healthcare. In: Tiwari A, Nordin AN (eds) Advanced biomaterials and biodevices. Wiley-Scrivener, New York, pp 243–288

Arfin T, Mohammad F (2016) Chemistry and structural aspects of chitosan towards biomedical applications. In: Ikram S, Ahmed S (ed), Natural polymers: derivatives, blends and composites, vol I, Nova Sciences Publishers, New York, pp. 265–280

Arfin T, Tarannum A (2017) Polymer materials: from the past to the future. In: Ahmed S, Annu IS (eds) Green polymeric materials: advances and sustainable development. Nova Science Publishers, New York, pp 73–88

Armentano I, Bitinis N, Fortunati E, Mattioli S, Rescignano N, Verdejo R, Lopez-Manchado MA, Kenny JM (2013) Multifunctional nanostructured PLA materials for packaging and tissue engineering. Prog Polym Sci 38(10):1720–1747

Athar S, Arfin T (2017) Commercial and prospective applications of gelatin. In: Ikram S, Ahmed S (eds) Natural polymers: derivatives, blends and composites, vol II. Nova Science Publishers, New York, pp 199–216

Avella M, De Vlieger JJ, Errico ME, Fischer S, Vacca P, Volpe MG (2005) Biodegradable starch/clay nanocomposite films for food packaging applications. Food Chem 93(3):467–474

Ayana B, Suin S, Khatua BB (2014) Highly exfoliated eco-friendly thermoplastic starch (TPS)/poly (lactic acid)(PLA)/clay nanocomposites using unmodified nanoclay. Carbohydr Polym 110:430–439

Azimi B, Nourpanah P, Rabiee M, Arbab S (2014) Poly (ε-caprolactone) fiber: an overview. J Eng Fibers Fabr 9(3):74–90

Cerqueira MA, Bourbon AI, Pinheiro AC, Martins JT, Souza BWS, Teixeira JA, Vicente AA (2011) Galactomannans use in the development of edible films/coatings for food applications. Trends Food Sci Technol 22(12):662–671

Chang PR, Jian R, Yu J, Ma X (2010) Starch-based composites reinforced with novel chitin nanoparticles. Carbohydr Polym 80(2):420–425

Choi JH, Choi WY, Cha DS, Chinnan MJ, Park HJ, Lee DS, Park JM (2005) Diffusivity of potassium sorbate in κ-carrageenan based antimicrobial film. LWT-Food Sci Technol 38 (4):417–423

da Silva FC, da Fonseca CR, de Alencar SM, Thomazini M, de Carvalho Balieiro JC, Pittia P, Favaro-Trindade CS (2013) Assessment of production efficiency, physicochemical properties and storage stability of spray-dried propolis, a natural food additive, using gum Arabic and OSA starch-based carrier systems. Food Bioprod Process 91(1):28–36

Gemili S, Yemenicioglu A, Altinkaya SA (2010) Development of antioxidant food packaging materials with controlled release properties. J Food Eng 96(3):325–332

Hansen NM, Plackett D (2008) Sustainable films and coatings from hemicelluloses: a review. Biomacromolecules 9(6):1493–1505

Izawa H, Nawaji M, Kaneko Y, Kadokawa JI (2009) Preparation of glycogen-based polysaccharide materials by phosphorylase-catalyzed chain elongation of glycogen. Macromol Biosci 9 (11):1098–1104

Jantanasakulwong K, Leksawasdi N, Seesuriyachan P, Wongsuriyasak S, Techapun C, Ougizawa T (2016) Reactive blending of thermoplastic starch, epoxidized natural rubber and chitosan. Eur Polym J 84:292–299

Li C, Fu X, Luo F, Huang Q (2013) Effects of maltose on stability and rheological properties of orange oil-in-water emulsion formed by OSA modified starch. Food Hydrocoll 32(1):79–86

Lim LT, Auras R, Rubino M (2008) Processing technologies for poly (lactic acid). Prog Polym Sci 33(8):820–852

Lo HY, Kuo HT, Huang YY (2010) Application of Polycaprolactone as an anti-adhesion biomaterial film. Artif Organs 34(8):648–653

Lopez O, Garcia MA, Villar MA, Gentili A, Rodriguez MS, Albertengo L (2014) Thermocompression of biodegradable thermoplastic corn starch films containing chitin and chitosan. LWT-Food Sci Technol 57(1):106–115

Lörcks J (1998) Properties and applications of compostable starch-based plastic material. Polym Degrad Stab 59(1–3):245–249

Marcos B, Aymerich T, Monfort JM, Garriga M (2007) Use of antimicrobial biodegradable packaging to control *Listeria monocytogenes* during storage of cooked ham. Int J Food Microbiol 120(1):152–158

Mogarkar PR, Arfin T (2017) Chemical and structural importance of starch based derivatives and its applications. In: Ikram S, Ahmed S (eds) Natural polymers: derivatives, blends and composites, vol II. Nova Science Publishers, New York, pp 73–88

Molyneux RJ (1993) Isolation, characterization and analysis of polyhydroxy alkaloids. Phytochem Anal 4(5):193–204

Muller J, González-Martínez C, Chiralt A (2017) Combination of poly (lactic) acid and starch for biodegradable food packaging. Materials 10(8):952

Nguyen VT, Gidley MJ, Dykes GA (2008) Potential of a nisin-containing bacterial cellulose film to inhibit *Listeria monocytogenes* on processed meats. Food Microbiol 25(3):471–478

Nolas GS, Morelli DT, Tritt TM (1999) Skutterudites: a phonon-glass-electron crystal approach to advanced thermoelectric energy conversion applications. Annu Rev Mater Sci 29(1):89–116

Park HM, Li X, Jin CZ, Park CY, Cho WJ, Ha CS (2002) Preparation and properties of biodegradable thermoplastic starch/clay hybrids. Macromol Mater Eng 287(8):553–558

Pavlath AE, Voisin A, Robertson GH (1999) Pectin-based biogradable water insoluble films application of polymers in foods. In Die Makromolekulare Chemie, Macromolecular, Symposium, Dallas, 1999

Reis KC, Pereira J, Smith AC, Carvalho CWP, Wellner N, Yakimets I (2008) Characterization of polyhydroxybutyrate-hydroxyvalerate (PHB-HV)/maize starch blend films. J Food Eng 89 (4):361–369

Rui L, Xie M, Hu B, Zhou L, Yin D, Zeng X (2017) A comparative study on chitosan/gelatin composite films with conjugated or incorporated gallic acid. Carbohydr Polym 173:473–481

Shen XL, Wu JM, Chen Y, Zhao G (2010) Antimicrobial and physical properties of sweet potato starch films incorporated with potassium sorbate or chitosan. Food Hydrocoll 24(4):285–290

Sheth M, Kumar RA, Davé V, Gross RA, McCarthy SP (1997) Biodegradable polymer blends of poly (lactic acid) and poly (ethylene glycol). J Appl Polym Sci 66(8):1495–1505

Siracusa V, Rocculi P, Romani S, Dalla MR (2008) Biodegradable polymers for food packaging: a review. Trends Food Sci Technol 19(12):634–643

Taravel MN, Domard A (1993) Relation between the physicochemical characteristics of collagen and its interactions with chitosan: I. Biomaterials 14(12):930–938

Wan YZ, Luo H, He F, Liang H, Huang Y, Li XL (2009) Mechanical, moisture absorption, and biodegradation behaviours of bacterial cellulose fibre-reinforced starch biocomposites. Compos Sci Technol 69(7):1212–1217

Yen MT, Yang JH, Mau JL (2008) Antioxidant properties of chitosan from crab shells. Carbohydr Polym 74(4):840–844

Zhou W, Ma G, Su Z (2013) Microspheres for cell culture. In: Ma G, Su Z (eds) Microspheres and microcapsules in biotechnology: design, preparation, and applications. Pan Stanford Publishing, Boca Raton, pp 49–73

Interaction Phenomena Between Packaging and Product

S. K. Vimala Bharathi, J. A. Moses, and C. Anandharamakrishnan

Abstract Food packaging plays a vital role in food processing and preservation. The growing concern of consumers in food safety and quality has imparted a great deal of attention in understanding the interaction of food and packaging materials, particularly polymers. This chapter presents a detailed note on the prospective health effects of polymeric materials and their interaction phenomena including migration (from package to product), permeation (from environment through package) and sorption (from product to package), highlighting appropriate illustrations from recent research works to provide an in-depth understanding of the subject. Factors such as physicochemical characteristics of the food, package and additives used in the package, as well as environmental and storage factors like temperature and duration, and their implications are presented. The chapter also describes the mechanism and mathematical models involved in interaction phenomena. Finally, food simulants for testing and testing methods, along with legislations to be adopted, are also discussed.

Keywords Food packaging · Migration · Permeation · Sorption · Food simulants

1 Introduction

Packaging forms an integral part of food processing and handling sector. According to Coles (2003), 'Packaging is defined as a means of safely and cost effectively delivering products to the consumer in accordance with the marketing strategy of the organisation'. The basic functions of packaging are containment, protection, presentation, market promotion, preservation, communication and so on (Coles 2003). In general, the packaging material used can be primary, secondary, tertiary and quaternary based on the nearness level to the packaged product (Robertson 2013).

S. K. Vimala Bharathi · J. A. Moses · C. Anandharamakrishnan (✉)
Indian Institute of Food Processing Technology (IIFPT), Ministry of Food Processing Industries, Government of India, Thanjavur, India
e-mail: director@iifpt.edu.in

S. Ahmed (ed.), *Bio-based Materials for Food Packaging*,
https://doi.org/10.1007/978-981-13-1909-9_2

Primary packaging is of prime concern as it is in direct contact with the food material. Paper, glass, metals and natural substances are used as primary packaging material since long time. Polymerization of monomers like ethylene, propylene, vinyl alcohol, etc. has made the packaging sector versatile and flexible. Addition of some external agents like additives, plasticizers and colouring agents is done to improve the property of polymers, and polymers with those external agents are called plastics. However, these external agents on contact with the food material can have negative impact on the food as well as the health of the consumer.

The packaging sector is focussed more as the processed food market is growing tremendously. Also, the need and design of packaging material change with change in consumer behaviour. Packaging materials like nanocomposite, bio-based polymers are being explored mainly because of the negative impact of polymer-based packaging material on the environment. Biopolymers are biodegradable polymeric materials that consist of covalently bonded monomeric units. Under suitable conditions, these polymers are degraded by the action of naturally occurring microorganisms leaving no toxic residues into the environment. Biopolymers can be obtained from renewable sources or petroleum products (Vieira et al. 2011; Othman 2014). And, polymer composites are composed of a continuous phase (polymer matrix) and a dispersed phase (reinforced material) (Youssef 2013). Polymer nanocomposite is a composite material in which at least one dimension of the material (usually reinforcements/ fillers) is of nanoscale size, but the final composite material can be either of micro- or macro-scale (Azeredo et al. 2011). Despite nanocomposites being able to protect the packaged food and extend its shelf life, the possibility of migration of nanosized materials into food possesses a serious threat (Song et al. 2011). Thus, the knowledge on interaction and the factors affecting these phenomena are very essential.

2 Interaction

Interaction is a major concern mainly in polymer-based packaging materials. Nearly 42% of the polymers produced are used as packaging materials, and about half of them are used for food packaging applications (Silvestre et al. 2011; Rhim et al. 2013). Polymers are preferred for food packaging applications due to their flexibility, strength, stability, low density, ease of processing, reduced cost and controlled hydrophobicity. But packing of food product in polymers can induce undesirable quality changes in packed food. For instance, degree of browning and ascorbic acid degradation of orange and grapefruit juice was found to be high when packed in polyethylene-laminated cartons than in glass (Mannheim et al. 1987). Interaction is the mass transfer phenomena involving migration (from packaging material to product), permeation (from the outside environment to food contact environment through packaging material) and sorption (adsorption/absorption of food components on the packaging material), either in combination or individually. Interaction of packaging material with the food and the environment plays a major role on quality of the product as well as integrity of the package. These effects have impact on market value, goodwill of the manufacturer, sensory attributes of products and health effect of consumers. Moreover, strict legislations are set for 'zero tolerance' of carcinogenic migrants (Gilbert et al. 1980), and also the compounds that migrate into food are considered as indirect food additive by FDA. Hence, knowledge on factors affecting interaction phenomena and its effect are of high importance. In this section, interaction phenomena like migration, permeation and sorption are discussed in detail.

2.1 *Migration*

Migration mainly denotes diffusion of chemical compounds or polymer residues from the food contact material to the product. The process of migration can be described in four steps: diffusion of substances through polymers, desorption from the surface of polymers, sorption at polymer-food interface and desorption into the food material (Ferrara et al. 2001). According to European and American regulations, migration can be global or specific. The former denotes the overall migrants migrated from the packaging material into the food, whereas the latter denotes the specific substance migrated (Abbes et al. 2015). In general, migration not only happens during packaging but also has the possibility of occurrence during other operations like handling, processing, storage, transportation, preparation and consumption, where the food comes in direct contact with various materials (Castle 2007). Goodson et al. (2004) studied the migration of bisphenol A from the can coating by considering various parameters like type of food (soup, minced beef, evaporated milk, carrot and a food simulant (10% ethanol)), can damage, storage conditions (5 °C (chilled) and 20 °C (ambient) for 9 months and also 40 °C (accelerated) for 3 months). The study concluded that maximum amount (80–100%) of BPA migrated during processing from can coating. And, further migration did not occur in all the kinds of foods stored for the study period. However, migration of BPA (68.3 $\pm$ 9 μg kg^{-1}) into food simulant was comparatively higher than that of other foods tested (soup, 45.7 $\pm$ 5 μg kg^{-1}; minced beef, 53.8 $\pm$ 7.6 μg kg^{-1}; evaporated milk, 49.8 $\pm$ 10.9 μg kg^{-1}; carrot, 47.2 $\pm$ 5.1 μg kg^{-1}). Moreover, damage to the can showed little/no effect on migration of BPA (Goodson et al. 2004).

Migration of compounds from packaging material in small quantities possesses effect on sensory properties of the food, whereas migration in large scale can impose health effects (Risch and Hotchkiss 1991). Effects on sensory properties include mainly colour, taints and off flavours. This can be detected by trained panellist or by the use of instruments like gas chromatography-mass spectrometry (GC-MS) and electronic nose (Tice 1996). Compounds like vinyl polymer monomer used in polyvinyl chloride (PVC) synthesis, monomers of styrene and isocyanates in polyurethane have toxic effects. The level of migration is influenced by many factors such as amount as well as type of additives and residual monomers present, nature of the food, surrounding environment and so on. Table 1 explains some of the potential migrants and their migration at given condition. Several researchers have identified the presence of migrated compounds in food product. For instance, a European survey on the migration of epoxidized soybean oil (ESBO) from PVC gaskets into baby food found that out of 248 samples, 95 samples contained ESBO in the range of 1.5–135.2 mg kg^{-1}. Among these, 14.9% contained ESBO above 31.5 mg kg^{-1} of food, the level calculated on the basis of 1 mg kg^{-1} of body weight (tolerable daily intake recommended by Scientific Committee on Food) (Hammarling et al. 1998; Fantoni and Simoneau 2003). Similarly, Simoneau et al. (2012) conducted a study on migration of chemicals from plastic baby bottles and found that under hot-fill conditions of 2 h at 70 °C, diisobutyl phthalate and dibutyl phthalate migration levels were from 50 to 150 μg kg^{-1}, whereas diethylhexyl phthalate migration level was from 25 to 50 μg kg^{-1} (Simoneau et al. 2012).

Migration can be advantageous in case of active and intelligent packaging (Brown and Williams 2003). For instance, migration of antioxidants, like carvacrol, thymol, astaxanthin, etc., from polymers was reported to enhance the shelf life of packed food products (Colín-Chávez et al. 2013; Ramos et al. 2014). Similarly, migration of antimicrobial agents from polymers was able to protect the food products from microbial contamination and deterioration (Duran et al. 2016).

In general, migration occurs due to difference in concentration of the compounds present in the polymer. Due to high concentration difference, diffusion of compounds from the polymers occurs. Fick's law of diffusion governs the mass diffusion. It is analogous to Fourier's law of heat transfer by conduction. Fick's first law of diffusion states that at steady state, rate of movement of the diffusing compound through a unit cross-sectional area is proportional to the concentration gradient normal to the surface.

Mathematically,

Steady state flux along x:

$$N_A = -D.\frac{\partial C_p}{\partial x} \tag{1}$$

where:

C_p is the concentration of migrating compound present in the polymer.
D is diffusion coefficient.

Negative sign indicates that the diffusion process occurs in the direction opposite to that of concentration gradient (Bhunia et al. 2013; Abbes et al. 2015). Practically, diffusion coefficient is considered to be constant, and in polymers, diffusion coefficient depends on concentration of the compounds, temperature, glass transition temperature, polymer morphology, melting temperature and molecular mass of the migrant (Ferrara et al. 2001).

In general, the food and package interaction occurs in unsteady state condition (Bhunia et al. 2013). Fick's second law is based on unsteady state of diffusion. Mathematically, change in concentration C_p with respect to time can be written as:

Table 1 Potential migrants and their migration at the given condition

Migrants	Polymer	End-use material	Amount migrated	Condition	References
Dibutyl phthalate	Textile packaging	Meat products	11.11 mg kg^{-1}	28 days of storage at 4 °C	Bogdanovičová and Jarošová (2015)
Di-(2-ethylhexyl) phthalate			28.20 mg kg^{-1}		
Diisobutyl phthalate and Dibutyl phthalate	Silicone	Baby bottles	50–150 μg kg^{-1}	Hot-fill condition (70 °C for 2 h)	Simoneau et al. (2012)
Diethylhexyl phthalate			25–50 μg kg^{-1}		
Bisphenol A	Polyamide		0.5–100 μg kg^{-1}		
Toluene	Polyethylene terephthalate (PET) bottle	Soft drinks and juices	26.8–146.7 μg kg^{-1}	Spiked bottle with initial concentration of 297.5 ± 1.2 mg kg^{-1} at 30 days of storage	Franz and Welle (2008)
Chlorobenzene			57–183 μg kg^{-1}	Spiked bottle with initial concentration of 357.8 ± 1.5 mg kg^{-1} at 30 days of storage	
Dibutyl phthalate	PET	Soft drinks preserved with orthophosphoric acid	12.89 μg L^{-1}	30 days of storage at 22 °C	Bosnir et al. (2007)
		Soft drinks preserved with Na-benzoate	21.29 μg L^{-1}		
		Soft drinks preserved with K-sorbate	9 μg L^{-1}		
		Soft drinks preserved with Na-benzoate and K-sorbate	26.75 μg L^{-1}		
		Mineral water	11.33 μg L^{-1}		
Diethylhexyl phthalate		Soft drinks preserved with orthophosphoric acid	17.11 μg L^{-1}		
		Soft drinks preserved with Na-benzoate	15.93 μg L^{-1}		

(continued)

Table 1 (continued)

Migrants	Polymer	End-use material	Amount migrated	Condition	References
		Soft drinks preserved with K-sorbate	36. 60 μg L^{-1}		
		Soft drinks preserved with Na-benzoate and K-sorbate	15 μg L^{-1}		
		Mineral water	8. 78 μg L^{-1}		
Di-(2-ethylhexylexyl) adipate	PVC	Kefalotyri cheese	216.6 ± 11.3 mg kg^{-1}	1 day of storage at 5 °C	Goulas et al. (2000)
			345.4 ± 13.5 mg kg^{-1}	10 days of storage at 5 °C	
		Edam cheese	172.4 ± 10.3 mg kg^{-1}	1 day of storage at 5 °C	
			222.5 ± 6.9 mg kg^{-1}	10 days of storage at 5 °C	
		Feta cheese	49.2 ± 1.8 mg kg^{-1}	1 day of storage at 5 °C	
			133.9 ± 6.1 mg kg^{-1}	10 days of storage at 5 °C	
PET oligomers	PET tray	Lasagne	0.78–1.47 mg kg^{-1}	Cooking for 80 min at 204 °C	Castle et al. (1989)
			0.07–0.18 mg kg^{-1}	Microwave cooking for 3 min at a power level of 600 W	
		French fries	0.27–0.35 mg kg^{-1}	Cooking for 40 min at 204 °C	
	PET roasting bag	Roast beef	0.79–0.82 mg kg^{-1}	Cooking for 90 min at 204 °C	
	Susceptor carton	French fries	1.47–2.73 mg kg^{-1}	Microwave cooking for 3 min at a power level of 600 W	
	Susceptor pad	Pizza	0.43–0.96 mg kg^{-1}	Microwave cooking for 1. 5 min at a power level of 600 W	

$$\frac{\partial C_p}{\partial t} = \frac{\partial}{\partial x}\left(D\frac{\partial C_p}{\partial x}\right) \tag{2}$$

Considering diffusion coefficient to be constant, (2) becomes:

$$\frac{\partial C_p}{\partial t} = D\frac{\partial^2 C_p}{\partial x^2} \tag{3}$$

Similar to diffusion coefficient, partition coefficient (k_p) plays a vital role in polymer-food interface:

$$k_p = \frac{C_s}{C_p} \tag{4}$$

where:

C_S denotes the concentration of migrating compound present in the food.

Partition coefficient values vary with the polarity of the food material that is in contact with the polymer. For fatty foods, k_p is very low (≤ 1), whereas for pure water, k_p is very high ($\gg 1000$) (Piringer 2007). For practical applications, similar to diffusion coefficient, partition coefficient is also considered to be constant. Partition coefficient also depends on the temperature of the food and polymer. Further, temperature also influences diffusion process. Arrhenius-type equation governs the systems that are influenced by temperature (Ferrara et al. 2001). It gives the relationship between temperature and migrant. The equation is as follows (Hernandez and Giacin 1998):

$$D = D_0 \exp\left(-\frac{E_d}{\mathrm{RT}}\right) \tag{5}$$

where:

D_0 is a constant (pre-exponential factor).
E_d represents activation energy for diffusion.
R represents universal gas constant.
T represents absolute temperature.

2.2 *Permeation*

Permeation involves movement of volatile and aroma components, gases, moisture and other low molecular weight substances from the outside environment into food through the packaging material or vice versa. In general, permeation involves three steps: adsorption of the permeant on the polymer surface, diffusion through the polymer and desorption of the permeant from the polymer surface. It mainly causes unbalanced flavour profile, leading to change in sensory properties of the food product. Further, packing of food products with packaging material of effective barrier properties can extend the shelf life of the product stored. For instance, aluminium foil has better barrier properties that can protect the foods packed in them for longer time than other polymers (Lamberti and Escher 2007). Willige et al. (2002) studied the relationship between flavour absorption and oxygen permeability of low-density polyethylene (LDPE), polypropylene (PP), polycarbonate (PC) and polyethylene terephthalate (PET). The study found that flavour absorption had linear positive effect on oxygen permeability of LDPE and PP. It was because of change in network of polymeric material due to swelling after flavour absorption. In case of PC, flavour absorption had linear negative effect on oxygen permeability, due to blockage of microcavities. Moreover, PET did not have any significant effect on oxygen permeability due to flavour absorption (Van Willige et al. 2002).

Similarly, moisture and gas permeation also plays a major role in selection of packaging material for a particular food product. For instance, carbonated beverage packaging requires an efficient gas barrier packaging material. Further, gas permeation is of high importance in modified and controlled atmospheric packaging. In case of fresh fruits and vegetables packing, due to product's respiration and metabolism, the oxygen percentage present inside the pack decreases, leading to anaerobic decomposition. This can be prevented by using permeable membrane. The permeability rate varies with gas composition, temperature and relative humidity. For polymers, oxygen permeates four times slower than that of carbon dioxide (Del-Valle et al. 2003). Also, the permeability of gases increases with increase in temperature. Galić and Ciković (2001) reported the permeability of gases at temperatures from 20 to 60 °C. Further, to investigate the effect of gas permeability with respect to relative humidity of ambient and water activity of polymer, the polymers were subjected to simulants (15% (v/v) ethanol and 3% (v/v) acetic acid solution) prior to analysis. The results found that the permeability of polymers treated with simulants showed higher permeability than that of dry polymers. Zhang et al. (2001) investigated the effect of oxygen and water vapour permeability of ethylene vinyl alcohol (EVOH) copolymers with respect to relative humidity and found that beyond 75% RH, the permeabilities of the film increased.

Permeability of a substance is a quantitative representation of permeant passing through a polymer, and it depends on two main factors: diffusion factors and permeant solubility (Siracusa 2012; Abbes et al. 2015). Mathematically,

Permeability coefficient,

$$P = D.S \tag{6}$$

where D and S represent the diffusion coefficient and solubility coefficient, respectively.

Similar to diffusion process, permeation is also influenced by temperature. Hence, Arrhenius-type equation can be used to represent the relationship (Siracusa 2012). Mathematically:

$$P = P_0 \exp\left(-\frac{E_p}{RT}\right) \quad (7)$$

Here, P_0 represents a constant (pre-exponential factor), and E_p represents activation energy for permeation.

2.3 *Sorption*

Sorption can be adsorption and absorption. Adsorption involves transfer of mass from the food to the surface of the packaging material, resulting in increased concentration of components at the interface (Arvanitoyannis and Kotsanopoulos 2014). In absorption, initially, adsorption takes place, followed by permeation of components into the polymer matrix. In general, sorption of fat and flavour is a major area of concern. As sorption of fat has the possibility to enhance the migration of food components into package as well as chemical compounds from package into food (Lebossé et al. 1997).

Sorption of aromatic components present in food by polymers can result in degradation of quality. This phenomenon of loss in quality of food product by absorption of flavour from food by polymer or vice versa is known as 'scalping' (Sajilata et al. 2007). Scalping of flavour is a major problem to be addressed in beverage industry. Nevertheless, the mere presence of volatile flavours does not lead to scalping. For scalping to occur, a thermodynamically favourable condition must exist (Gnanasekharan and Floros 1997). But the major concern of flavour sorption is that loss in very small amount of flavour has significant effect on quality of the stored food product depending on the component sorbed (Gnanasekharan and Floros 1997). A study on packing of orange juice in glass bottles and polyethylene-laminated cardboard packages found that after 24 weeks of storage at 4 °C, up to 50% of d-limonene and little aldehydes and alcohols were absorbed into polyethylene-laminated cardboard packages. But, this had little/no effect on the sensory properties of the orange juice (Pieper et al. 1992).

At times, sorption can result in swelling of the packaging material, leading to increased migration and permeation. Further, it can reduce the mechanical properties of the polymer (Helmroth et al. 2002; Piringer 2007). For instance, the absorption of limonene on LDPE increased the oxygen permeability of the polymer and that was proportional to the concentration of limonene (Sadler and Braddock 1990; Lebossé et al. 1997).

Flavour absorption depends on characteristics of polymer (polarity, crystallinity, chain stiffness), flavour compounds (concentration, chemical composition, polarity) and environmental and external factors like temperature, relative humidity, duration of storage and composition of food matrix (Caner 2011). For liquid foods, the sorption is mainly affected by partition coefficient of flavour components, whereas for solid foods, the sorption is affected by solubility and vapour pressure of components in the polymer (Kwapong and Hotchkiss 1987; Lebossé et al. 1997).

Similar to migration, sorption is also a diffusion process. Hence, Fick's law governs sorption as well (Gnanasekharan and Floros 1997). Further, sorption process is also influenced by temperature. In general, sorption and temperature are positively related. Exceptionally, sorption and temperature are negatively correlated in few cases. For instance, sorption of vinyl chloride by dry casein particles was found to reduce with increasing temperature (Biran et al. 1979). In sorption process, the quantity of volatile component sorbed by the polymer can be measured by the parameter 'solubility'. Solubility 'S' can be given by the following equation (Siracusa, 2012):

$$S = S_0 \exp\left(-\frac{H_s}{RT}\right) \tag{8}$$

Here,

S_o denotes pre-exponential factor.
H_s represents the heat of sorption.

In other words, solubility coefficient is the amount of volatile component sorbed by a polymer of unit mass under unit partial pressure (Caner 2011).

Mathematically:

$$S = \frac{M}{v.p} \tag{9}$$

where M is the total amount of vapour absorbed; p and v are pressure and volume of the polymer, respectively.

3 Factors Affecting Interaction Phenomena

The extent of the abovementioned interaction phenomena and their impact on the quality of the food material as well as the integrity of the package depends on various factors like characteristics of food material, nature of the packaging material and the additive used, as well as on the storage and handling environment of the package.

3.1 Nature of the Food Material

The effect of interaction varies with the nature of food material, and hence, in most of the studies involving migration and permeation, various food-simulating agents like water, oil, acetic acid, ethanol and hexane (Arvanitoyannis and Kotsanopoulos 2014) are used. A study on migration of nanosilver into various food simulants for a period of 15 days at 50 °C revealed that highest migration of nanosilver was found in hexane, followed by 4% acetic acid (v/v) and ultrapure water. The migration into 95% ethanol (v/v) was found to be the least (Huang et al. 2011). The composition of food matrix also shows significant effect on the absorption of volatiles into packaging material. For instance, Van Willige et al. (2000) studied the effect of various food matrices on the absorption of flavour compounds like limonene, decanal, linalool and ethyl 2-methyl butyrate into linear low-density polyethylene (LLDPE) and proved that composition of food matrix influences the absorption of volatiles. The composition containing fat/oil had major effect on flavour interaction, followed by proteins and polysaccharides and then by disaccharides (van Willige et al. 2000). The interaction of packaging and product also depends on the pH of the food. A study on effect of pH on the migration of toxic metals like lead, cobalt, nickel and zinc from ceramic food packaging materials showed that the migration rate increased with decrease in pH value (Dong et al. 2014)

Further, the nature and concentration of volatile materials present in food also has considerable impact on the interaction. The higher the concentration of the migrating/penetrating/diffusing material, the higher will be the rate of interaction. Also, the presence of similar flavour compounds also influences the interaction (especially sorption) of particular compound because of reduced availability of free site in the polymer. For instance, the degree at which limonene is sorbed into the polymer has comparatively reduced when it is present in the blend of flavoured compounds than individual presence (Halek and Luttmann 1991; Caner 2011). At times, the compatibility of the food material with the polymer also plays a significant role on interaction. For example, when acidic foods come in contact with metal containers, the erosion and, hence, leaching of metals from the container into the food material are accelerated (Castle 2007). Thus, the selection of the packaging material used for packing a product should be based on the nature of the food material. Apart from the factors discussed above, factors like molecular weight of the interacting molecule, its carbon chain length, structure, functional group and polarity also play important roles in determining the rate and extent of interaction (Caner 2011).

3.2 *Nature of the Packaging Material*

Similar to the characteristics of food packed, the nature of the packaging material also has considerable impact on the interaction. The packaging material used can be impermeable, permeable or porous. In case of impermeable materials like glass and metals, the migration occurs only at the surface of the material, whereas in case of permeable materials like plastics and rubbers, the migration occurs both on surface and from the inner side of the material. But these materials offer some amount of resistance. Porous materials including paper and boards facilitate rapid migration (Castle 2007). The rate of interaction between polymers and the food material depends on factors like molecular weight, density, additive present, crystallinity, glass transition temperature and so on. An experiment on migration of carbon black found that the migration rates in LDPE and polystyrene (PS) decreased by a factor of

10 (approx.), when the molecular weights were doubled (Bott et al. 2014). In case of PP, the migration rate increases with decrease in degree of crystallinity (Alin and Hakkarainen 2010). And highly crystalline polymers like HDPE and PP show comparatively lesser absorptivity than that of other polymers like LDPE and surlyn (Charara et al. 1992). Besides, the polymers with lower glass transition temperatures show higher permeability as well as high diffusion coefficients. The affinity of the polymer toward the interacting molecule also influences interaction. For instance, packing of non-polar materials like oils and fats in polyolefins is not advisable because of its lipophilic nature (Caner 2011). Similarly, the rate of sorption also decreases with increase in density of the polymer.

Other than the polymer itself, the nature of additives like colourants, plasticizers, stabilizers, fillers, blowing agents, antioxidants, antimicrobials, slip compounds and printing ink also migrates into the food material, and their characteristics impact the rate of migration (Guart et al. 2011).

3.3 *Environmental and Storage Factors*

The environment in which the product along with the package is placed has a major impact on the interaction phenomena. As discussed in previous sections, the interaction is influenced due to temperature, and thus according to Arrhenius-type equations with increase in temperature, the rate of interaction increases. Owing to increase in temperature, free volume, mobility of the interacting material present in the food as well as the polymer, degree of swelling of the polymer increases, whereas degree of crystallinity decreases, and also relaxation of polymer chain occurs (Caner 2011). A study was conducted on the effect of temperature during processing and temperature and time of storage on bisphenol A migration from can containing organosol and epoxy coatings to a fatty food simulant (sunflower oil) and tuna in oil. The study found that maximum migration level of BPA from organosol can coating into fatty food simulant (403.6 ± 52.9 μg/kg) and tuna in oil (27.4 ± 11.3 μg/kg) occurred at heat processed cans stored 25 °C for a period of 160 days. The rate of migration gradually decreased with increase in storage time for heat processed cans. Also, non-heat processed cans (186.1 ± 32.2 μg/kg) showed slower migration rate than that of heat processed cans (403.6 ± 52.9 μg/kg) stored at same temperature (25 °C) and time (160 days) (Munguia-Lopez et al. 2005). Similarly, a study on migration of nanosilver into different food-simulating solutions (water, hexane, 95% ethanol (v/v) and 4% acetic acid (v/v)) found that irrespective of the simulating solution adopted, the amount of nanosilver migration increased gradually with time and temperature (Huang et al. 2011).

Another important factor to be considered is relative humidity. The effect of humidity on the interaction varies with the polymer used. For instance, absorption of water in polymers like EVOH increases the permeability, whereas in case of polymers like PET and nylon, it decreases oxygen permeability. However, in a few cases like polyolefins, humidity has no significant effect on the permeability of the polymer (Linssen et al. 2003). A study on the effect of relative humidity on the release of antimicrobial agent (carvacrol) from EVOH coating on PP films found that diffusion coefficient increased with increase in humidity (Cerisuelo et al. 2012).

Sometimes, effects of interaction phenomena are inter-dependent. For instance, LDPE and PP showed increased oxygen permeability by 21% and 130%, respectively, after 8 h of exposure to flavours, whereas in the case of PC, the oxygen permeability decreased by 11% after 21 days of exposure. In case of PET, no significant effect was observed. Thus, the study proved that flavour absorption had effect on oxygen permeability of some polymers (Van Willige et al. 2002).

4 Effects of Interaction

In general, interaction phenomena result in mass transfer of any of the component either from the packaging material to food or vice versa. It also occurs between the outside environment and food through packaging material. Hence, there occurs a change in composition of the food material, the packaging material and the environment. However, change in environment is negligible, and it is not considered. This section deals about the effects of interaction phenomena with the packaging material as well as the product and the health effects due to consumption of interacted food.

4.1 On Polymers

The main effect of interaction on the polymeric material leads to change in structure and nature of the polymeric material. This may occur in one or many of the following ways:

- Migration of residual monomers or oligomers into the food material
- Migration of additives into the food material
- Swelling of polymeric material due to absorption of components
- Alteration in polymeric network due to absorption of components
- Change in permeability behaviour of polymeric material
- Development of voids due to migration
- Filling of microcavities or void spaces in the polymer by the sorbed components
- Relaxation of polymer linkages
- Change in crystallinity of the polymeric material due to migration
- Shift in glass transition temperature of the polymeric material due to migration

4.2 On Product

Change in sensory properties of the food material is one of the main problems associated with the interaction. Sensory changes may occur due to direct absorption of volatile compounds into the polymeric material or through indirect changes which occurred due to chemical reactions like oxidation and hydrolysis that has occurred because of increased permeability and so on (Duncan and Webster 2009). Flavour scalping in small amount has significant effect on the organoleptic properties of food. However, the nature of flavour absorbed plays a major role in alteration of sensory properties. For instance, d-limonene does not have much effect on the flavour of orange juice. Hence, the absorption of d-limonene to a level of 40% did not significantly affect the orange juice quality. Additionally, d-limonene absorption leads to prevent the formation of off-flavour compounds like α-terpineol (Durr et al. 1981). Another study conducted on absorption of aroma compounds like lemon and orange into LDPE found that sensory panellists were able to identify the significant difference in intensity of aroma (Kwapong and Hotchkiss 1987). To balance the lost favour due to scalping, manufacturers often add additional flavour, so that the end product reaching consumers is acceptable in terms of flavour (Lebossé et al. 1997).

Apart from organoleptic changes, the interaction of food components with packaging material and environment may lead to detrimental effect on the quality of the food. For instance, enhanced permeation may result in oxidation of food, thus reducing the shelf life of the packed food. Mannheim and his team found that citrus juices packed in laminated cartons had limited shelf life than those packed in glass jars (Mannheim et al. 1987).

In cases like packaging material containing antimicrobial or antioxidant agents, interaction can be beneficial. A study on migration of antioxidant (α-tocopherol) from a multilayer packaging material containing high-density polyethylene (HDPE), EVOH, LDPE layers and α-tocopherol into whole milk powder found that antioxidant delivery reduced the rate of lipid oxidation and, hence, extended the shelf life of whole milk powder (Granda-Restrepo et al. 2009).

4.3 Health Effects

Though packaging effectively maintains the safety and nutrition of the food, the use of polymers can lead to negative impact on the health of the consumers. This may be because of migration of undesirable and hazardous components from the polymers into the food material. The effects can be due to exposure for a longer period of time or shorter time period, depending on the component. Hence, the toxicity of the migrating substance is to be considered. Based on the extent of harmfulness, several legislations have set the permissible limit for the migrating substances to be present. Some compounds migrated into food are extremely toxic to consumers. For instance, long-term exposure to BPA (Fig. 1) is dangerous as it mimics female hormone

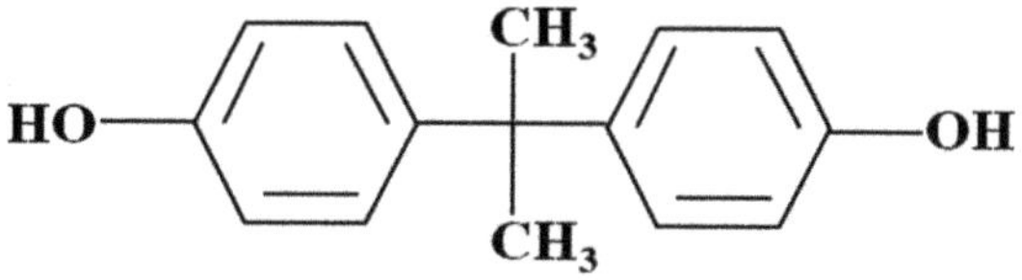

Fig. 1 Chemical structure of Bisphenol A

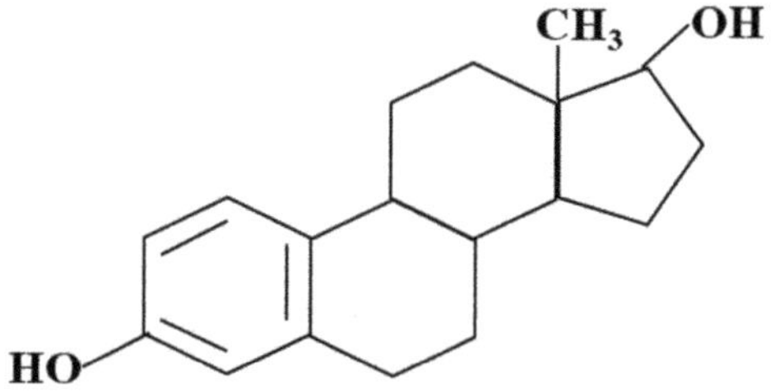

Fig. 2 Chemical structure of oestradiol (an oestrogen)

oestrogens (Fig. 2) and it is a potential endocrine disrupter (Thomson and Grounds 2005). Also, fluorinated organic chemicals like perfluorooctanesulfonate (PFOS) and perfluorocarboxylates (PFCs) were found to cause toxicological effect on monkeys and rodents (Kumar 2005).

5 Testing Methods

Testing the quality as well as the quantity of the interacting substance is very important for the manufacturer to ensure that the material meets the standard and also to make sure that packaging material has negligible effect on the final product reaching the consumer and it does not cause any negative impact on health after consumption of the packed food material. To detect and analyse the amount of components present, simulants, mathematical and predictive models and chromatographic techniques are used.

5.1 Simulants

In general, foods are classified into aqueous, acidic, alcoholic, fat-based and dry foods (Castle 2007). Hence, based on the nature of the food material, various simulants are used to mimic the environment. The simulants recommended by US Food and Drug Administration (US FDA) and European Union (EU) for different types of food are illustrated in Table 2 (Castle 2007; Bhunia et al. 2013). These simulants are selected accordingly, and then overall migration or specific migration is calculated by exposing the packaging material to the simulants for the recommended time and temperature (Table 3) (FDA 2007). Overall migration is calculated by evaporating the simulants and then calculating the weight of the

Table 2 Types of food and food simulants recommended by FDA and EU

Type of food	Simulant: Recommended by US FDA	Simulant: Recommended by EU
Aqueous	10% ethanol (v/v)	10% ethanol (v/v)
Acidic		3% acetic acid (w/v)
Low alcohol		20% ethanol (v/v)
High alcohol	50% ethanol (v/v)	
Fat-based	Food oil, HB307 (a product containing mixture of synthetic triglycerides), Miglyol 812 (coconut oil derivative)	50% ethanol and vegetable oil
Dry	Tenax (modified polyphenylene oxide)	Tenax

Table 3 Migration protocols for short-term accelerated testing as recommended by FDA

Intended end-use condition	Simulated condition
High temperature, heat sterilized or retorted	121 °C for 2 h, followed by 40 °C for 238 h
Boiling water sterilized	100 °C for 2 h, followed by 40 °C for 238 h
Hot-filled or pasteurized above 66 °C	100 °C for 30 min, followed by 40 °C for 10 days (or) 66 °C for 2 h, followed by 40 °C for 238 h
Hot-filled or pasteurized below 66 °C	66 °C for 30 min, followed by 40 °C for 10 days (240 days)
Room temperature filled and stored	40 °C for 10 days
Refrigerated storage	20 °C for 10 days
Frozen storage	20 °C for 5 days

remaining residues. This method is limited in case of fatty simulants, because of difficulty in vaporization of simulant. Hence, the loss in weight of the packaging material, followed by subtracting with the amount of fat sorbed gives the overall migration. Specific migration is calculated by mathematical models or analytical techniques. In case, if overall migration value or the total amount of the particular component present in the polymer is less than the maximum permissible limit, then the need for specific migration test is neglectable (Veraart and Coulier 2007).

Unlike other components, the interaction of a flavour with polymers cannot be studied by food simulants. In general, diluted aqueous solution of particular flavour is used to simulate the environment. For instance, to mimic the flavour of citrus products, d-limonene is used (Gnanasekharan and Floros 1997). To test the transfer of odour and taste from packaging materials, ASTM (American Society for Testing and Materials) standard test methods are adopted. According to the ASTM standard, 0.9 m^2 of test material is placed in a required environment for a minimum of 20 h (Tice 1996).

5.2 Analytical Instruments

The quantification of interacted compounds can be done through mathematical models as well as chromatographic techniques. Application of chromatographic techniques on interaction phenomena directly involving food stuffs is limited than that related to simulants and packaging material. Though some studies have been carried out on the detection of interacted components, there are no standard methods available for analysis of the most of the migrating substances. Also, some migrants like perfluorochemicals cannot be detected by conventional techniques like GC-MS or LC-UV. Begley et al. (2005) adopted LC-MS technique to characterize perfluorochemicals. Table 4 describes some of the commonly found interacting components and the method of detection adopted.

5.3 Mathematical Models

As the experiments are time-consuming and laborious and involve high cost, several mathematical models have been developed to determine diffusion coefficients as well as partition coefficients and also to predict interacting substance concentration (Lau and Wong 1997; Mousavi et al. 1998; Helmroth et al. 2002; Samsudin et al. 2018). A model was developed for the design optimization of the package containing antimicrobial agents. The objective of the model was to make sure the presence of the optimum concentration of the antimicrobial agent on the headspace, so as to prevent the growth of selected microorganisms. The study also claimed that the same model can be used to similar packaging systems also (Cerisuelo et al. 2012). Moreover, Price and Chaudhry (2014) developed a computational model to estimate the toxicity of compounds migrated from the polymer into food material (Price and Chaudhry 2014).

Apart from the testing methods described above, permeation of a particular packaging material can be calculated in terms of gas and moisture transmission rate. The permeability properties of the material depend on film characteristics like thickness, as well as on the surrounding environment. Gas transmission rate is tested by placing the packaging material in between a low-pressure and high-pressure environment, followed by measurement of pressure difference in case of manometric method and volume difference in case of volumetric method. Gas transmission rate is expressed in cc of specific gas permeated per unit area per unit time. Similarly, moisture transmission rate of a packaging material is tested by placing the material in between an environment of two different humidities. Due to vapour pressure difference, permeation of water vapour takes place, and this is calculated in g of moisture permeated m^{-2} day^{-1} (Huang and Qian 2008).

Table 4 Techniques adopted for the detection of selected interacting components

Interacting component	Polymer	Technique adopted	References
Erucamide, Irgafos 168, Oxidized Irgafos 168, Irganox 1076 and Irganox 1010	Multilayered packaging	UHPLC-VWD-(TOF) MS	Moreta and Tena (2015)
Irganox 1330, 1010, 3114 and 1076, Tinuvin 326, 327 and 328, Erucamide	Polypropylene, poly-cycloolefin, copolyester	UHPLC-MS/MS	(Pouech et al. (2014)
Antioxidants agents (carvacrol and thymol)	Polypropylene	GC-MS, HPLC-UV	Ramos et al. (2014)
Adhesives (hot melt, vinyl acetate, starch, polyvinyl acetate and acrylic)	Multilayered laminates	GC-O-MS	Vera et al. (2014)
Perfluorinated compounds	Paper, paperboard, aluminium foil	LC-MS/MS	Zafeiraki et al. (2014)
Perfluorochemicals	Polytetrafluoroethylene	LC-MS	Begley et al. (2005)
Phenol, Bisphenol A, 2,4-di-tert-butylphenol, Cyasorb UV511, bis (2-ethylhexylphthalate), Irganox 1076, Irgafos 168	Polycarbonate containers	HPLC-UV-fluorescence	Nerín et al. (2003)
Irganox 245, Irganox 1035, Chimmasorb 81, Uvitex OB	Blends of food contact polymers like polystyrenes, polyolefins	HPLC-UV-Fluorescence	Quinto-Fernández et al. (2003)
Anthracene, Benzophenone, Methyl stearate, Pentachlorophenol	From recycled paperboard through polypropylene	GC-FID	Song et al. (2003)
Caprolactam and its cyclic oligomers	Food grade nylon 6	HPLC-UV (Normal and reverse phase)	Barkby and Lawson (1993)
Irganox 1010	Polypropylene	H-NMR[a]	Ehret-Henry et al. (1992)
ε-Caprolactam and its dimer-monomer separation	–	HPLC-RI	Bonifaci et al. (1991)

[a]H-NMR Proton nuclear magnetic resonance; *GC-FID* gas chromatography with flame ionization detector, *GC-MS* gas chromatography-mass spectrometry, *GC-O-MS* gas chromatography-olfactometry-mass spectrometry, *HPLC-RI* high-performance liquid chromatography with refractive index detectors, *HPLC-UV* high-performance liquid chromatography with ultraviolet detectors, *HPLC-UV-Fluorescence* high-performance liquid chromatography with both ultraviolet and fluorescence detectors, *LC-MS* liquid chromatography-mass spectrometry, *LC-MS/MS* liquid chromatography-tandem mass spectrometry, *UHPLC-MS/MS* ultra-high-performance liquid chromatography coupled to a tandem mass spectrometer, *UHPLC-VWD-(TOF) MS* ultra-high-performance liquid chromatography coupled to variable wavelength and time of flight mass spectrometry

6 Regulations

The selection of packaging material for a food should be in such a way that the migration is limited and within the limits set by the legislative bodies. In general, FDA and European Commission (EC) bodies implement threshold policy as well as EC directives, respectively, to set the limits in a way that the compounds on consumption do not create any harm to the consumer (Arvanitoyannis and Kotsanopoulos 2014). The threshold level is set considering the dietary concentration as well as the toxic potency of the harmful component. Regulations are set for labelling as well to mention the possibility of migration of allergens like latex from cold seal adhesives. This is because these allergens may cause serious health hazards in certain individuals (Castle 2007). In case of FDA, the food contact materials should be GRAS (generally recognized as safe) approved. If not, the material and/or the additive migrated from the material should prove to be safe. The level of migration limit varies with the nature of the additive. For instance, according to FDA, carcinogenic compounds if present less than 0.5 $\mu g\ kg^{-1}$a possess a risk of less than one in a million lives (Arvanitoyannis and Kotsanopoulos 2014). Thus, the migration of the compounds from the packaging material at their intended use condition should be less than the maximum permissible limit set by these governing bodies, for the safety of the consumers.

7 Conclusion

In summary, as different kinds of packaging materials have evolved for food applications, interaction of the package and the product is one of the most important properties to be considered. The knowledge on these interaction phenomena is of significant importance in all the stages involving food and the packaging material, as it can lead to unsafe and below standard food product to the consumer. Further, clear understanding on the role of the storage and handling environment as well as the nature of the food and packaging material on the interaction enables to ensure the safety of the product. Besides, the testing of the materials in the intended environment gives the level of interaction, and the manufacturers and handlers should ensure that the level does not exceed the maximum permissible limit set by the legislative bodies. Currently, the legislative bodies like FDA and EC are exploring more on the toxicity and harmfulness of the unintentional additives in the packaging material.

References

Abbes B, Abbes F, Guo Y-Q (2015) Interaction phenomena between packaging and product. In: Alavi S, Thomas S, Sandeep K et al (eds) Polymers for packaging applications, 1st edn. Apple Academic Press, New Jersey, pp 39–70

Alin J, Hakkarainen M (2010) Type of polypropylene material significantly influences the migration of antioxidants from polymer packaging to food simulants during microwave heating. J Appl Polym Sci n/a-n/a. https://doi.org/10.1002/app.32472
Arvanitoyannis IS, Kotsanopoulos KV (2014) Migration phenomenon in food packaging. Food-package interactions, mechanisms, types of migrants, testing and relative legislation-a review. Food Bioprocess Technol 7:21–36. https://doi.org/10.1007/s11947-013-1106-8
Azeredo H, Mattoso L, McHugh T (2011) Nanocomposites in food packaging – a review. In: Reddy B (ed) Advances in diverse industrial applications of nanocomposites. InTech, p 57–78
Barkby CT, Lawson G (1993) Analysis of migrants from nylon 6 packaging films into boiling water. Food Addit Contam 10:541–553. https://doi.org/10.1080/02652039309374177
Begley TH, White K, Honigfort P et al (2005) Perfluorochemicals: potential sources of and migration from food packaging. Food Addit Contam 22:1023–1031. https://doi.org/10.1080/02652030500183474
Bhunia K, Sablani SS, Tang J, Rasco B (2013) Migration of chemical compounds from packaging polymers during microwave , conventional heat treatment, and storage. Compr Rev Food Sci Food Saf 12:523–545. https://doi.org/10.1111/1541-4337.12028
Biran D, Giacin JR, Hayakawa K, Gilbert SG (1979) Vinylchloride sorption by dry casein particles: mechanistic considerations. J Food Sci 44:59–61
Bogdanovičová S, Jarošová A (2015) Phthalate migration from packaging materials into food. Potravin Slovak J Food Sci 9:275–279
Bonifaci L, Frezzotti D, Cavalca G et al (1991) Analysis of epsilon-caprolactam and its cyclic oligomers by high-performance liquid chromatography. J Chromatogr 585:333–336
Bosnir J, Puntaric D, Galic A et al (2007) Migration of phthalates from plastic containers into soft drinks and mineral water. Food Technol Biotechnol 45:91–95
Bott J, Störmer A, Franz R (2014) Migration of nanoparticles from plastic packaging materials containing carbon black into foodstuffs. Food Addit Contam A 31:1769–1782. https://doi.org/10.1080/19440049.2014.952786
Brown H, Williams J (2003) Packaged product quality and shelf life. In: Richard C, McDowell D, Kirwan MJ (eds) Food packaging technology. Blackwell Publishing Ltd, Hoboken, pp 65–94
Caner C (2011) Sorption phenomena in packaged foods: factors affecting sorption processes in package-product systems. Packag Technol Sci 24:259–270. https://doi.org/10.1002/pts
Castle L (2007) Chemical migration into food: an overview. In: Barnes KA, Sinclair CR, Watson DH (eds) Chemical migration and food contact materials, 1st edn. CRC Press, New York, pp 1–13
Castle L, Mayo A, Crews C, Gilbert J (1989) Migration of poly (ethylene terephthalate) (PET) oligomers from PET plastics into foods during microwave and conventional cooking and into bottled beverages. J Food Prot 52:337–342
Cerisuelo JP, Muriel-galet V, Bermúdez JM et al (2012) Mathematical model to describe the release of an antimicrobial agent from an active package constituted by carvacrol in a hydrophilic EVOH coating on a PP film. J Food Eng 110:26–37. https://doi.org/10.1016/j.jfoodeng.2011.12.013
Charara ZN, Williams JW, Schmidt RH, Marshall MR (1992) Orange flavor absorption into various polymeric packaging materials. J Food Sci 57:963–968. https://doi.org/10.1111/j.1365-2621.1992.tb14334.x
Coles R (2003) Introduction. In: Richard C, McDowell D, Kirwan MJ (eds) Food packaging technology. Blackwell Publishing Ltd, Hoboken, pp 1–31
Colín-Chávez C, Soto-Valdez H, Peralta E et al (2013) Diffusion of natural astaxanthin from polyethylene active packaging films into a fatty food simulant. Food Res Int 54:873–880. https://doi.org/10.1016/j.foodres.2013.08.021
Del-Valle V, Almenar E, Lagarón JM et al (2003) Modelling permeation through porous polymeric films for modified atmosphere packaging. Food Addit Contam 20:170–179. https://doi.org/10.1080/0265203021000042869

Dong Z, Lu L, Liu Z et al (2014) Migration of toxic metals from ceramic food contact materials into acid food. Math Probl Eng:1–7. https://doi.org/10.1002/pts.2122

Duncan SE, Webster JB (2009) Sensory impacts of food-packaging interactions. In: Taylor SL (ed) Advances in Food and Nutrition Research, Volume 56. Elsevier, Amsterdam, pp 17–64

Duran M, Aday MS, Zorba NND et al (2016) Potential of antimicrobial active packaging "containing natamycin, nisin, pomegranate and grape seed extract in chitosan coating" to extend shelf life of fresh strawberry. Food Bioprod Process 98:354–363. https://doi.org/10.1016/j.fbp.2016.01.007

Durr P, Schobinger U, Waldvogel R (1981) Aroma quality of orange juice after filling and storage in soft packages and glass bottles. Alimenta 20:91

Ehret-Henry J, Bouquant J, Scholler D et al (1992) 1H-NMR for the safety control of food packaging materials: analysis of extracts from polyolefin samples. Food Addit Contam 9:303–314. https://doi.org/10.1080/02652039209374076

Fantoni L, Simoneau C (2003) European survey of contamination of homogenized baby food by epoxidized soybean oil migration from plasticised PVC gaskets. Food Addit Contam 20 (11):1087–1096. https://doi.org/10.1080/02652030310001615186

FDA (2007) U.S. food and drug administration. https://www.fda.gov/Food/GuidanceRegulation/ucm081818.htm. Accessed 13 Dec 2017

Ferrara G, Bertoldo M, Scoponi M, Ciardelli F (2001) Diffusion coefficient and activation energy of Irganox 1010 in poly(propylene-co-ethylene) copolymers. Polym Degrad Stab 73:411–416. https://doi.org/10.1016/S0141-3910(01)00121-5

Franz R, Welle F (2008) Migration measurement and modelling from poly (ethylene terephthalate) (PET) into soft drinks and fruit juices in comparison with food simulants. Food Addit Contam Part A 25:1033–1046. https://doi.org/10.1080/02652030701837381

Galić K, Ciković N (2001) Permeability characterisation of solvent treated polymer materials. Polym Test 20:599–606. https://doi.org/10.1016/S0142-9418(00)00081-7

Gilbert SG, Miltz J, Giacin JR (1980) Transport considerations of potential migrants from food packaging materials. J Food Process Preserv 4:27–49. https://doi.org/10.1111/j.1745-4549.1980.tb00594.x

Gnanasekharan V, Floros JD (1997) Migration and sorption phenomena in packaged foods migration and sorption phenomena in packaged foods. Crit Rev Food Sci Nutr 37:519–559

Goodson A, Robin H, Summerfield W, Cooper I (2004) Migration of bisphenol a from can coatings – effects of damage, storage conditions and heating. Food Addit Contam 21:1015–1026. https://doi.org/10.1080/02652030400011387

Goulas AE, Anifantaki KI, Kolioulis DG, Kontominas MG (2000) Migration of di- (2-ethylhexylexyl) Adipate plasticizer from food-grade polyvinyl chloride film into hard and soft cheeses. J Dairy Sci 83:1712–1718. https://doi.org/10.3168/jds.S0022-0302(00)75040-5

Granda-Restrepo DM, Soto-Valdez H, Peralta E et al (2009) Migration of α-tocopherol from an active multilayer film into whole milk powder. Food Res Int 42:1396–1402. https://doi.org/10.1016/j.foodres.2009.07.007

Guart A, Bono-Blay F, Borrell A, Lacorte S (2011) Migration of plasticizers, phthalates, bisphenol A and alkylphenols from plastic containers and evaluation of risk. Food Addit Contam A 28:676–685. https://doi.org/10.1080/19440049.2011.555845

Halek GW, Luttmann JP (1991) Sorption behaviour of citrus-flavor compounds in polyethylenes and polypropylenes. In: Risch SJ, Hotchkiss JH (eds) Food and packaging interactions II. American Chemical Society, Washington, DC, pp 212–226

Hammarling L, Gustavsson H, Svensson K et al (1998) Migration of epoxidized soya bean oil from plasticized pvc gaskets into baby food. Food Addit Contam 15:203–208. https://doi.org/10.1080/02652039809374631

Helmroth E, Rijk R, Dekker M, Jongen W (2002) Predictive modelling of migration from packaging materials into food products for regulatory purposes. Trends Food Sci Technol 13:102–109

Hernandez RJ, Giacin JR (1998) Factors affecting permeation, sorption, and migration processes in package-product systems. In: Taub IA, Singh RP (eds) Food storage stability, 1st edn. CRC Press, New York

Huang J, Qian X (2008) Comparison of test methods for measuring water vapor permeability of fabrics. Text Res J 78:342–352. https://doi.org/10.1177/0040517508090494

Huang Y, Chen S, Bing X et al (2011) Nanosilver migrated into food – simulating solutions from commercially available food fresh containers. Packag Technol Sci 24:291–297. https://doi.org/10.1002/pts

Kumar KS (2005) Fluorinated organic chemicals: a review. Res J Chem Environ 9:50–79

Kwapong OY, Hotchkiss JH (1987) Comparative sorption of aroma compounds by polyethylene and lonomer food-contact plastics sensory evaluation. J Food Sci 52:761–763

Lamberti M, Escher F (2007) Aluminium foil as a food packaging material in comparison with other materials. Food Rev Int 23:407–433. https://doi.org/10.1080/87559120701593830

Lau O-W, Wong S-K (1997) Mathematical model for the migration of plasticisers from food contact materials into solid food. Anal Chim Acta 347:249–256. https://doi.org/10.1016/S0003-2670(97)00175-X

Lebossé R, Ducruet V, Feigenbaum A (1997) Interactions between reactive aroma compounds from model citrus juice with polypropylene packaging film. J Agric Food Chem 45:2836–2842. https://doi.org/10.1021/jf960957e

Linssen JP, van Willige RW, Dekker M (2003) Packaging-flavour interactions. In: Ahvenainen R (ed) Novel food packaging techniques, 1st edn. CRC Press, New York, pp 144–171

Mannheim CH, Miltz J, Letzter A (1987) Interaction between polyethylene laminated cartons and aseptically packed citrus juices. J Food Sci 52:737–740

Moreta C, Tena M (2015) Determination of plastic additives in packaging by liquid chromatography coupled to high resolution mass spectrometry. J Chromatogr A 1414:77–87. https://doi.org/10.1016/j.chroma.2015.08.030

Mousavi SM, Desobry S, Hardy J (1998) Mathematical modelling of migration of volatile compounds into packaged food via package free space. Part I: cylindrical shaped food. J Food Eng 36:453–472. https://doi.org/10.1016/S0260-8774(98)00059-4

Munguia-Lopez EM, Gerardo-Lugo S, Peralta E et al (2005) Migration of bisphenol A (BPA) from can coatings into a fatty-food simulant and tuna fish. Food Addit Contam 22:892–898. https://doi.org/10.1080/02652030500163674

Nerín C, Fernández C, Domeño C, Salafranca J (2003) Determination of potential migrants in polycarbonate containers used for microwave ovens by high-performance liquid chromatography with ultraviolet and fluorescence detection. J Agric Food Chem 51:5647–5653. https://doi.org/10.1021/jf034330p

Othman SH (2014) Bio-nanocomposite materials for food packaging applications: types of biopolymer and nano-sized filler. Agric Agric Sci Proc 2:296–303. https://doi.org/10.1016/j.aaspro.2014.11.042

Pieper G, Borgudd L, Ackermann P, Fellers P (1992) Absorption of aroma volatiles of orange juice into laminated carton packages did not affect sensory quality. J Food Sci 57:1408–1411. https://doi.org/10.1111/j.1365-2621.1992.tb06870.x

Piringer O (2007) Mathematical modelling of chemical migration from food contact materials. In: Barnes KA, Sinclair CR, Watson DH (eds) Chemical engineering journal, 1st edn. CRC Press, New York, pp 180–202

Pouech C, Lafay F, Wiest L et al (2014) Monitoring the extraction of additives and additive degradation products from polymer packaging into solutions by multi-residue method including solid phase extraction and ultra-high performance liquid chromatography-tandem mass spectrometry analysis. Anal Bioanal Chem 406:1493–1507. https://doi.org/10.1007/s00216-013-7551-4

Price N, Chaudhry Q (2014) Application of in silico modelling to estimate toxicity of migrating substances from food packaging. Food Chem Toxicol 71:136–141. https://doi.org/10.1016/j.fct.2014.05.022

Quinto-Fernández EJ, Pérez-Lamela C, Simal-Gándara J (2003) Analytical methods for food-contact materials additives in olive oil simulant at sub-mg kg-1 level. Food Addit Contam 20:678–683. https://doi.org/10.1080/0265203031000114219

Ramos M, Beltrán A, Peltzer M et al (2014) Release and antioxidant activity of carvacrol and thymol from polypropylene active packaging films. LWT – Food Sci Technol 58:470–477. https://doi.org/10.1016/j.lwt.2014.04.019

Rhim J, Park H, Ha C (2013) Bio-nanocomposites for food packaging applications. Prog Polym Sci 38:1629–1652. https://doi.org/10.1016/j.progpolymsci.2013.05.008

Risch SJ, Hotchkiss JH (eds) (1991) Food and packaging interactions II, ACS symposium. American Chemical Society, Washington, DC

Robertson GL (2013) Food packaging: principles and practice, 3rd edn. CRC Press, New York

Sadler GD, Braddock RJ (1990) Oxygen permeability of low density polyethylene as a function of limonene absorption: an approach to modeling flavor “Scalping”. J Food Sci 55:587–588. https://doi.org/10.1111/j.1365-2621.1990.tb06826.x

Sajilata MG, Savitha K, Singhal RS, Kanetkar VR (2007) Scalping of flavors in packaged foods. Compr Rev Food Sci Food Saf 6:17

Samsudin H, Auras R, Mishra D et al (2018) Migration of antioxidants from polylactic acid films: a parameter estimation approach and an overview of the current mass transfer models. Food Res Int 103:515–528

Silvestre C, Duraccio D, Cimmino S (2011) Food packaging based on polymer nanomaterials. Prog Polym Sci 36:1766–1782. https://doi.org/10.1016/j.progpolymsci.2011.02.003

Simoneau C, Van den Eede L, Valzacchi S (2012) Identification and quantification of the migration of chemicals from plastic baby bottles used as substitutes for polycarbonate. Food Addit Contam – Part A Chem Anal Control Expo Risk Assess 29:469–480. https://doi.org/10.1080/19440049.2011.644588

Siracusa V (2012) Food packaging permeability behaviour: a report. Int J Polym Sci. https://doi.org/10.1155/2012/302029

Song YS, Begley T, Paquette K, Komolprasert V (2003) Effectiveness of polypropylene film as a barrier to migration from recycled paperboard packaging to fatty and high-moisture food. Food Addit Contam 20:875–883. https://doi.org/10.1080/02652030310001597592

Song H, Li B, Lin QB et al (2011) Migration of silver from nanosilver-polyethylene composite packaging into food simulants. Food Addit Contam – Part A 28:1758–1762. https://doi.org/10.1080/19440049.2011.603705

Thomson BM, Grounds PR (2005) Bisphenol A in canned foods in New Zealand: an exposure assessment. Food Addit Contam 22:65–72. https://doi.org/10.1080/02652030400027920

Tice P (1996) Packaging materials as a source of taints. In: Saxby MJ (ed) Food taints and off-flavours. Springer Science and Businees Media, Berlin, pp 226–263

van Willige RW, Linssen JP, Voragen AG (2000) Influence of food matrix on absorption of flavour compounds by linear low-density polyethylene: proteins and carbohydrates. J Sci Food Agric 80:1779–1789. https://doi.org/10.1002/1097-0010(20000915)80:12<1779::AID-JSFA726>3.0.CO;2-F

Van Willige RWG, Linssen JPH, Meinders MBJ et al (2002) Influence of flavour absorption on oxygen permeation through LDPE, PP, PC and PET plastics food packaging. Food Addit Contam 19:303–313. https://doi.org/10.1080/02652030110081146

Vera P, Canellas E, Nerín C (2014) Migration of odorous compounds from adhesives used in market samples of food packaging materials by chromatography olfactometry and mass spectrometry (GC – O – MS). Food Chem 145:237–244

Veraart R, Coulier L (2007) Compliance testing of chemical migration from food contact materials. In: Barnes KA, Sinclair CR, Watson DH (eds) Chemical migration and food contact materials. CRC Press, New York, pp 87–121

Vieira MGA, Da Silva MA, Dos Santos LO, Beppu MM (2011) Natural-based plasticizers and biopolymer films: a review. Eur Polym J 47:254–263. https://doi.org/10.1016/j.eurpolymj.2010.12.011

Youssef AM (2013) Polymer nanocomposites as a new trend for packaging applications. Polym Plast Technol Eng 52:635–660. https://doi.org/10.1080/03602559.2012.762673

Zafeiraki E, Costopoulou D, Vassiliadou I et al (2014) Determination of perfluorinated compounds (PFCs) in various foodstuff packaging materials used in the Greek market. Chemosphere 94:169–176

Zhang Z, Britt IJ, Tung MA (2001) Permeation of oxygen and water vapor through EVOH films as influenced by relative humidity. J Appl Polym Sci 82:1866–1872. https://doi.org/10.1002/app.2030

Testing Methods for Packaging Materials

Heera Jayan, J. A. Moses, and C. Anandharamakrishnan

Abstract An extensive knowledge about the packaging material is crucial in designing of packaging material for particular products. Properties of packaging materials such as mechanical strength, gas permeability, and sealing ability determine the suitability of packaging material for intended purpose. Thus, it is necessary to determine the properties of packaging material to select the suitable material for any product. Testing of packaging materials generally involves determination of physical properties. But it is also important to study the chemical stability of a packaging material, particularly when it would be used for food-based applications, so as to ensure the safety of the product. There are several standard testing protocols for packaging materials, and this chapter provides detailed information about testing methods that are in use to evaluate various properties of packaging materials. Further, some methods are specific to the type of packaging material, and these are also covered in this chapter.

Keywords Properties of packaging material · Testing methods · Food packaging · Testing standard

1 Introduction

Food packaging is a major business sector, and numerous innovations in packaging of food products are launched in the market every day. Owing to the constantly changing behavior of consumers, innovative packaging technologies have been developed recently. Developed material should meet the standards to be successful in the market. Packaging material provides several functions including protection, containment, convenience, and communication to the product (Robertson 2016). Wide range of materials are available for packaging of food products including

H. Jayan · J. A. Moses · C. Anandharamakrishnan (✉)
Indian Institute of Food Processing Technology (IIFPT), Ministry of Food Processing Industries, Government of India, Thanjavur, India
e-mail: director@iifpt.edu.in

S. Ahmed (ed.), *Bio-based Materials for Food Packaging*,
https://doi.org/10.1007/978-981-13-1909-9_3

materials made of paper, glass, plastics, biodegradable material, metals, ceramics, and wood. The packaging material should be cost effective or both consumers and manufacturers. It should also protect the food material inside during storage, transportation, and distribution.

Each packaging material is different in terms of physical, chemical, and functional properties. The various properties offer different aspects as the packaging of food material is concerned. The properties of packaging material serve as a basis for selection of material for packing of food products. The basic material properties of packaging material that influence the quality and safety are (Kim et al. 2014)

- Barrier properties
- Mechanical properties
- Chemical reactivity
- Migration properties

A broad knowledge about the properties of packaging material is crucial in designing of packaging material for particular products. Properties of packaging materials such as mechanical strength, gas permeability, and sealing ability determine the suitability of packaging material for intended purpose (Marsh and Bugusu 2007). It is necessary to determine the properties of packaging material to select the suitable material so that it will maintain the safety and quality of packaging material.

Testing of packaging materials generally involves determination of physical properties such as mechanical strength and permeability. But it is also important to study the chemical stability of a packaging material, particularly when it would be used for food-based applications. Determination of properties also helps to predict the performance of packaging material when the packages are in the real environmental conditions (Arvanitoyannis and Bosnea 2004). A series of testing methods are required to confirm the packaging material possesses required properties. There are many crucial factors to be considered while determining properties of packaging material. The material should be tested in the same environment as that of the product is intended to perform, because almost all the properties changes as the environment around the package changes. Also, the performance of the packaging material deviates from the predicted performance. So it is important to study the property and the behavior of packaging materials in different environmental conditions.

There are several standard testing protocols for packaging materials, and this chapter provides detailed information about testing methods that are in use to evaluate various properties of packaging materials.

2 Regulation in Testing Methods

When packaging materials are used for particular food product, it should meet certain standards. These standards are provided by various governmental and nongovernmental organizers. The standards are set for each testing method, to determine the properties of packaging material. Several packaging materials are available in the market which are intended for packaging of food products; they should meet the requirements of the packaging materials that are provided by the various organizers to ensure the safety and quality. There are various agencies that provide standards or testing methods of packaging materials.

2.1 *ASTM International*

ASTM International is a globally recognized organizer which was established in 1989 for the development and delivery of voluntary standards. Globally, over 12,000 ASTM standards are used to improve product quality, strengthen market trade, consumer confidence, and enhance health and safety. ASTM International has 140 standard writing committees which revises each method annually and publish book for all the developed standards. ASTM committee D10 develops standards and testing procedures regarding packaging materials (ASTM 2017). ASTM standards are more flexible for simulation of the various distribution environment compared to other standards. It also considers the economical value of the product that is going to be packed in the packaging material, and the testing methods can be altered accordingly.

2.2 *International Organization for Standardization (ISO)*

ISO is a nongovernmental international organization which was established in 1947. ISO is the world's largest organization that develops voluntary standards. The organization has 162 members all around the world, and the technical committee will work on development of new standard and updating existing standards. ISO published 21980 International Standards covering almost every industry, including food safety, agriculture, and healthcare. ISO standards respond to a need in the market based on global expert opinion. TC-122 provides standards for terminology and definitions, packaging dimensions, performance requirements, and tests (ISO 2017).

2.3 *International Safe Transit Association (ISTA)*

ISTA established in 1948 which will provide standard for testing of packaging materials. ISTA packaging standards are used to assess the safety of product during transport with the help of vibration, impact, drop, and temperature testing. All tests simulate the transportation conditions in which the product will experience during transit. ISTA gives five series of standards including (ISTA 2017):

- 1 Series: Non-Simulation Integrity Performance Tests
- 2 Series: Partial Simulation Performance Tests
- 3 Series: General Simulation Performance Tests
- 4 Series: Enhanced Simulation Performance Tests
- 5 Series: Focused Simulation Guides

2.4 Technical Association of the Pulp and Paper Industry (TAPPI)

TAPPI is international nongovernmental organizations which consist of 14,000 members which work in the area of pulp and paper. TAPPI provides standards and technical information about paper to measure and evaluate paper and paper products. They also provide standards for such as corrugated fiberboard, flexible packaging, lamination, adhesives, coatings, and extrusion packaging. All the TAPPI standards are ANSI (American National Standards) certified. T 400–500 Series gives standards for testing of paper and paperboard.

3 Testing Methods for Packaging Materials

3.1 Analysis of Mechanical Properties

3.1.1 Compression Test

Compression testing of packaging materials is done to determine the compression strength of materials such as corrugated box, paper boards, and cans. Compression strength is usually measured to study the characteristics of packaging while stacking. During stacking of containers, the container at bottom experience more force than at the top. The container should withstand the force during transportation, distribution, and handling in bulk (Horvath et al. 2017). So it is important to study the compression strength of packaging material. The compression tester consists of a fixed head, where the container to be tested is placed, and a movable head which help to apply a fixed amount of force to the container as shown in Fig. 1. The test generally gives the deformation characteristics of material under compressive force. Compressive properties of rigid boxes or cans include modulus of elasticity, yield stress, deformation beyond yield point, and compressive strength (Frank 2014). During compression testing, some plastic containers or cans continue to deform until a flat disk is obtained. As a result the compressive strength will be steadily increasing without any fracture point, which indicates that there is no meaning for compression test.

There are several test methods according to ASTM for testing the compressive strength. The procedure can be selected according to the need of the study and packaging material. The type of force transfer differs in each method. In the test

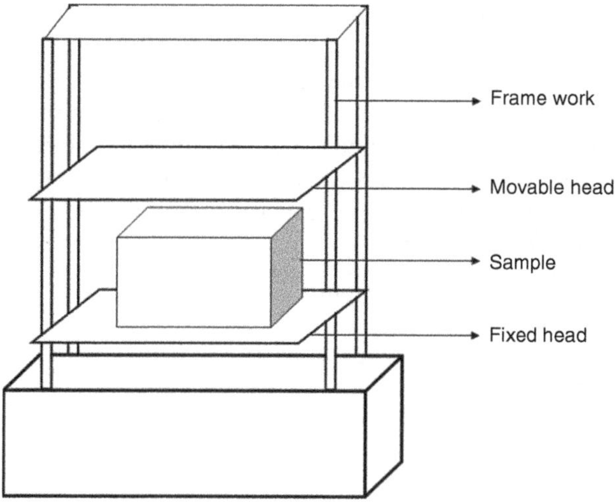

Fig. 1 Schematic representation of compression tester

method, ASTM D6641/D6641M compressive force is transmitted by combined shear and end loading, D5467/D5467M compressive force is transmitted by subjecting a honeycomb core sandwich beam with thin skins to four-point bending, and D695 compressive force is transmitted into the specimen by end loading.

Bucci et al. (2005) conducted dynamic compression test on PHB containers produced by injection molding. Resistance to dynamic compression test was carried out in accordance with ASTM 2659-95 (Standard Test Method for Column Crush Properties of Blown Thermoplastic Containers). The properties include crushing yield load, deflection at crushing yield load, crushing yield load at failure, and apparent crushing stiffness. PHB jar was more elastic than PP jar in terms of deformation value. They also stated that upper compression plate used was not as the same dimension as that of jar, so the compression applied was concentrated on the center, and the result was not valid for determining the stacking strength of prepared PHB jar. The following are the standard methods adopted for testing of compression strength of packaging materials:

- ASTM D695 – 15 – Standard Test Method for Compressive Properties of Rigid Plastics
- ASTM D3410/D3410M – 16 – Standard Test Method for Compressive Properties of Polymer Matrix Composite Materials with Unsupported Gage Section by Shear Loading
- ASTM D6641/D6641M – 16e1 – Standard Test Method for Compressive Properties of Polymer Matrix Composite Materials Using a Combined Loading Compression (CLC) Test Fixture
- ISO 604:2002 – Plastics – Determination of compressive properties

3.1.2 Tensile Test

Tensile test measures the tensile properties of packaging material by applying controlled tension until failure. Tensile test is carried out to determine maximum load that can be applied to a material before it ruptures. Tensile properties may vary with specimen thickness, method of preparation, speed of testing, type of grips used, and manner of measuring extension. The properties measured during tensile test include (ASTM 2014):

- Ultimate tensile strength
- Ultimate tensile strain
- Modulus of elasticity
- Poisson's ratio
- Transition strain

The following are the standard methods adopted for testing of tensile properties of packaging materials:

- ASTM D638 – 14 Standard Test Method for Tensile Properties of Plastics
- ASTM D882 – Tensile Properties of Thin Plastic Sheeting
- ASTM D3039 – Standard Test Method for Tensile Properties of Polymer Matrix Composite Materials
- ASTM D1708 – 13 – Standard Test Method for Tensile Properties of Plastics by Use of Microtensile Specimens
- ISO 527-1:2012 – Plastics – Determination of tensile properties – Part 1: General principles

Generally, the tensile strength is measured in universal testing machines (Zhang et al. 2015). The film is place between two jaws which are drawn apart at a constant rate. The upper jaw is movable and the jaws are supported by a frame. The load which is applied on the material can be varied according to the nature and indented application of material. The materials are drawn and a load vs. deformation curve is drawn as shown in figure (dgjhdg). The graph can be used to identify the tensile properties of the material. The tensile properties of the material decide how the material behaves under various forces. The stress-strain characteristics of packaging material depend upon rate of application of stress, temperature, and condition of the specimen. The tensile strength obtained for packaging material by above test methods cannot be considered valid for packaging material for application under different environments.

Auras et al. (2005) used universal testing machine to compare the tensile properties of oriented Poly (lactide) polymers with existing PET and oriented PS. Conditioning of packaging films prior to the test is important to avoid variability.

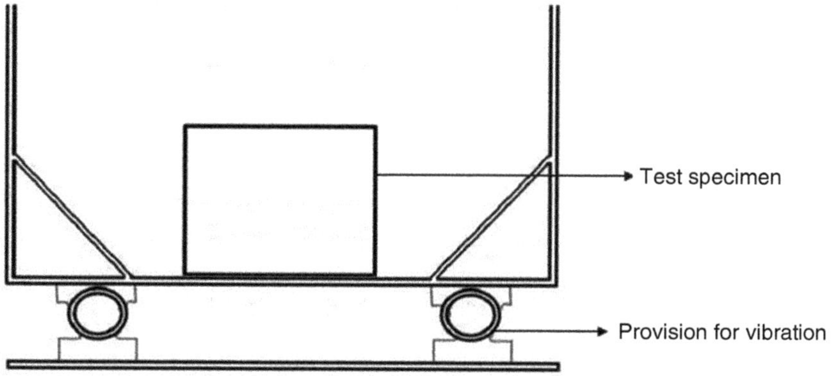

Fig. 2 Schematic representation of vibration tester

3.1.3 Vibration Test

A product may get damaged during transportation when the vibration frequency is not within the resonant frequency of the product. This kind of damage is common in fruit packaging where the skin of the fruit get damaged due to vibration which eventually leads to spoilage. So it is important to determine the resonant frequency of the package so that adequate protection can be given to save the product from getting spoiled. It will also provide information about complex interactions between the components inside the product as they are subjected to vibration when they are kept inside the package (Twede and Harte 2003) (Fig. 2).

According to ASTM the vibration test can be done in two different methods.

- Sinusoidal vibration
- Random vibration

Sinusoidal test also known as sweep-dwell test. It includes a sweep test that determines the natural resonant frequency and a dwell test in which the product will be vibrated at its natural resonant frequency. The test offers the same vibration frequency and time or all kinds of products regardless of the transportation conditions and product characteristics.

Random vibration test will simulate the condition the product experiences while it is being transported through land, air, or sea. During the test the product experiences predefined vibration intensity or fixed time duration (ASTM 2015). The following are the standard methods adopted for vibration testing of packaging materials:

- ASTM D3580 – 95(2015) – Standard Test Methods for Vibration (Vertical Linear Motion) Test of Products
- ASTM D999 – 08(2015) – Standard Test Methods for Vibration Testing of Shipping Containers
- ISO 8318:2000 – Packaging – Complete, filled transport packages and unit loads – Sinusoidal vibration tests using a variable frequency

- ISO 2247:2000 – Packaging – Complete, filled transport packages and unit loads – Vibration tests at fixed low frequency

3.1.4 Drop Test

The materials get damaged when the package is dropped on the ground during storage or transportation. This kind of failure happens in packages which are handled manually. The packages are tested for its drop resistance before selection of packaging material, and the test conditions are simulated by drop tester. The test assesses the ability of the container to withstand the drop that occurred during the test and also to measure the ability of container to protect the product during all the conditions that occur during transportation. This test allows selecting suitable packaging material for particular packaging (Paine and Paine 2012). Drop test is generally carried out for loaded boxes, cylindrical containers, and flexible packages. The drop tester consists of an adjustable plane surface on which the containers to be tested are placed. The surface possesses a special release mechanism which can be controlled by a lever. The conditioned packages are filled with indented products or dummy products which have the same dimension and characteristics, and packages are dropped from a predetermined height on a rigid surface (ASTM D5276 – 98 2017). The following are the standard methods adopted for drop testing of packaging materials:

- ASTM D5276 – 98(2017) – Standard Test Method for Drop Test of Loaded Containers by Free Fall
- ISO 2248:1985 – Packaging – Complete, filled transport packages – Vertical impact test by dropping

ASTM standard protocol allows only testing of free falling on flat surface, but generally the dropping or free falling does not happen on flat surface during handling. ISO 2248:1985 allows conducting of drop test over different orientations of packaging materials.

- Edge drop: the maximum angle between impacting edge and horizontal surface should be 2°
- Corner drop: the angle between surface of the package and the horizontal surface should be ± 5°

The exact drop heights, drop orientations, and acceleration values are published in ASTM D4169 and also in various ISTA standards. Both the methods possess free falling of product under gravity. Much sophisticated instruments can be used to produce different shock or impact on the packaging material to study shock or impact strength of packaging material.

3.1.5 Impact Test

Large packages experience high amount of horizontal shock when the materials are being transported in trucks or carried through rails. The package should be selected in such a way that it should protect the product without causing any damage or spoilage. The impact strength of packaging material for different packaging material can be studied under various conditions of impact force and orientation with the help of impact tester. Impact testers are complicated instrument with programming device to produce different impact force and impact angle.

The equipment consist of an inclined path containing a carriage on which the packages are placed and a wall where the package collides at the end of the path. The inclination of the surface and the speed at which the package hit on the wall can be adjusted according to the product and packaging material (Natarajan et al. 2014). The following are the standard methods adopted for impact testing of packaging materials:

- ASTM D880, Standard Test Method for Incline Impact Testing for Shipping Containers and Systems
- ASTM D4003, Standard Test Methods for Programmable Horizontal Impact Testing for Shipping Containers and Systems
- ASTM D5277, Standard Test Method for Performing Programmed Horizontal Impacts Using an Inclined Impact Tester
- ISO 2244:2000 Packaging – Complete, Filled Transport Packages and Unit Loads – Horizontal Impact Tests
- ISO 2248:1985 – Packaging – Complete, filled transport packages – Vertical impact test by dropping
- ISO 13355:2016 – Packaging – Complete, filled transport packages and unit loads – Vertical random vibration test

There are two methods followed in ASTM D880 – 92 (2015), procedure A to test the ability of container to withstand the impact force and procedure B to test the ability of package to protect the product inside.

3.1.6 Heat Seal Integrity

The strength and efficiency of sealing in the packaged goods affect the predicted shelf life of food product. The sealing should be proper and it should hold the material inside without any leakage. The sealing of the packaged goods depends upon the conditions maintained for sealing, type of equipment used, material selected for packaging, and the environment in which the material is being stored. At a regular interval, the heat seal integrity is visually inspected in industries. Improper sealing and sealing integrity can be identified by analysis seal characteristics and defect. Process variation in sealing machine can be indicated by visual seal defects. ASTM F1886 provides standard method for determination of integrity of seals by visual method. This method allows the determination of channels in the package seal down to a width of 75 μm, but it can only be used for package with one transparent side so that the seal area may be completely visible. ASTM F1921/F1921M – 12e1 describes standards or testing seal strength of packaging material while they are still hot. During form fill sealing the packed material, the contents are dropped immediately after the horizontal bars have been opened. So the package experiences a force before actual time required for cooling is completed to obtain a strong seal. The test requires equipment which will seal and check the strength of the seal simultaneously. The following are the standard methods adopted for testing of heat seal integrity of packaging materials:

- ASTM F1886/F1886M – 16 – Standard Test Method for Determining Integrity of Seals for Flexible Packaging by Visual Inspection
- ASTM F1921/F1921M – 12e1 – Standard Test Methods for Hot Seal Strength (Hot Tack) of Thermoplastic Polymers and Blends Comprising the Sealing Surfaces of Flexible Webs
- ASTM F88/F88M – 15 – Standard Test Method for Seal Strength of Flexible Barrier Materials
- ASTM F2391 – 05 (2016) – Standard Test Method for Measuring Package and Seal Integrity Using Helium as the Tracer Gas

3.2 *Testing Methods for Physical Properties*

3.2.1 Thickness Measurement

The accurate measurement of thickness is crucial for characterization to check whether the film meet certain specifications for storing particular product. Film thickness changes as the process parameter changes. The thickness affects the permeability and mechanical strength of films, and it also affects the product shelf life. So it is important to measure the thickness. Thickness of packaging film is usually measured using a micrometer. Screw gauge or vernier calipers can be employed but the error will be more. Nowadays digital micrometers are used which can give extremely accurate measurement up to 0.001 mm. The film to be measured is placed between anvil and spindle (ASTM F2251-13 2013).

3.2.2 Gloss Detection

Gloss is associated with the appearance or surface shininess of packaging material. The measurement of gloss is important to predict consumer acceptance. Gloss is measured using a gloss meter by projecting a fixed intensity of light at a particular angle (60, 20, or 85°) onto the surface of the specimen and measuring the amount of reflected light at the opposite angle. Gloss ratings are obtained by comparing the reflectance from the surface of the specimen to that from a black glass standard which gloss value is assumed to be 100. The gloss of a particular film increases as the incident angle increases. Also the gloss value depends upon the refractive index of the specimen. The gloss value of film changes with the molding process parameters and aberrations on the surface of film. So the gloss of the film should be measured to study the consumer acceptance (Singh et al. 2017). The following are the standard methods adopted for testing of gloss value of packaging materials:

- ASTM D523 covers the measurement of the specular gloss of nonmetallic specimens for gloss meter geometries of 20, 60, and 85°.
- ASTM D2457 describes procedures for the measurement of gloss of plastic films and solid plastics, both opaque and transparent at 20, 45, 60, and 75°.

3.2.3 Leak Testing

Leaks in packages occur due to imperfections in the packaging material or improper sealing. Leakage will lead to spoilage of food product inside the packages by the entry of unwanted gases, harmful microorganisms, or other contaminants. So the detection of leaks in packed food product is important. There are several methods for detection of leak in food packages given by ASTM standards, and they can be selected according to the material being tested and the intended use of the packaging material. Heat seal strength along with leak testing provides information about integrity of packaging material (ASTM D3580-95 2015). The following are the standard methods adopted for testing of leakage of packaging materials:

- ASTM F2096 – 11 – Standard Test Method for Detecting Gross Leaks in Packaging by Internal Pressurization (Bubble Test)
- ASTM D3078 – 02 (2013) – Standard Test Method for Determination of Leaks in Flexible Packaging by Bubble Emission
- ASTM F2338 – 09 (2013) – Standard Test Method for Nondestructive Detection of Leaks in Packages by Vacuum Decay Method
- ASTM F3169 – 16 – Standard Test Method for Leak Detection in Blister Packaging by Vacuum Deflection Method by Laser Measurement
- ASTM F3039 – 15 – Standard Test Method for Detecting Leaks in Nonporous Packaging or Flexible Barrier Materials by Dye Penetration
- ASTM D4991 – 07 (2015) – Standard Test Method for Leakage Testing of Empty Rigid Containers by Vacuum Method

3.2.4 Water Vapor Transmission Rate

Water vapor transmission rate is the amount of water transferred trough packaging material per unit time trough unit area. The main function of packaging material is to protect the product from outer environment to reduce wastage due to spoilage. The package should not allow water vapor to transmit and enter into product. Suitable packaging material should be selected according to the characteristics of the product. The selected material should maintain the quality of the packaged foods, so it is important to know the water vapor transmission rate of packaging material before selecting the material for packaging. According to ASTM the WVTR is expressed in g/m^2 day. Both ASTM methods determine water vapor transmission rate and permeance of the film to water vapor. Additionally, ASTM F1249 determines water vapor permeability coefficient for homogeneous materials.

Gravimetric Method

In this method a known amount of desiccant is placed in an impermeable jar, and the mouth of the jar is covered with material to be tested. The mouth should be completely covered in such a way that the moisture transfer only happens through the membrane. The sample is placed in a controlled environment (25 ± 0.5 °C and 75 ± 2 RH for temperate region; 38 ± 0.5 °C and 90 ± 2 RH for tropic region), and the weight gained by the desiccant is measured for 2–14 days. Graph should be plotted as weight gain VS. time, which follows a straight line. The water vapor transmission rate can be found using the following formula (ASTM E96 1995):

$$\text{WVTR}\left(\text{g H}_2\text{O/day } m^2\right) = \frac{Q}{\text{At}} = \frac{\text{slope}}{\text{area.}}$$

The following are the standard methods adopted for testing of water vapor transmission rate of packaging materials:

- ISO 2528 – Sheet materials – Determination of water vapour transmission rate – Gravimetric (dish) method
- ASTM E96 – Standard Test Methods for Water Vapor Transmission of Materials
- ASTM D1653 – Standard Test Methods for Water Vapor Transmission of Organic Coating Films

Modulated Infrared Sensor

Since the gravimetric methods have several disadvantages including time consumption, low accuracy, and low detection limit, an alternative method using infrared sensor has been suggested by ASTM F1249. In this method the film is kept between a dry and wet chamber. The wet chamber is saturated with water vapor using a sponge containing HPLC grade water, and nitrogen gas is continuously circulated through dry chamber. The vapor transferred through the membrane from wet to dry chamber is carried by the nitrogen gas. The amount of water vapor in nitrogen gas is measured by passing infrared light at 6.3 μm, which is the critical wavelength at which moisture absorbs infrared light. Recently, ASTM starts to develop new test method for water vapor transmission rate through plastic film and sheeting using an electrolytic detection sensor. The sensor is based on coulometric P_2O_5. The following are the standard methods adopted for testing of water vapor transmission rate of packaging materials:

- ISO 15106-2 – Plastics – Film and sheeting – Determination of water vapour transmission rate – Part 2: Infrared detection sensor method
- ASTM F1249 – Standard Test Method for Water Vapor Transmission Rate Through Plastic Film and Sheeting Using a Modulated Infrared Sensor
- ASTM F372 – 99 – Standard Test Method for Water Vapor Transmission Rate of Flexible Barrier Materials Using an Infrared Detection Technique
- ASTM F1249 – 13 – *Standard Test Method for Water Vapor Transmission Rate Through Plastic Film and Sheeting Using a Modulated Infrared Sensor*

Relative Humidity Sensor

In this method the wet and dry chambers are maintained at high and low relative humidity, respectively. Both chambers are isolated by the material being tested. The moisture migration happens only through the film. The amount of moisture transferred is measured using a relative humidity sensor fixed in the dry chamber (Fig. 3). The performance of both infrared and relative humidity sensor are similar, but in later method nitrogen gas is not flushed through the dry chamber to remove moisture. This allows us to study the transfer of moisture from a package into the environment also.

The following are the standard methods adopted for testing of water vapor transmission rate of packaging materials:

- ISO 15106-1 – Plastics – Film and sheeting – Determination of water vapor transmission rate – Part 1: Humidity detection sensor method
- ASTM E398 – Standard Test Method for Water Vapor Transmission Rate of Sheet Materials Using Dynamic Relative Humidity Measurement

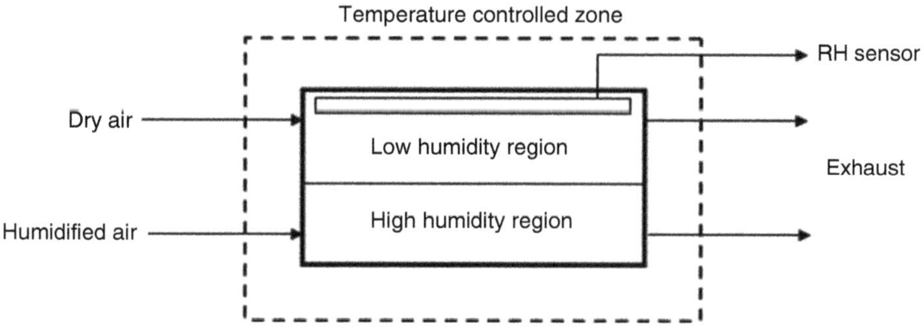

Fig 3 Schematic representation of relative humidity sensor (ASTM 2013)

3.2.5 Gas Permeability Testing

Most of the food materials are sensitive to oxygen, and the shelf life of certain foods depends on the gas permeability of packaging material. Gas permeability is an important property of a packaging material which determines its suitability for the application. It depends upon the solubility of gas in the polymer material and diffusivity of gas through the material. ASTM D1434 provides two procedures which determine gas transmission rate (GTR), permeance, and permeability (only for homogeneous material) for polymer material.

In this method the sample is sealed between two chambers in such a way that it acts as a semipermeable membrane. One chamber is filled with the gas being tested and kept at high pressure. The other chamber is kept at low pressure. The gases pass through the membrane and get accumulated in evacuated chamber. Based on the way in which the amount of gas is measured, the method varies. In manometric method, the lower pressure chamber is evacuated first, and the gas concentration is measured by increase in pressure. In volumetric method, the lower pressure chamber is maintained near atmospheric pressure, and the gas concentration is specified by increase in volume.

However, ASTM (F1307, D3985, F1927) also gives a coulometric sensor-based method for determination of transmission of oxygen through the packaging film. Other conventional method allows testing various gases, where coulometric sensor only detects oxygen transmission through the membranes. This test has the same setup as that of water vapor transmission rate (ASTM F1249). In this method one chamber is filled with oxygen and the other is flushed with nitrogen gas. The oxygen transferred through the membrane is carried by nitrogen gas to the coulometric sensor. The oxygen produces an electric current proportional to the concentration of oxygen present at the sensor per unit time. As time increases, the current generated also increases and the value is noted down when the value becomes constant. The oxygen transmission rate can be determined by the following formula (Singh et al. 2017):

$$\text{Oxygen transmission rate} = \frac{(E_e - E_o) * Q}{A * R_L}$$

where E_e = steady-state voltage level; E_o = zero voltage level; A = specimen area; Q = calibration constant; and R_L = value of load resistance.

The following are the standard methods adopted for testing of gas permeability of packaging materials:

- ASTM F1927 – 14 – Standard Test Method for Determination of Oxygen Gas Transmission Rate, Permeability and Permeance at Controlled Relative Humidity Through Barrier Materials Using a Coulometric Detector
- ASTM D3985 – 05(2010)e1 – Standard Test Method for Oxygen Gas Transmission Rate Through Plastic Film and Sheeting Using a Coulometric Sensor

- ASTM F1307 – 14 – Standard Test Method for Oxygen Transmission Rate Through Dry Packages Using a coulometric Sensor

3.3 Biodegradability Testing

Because of negative impact of synthetic polymer material in the environment, researches are focused on developing packaging material which can be degraded naturally. Biodegradable plastics are derived from plant- and animal-based materials or from other renewable resources. It is more important to prove the extent of biodegradability of prepared biodegradable film than determining the mechanical and barrier properties (Siracusa et al. 2008). If a material is claimed to be biodegradable which can be made into a film for packaging of food products, the material should be tested for its biodegradability to all under the legislative framework. A number of national, European, and international standards have been developed to determine the compostability of developed packaging material. The following are the standard methods adopted for testing of biodegradability of packaging materials:

- ASTM D6868 – 17 – Standard Specification for Labeling of End Items that Incorporate Plastics and Polymers as Coatings or Additives with Paper and Other Substrates Designed to be Aerobically Composted in Municipal or Industrial Facilities
- ASTM G21 – 15 – Standard Practice for Determining Resistance of Synthetic Polymeric Materials to Fungi
- ISO 14851:1999 – Determination of the ultimate aerobic biodegradability of plastic materials in an aqueous medium – Method by measuring the oxygen demand in a closed respirometer
- ISO 14853 Plastics – Determination of the ultimate anaerobic biodegradation of plastic materials in an aqueous system – Method by measurement of biogas production

Several methods have been developed or testing of compostability, but the selection of appropriate test is crucial to get a reliable results. The method should be chosen according to the chemical nature of film and condition of the environment in which the package indented to use. Biodegradation or compostability can be indicated by loss of weight, carbon dioxide production, change in tensile strength, dimensions, and chemical and physical properties (Singh and Sharma 2008).

All the test method can be categorized into field tests, simulation tests, and laboratory tests. In field test the biodegradable test material is buried in soil, lake, and river allowing it to undergo composting process. But in such methods, the environmental conditions such as temperature, pH, or humidity can't be controlled. Further, the analysis of residues after the degradation process is complicated because the environment itself is complicated for any further retrieval. In order to avoid all these limitations, various simulation techniques have been developed. The test is conducted in a reactor which will maintain required temperature, pH, and humidity.

The environment as similar to that of soil, compost, or seawater can be created using a simulation model in laboratory. This method is commonly employed or analysis of residue, intermediates, and CO_2 produced during degradation process. Some of the simulation test involves soil burial test (Kaur et al. 2009; Sharma et al. 2014), composting test (Yang et al. 2005), simulating landfills (Müller 2005; Shin et al. 1997), and aqueous aquarium tests (Puechner et al. 1995; Tosin et al. 2012). The laboratory test includes usage of special synthetic media for growth of microbes and inoculation culture. The inoculums can be mixed population which is taken from waste water or individual microbial strains. Among all the other methods, laboratory method produces reproducible result. The microbial strain can be selected according to the nature of polymer to be tested, and the degradation rate will be higher when compared to test conducted in natural environment.

3.3.1 Soil Burial Method

It is the most commonly employed method for determination of biodegradability which can be performed under laboratory conditions. The film to be tested is buried in soil at a specific depth after cutting the film into required dimension. Twenty to 40% humidity should be maintained by sprinkling water on the soil at regular interval. Care should be taken to avoid any stagnation of water, because it affects the natural degradation process. After a specified time usually 15 days once, the film is taken out, rinsed with distilled water, and then dried in a vacuum oven until constant weight is obtained. Degradation rate of the film is calculated by taking weight loss of the sample with time (Thakore et al. 2001).

3.3.2 Microbiological Degradation Test

In soil burial method, the degradation happens due to the combined effect of all kinds of microorganisms, but in pure culture method, the action of specific bacteria or fungi can be studied. The film to be tested is disinfected to ensure the action of degradation is only due to targeted microbes. The disinfected film is placed inside sterile culture medium and incubated for 24 h to ensure asepsis. The culture of specific microbes is inoculated into medium containing the film, and the sample is incubated at required temperature and time optimal for the selected microorganism. The weight of sample is taken after washing with ethanol and drying at lower temperature. The weight of the film is compared with the film which is incubated in uncultured medium (Lee et al. 1991). The surface morphologies of the film, before and after microorganism treatment, can be studied with the help of a microscope.

3.3.3 Compost Method

In this method, the pre-weighed biopolymers are subjected to a mixture of a mature compost and then incubated at optimum temperature and relative humidity. After the required incubation time, the amount of material carbon converted to carbon dioxide is measured. The test can be conducted at various temperature including −20 °C, 4 °C, and 20 °C. It was expected that the viable cell count in compost stored at −20 °C will be less compared to 4 °C and 20 °C because of the act that the microorganisms will be under stress at subzero temperature. But in contrary to the expectation, mesophilic bacteria were fewer in compost stored at 4 °C and 20 °C when compared with compost stored at −20 °C (Singh et al. 2017).

3.3.4 Enzymatic Degradation Test

Enzymes accelerate the reaction rate of cellular metabolism of living microorganisms. So it is important to study the contribution of enzymes in degradation of biomaterials. The research conducted on degradation of polyurethanes shows that the degree of degradation was increased by ten times in the presence of cholesterol esterase enzyme.

The enzymatic test involves the addition of dried and weighed samples into enzymatic reaction mixture containing alpha and beta amylase (1 ml of α-amylase and 4 ml of β-amylase). The sample should be completely immersed in the reaction mixture, and they are incubated in shaker at required condition. After the period of incubation, the samples were taken out and rinsed with distilled water to remove enzymes. The samples are then dried under vacuum to remove water absorbed and weighed. The degree of enzymatic degradation (DED) was calculated as (Guohua et al. 2006)

$$\text{DED}\ (\%) = ((W_I - W_H)/W_I)\ 100$$

where W_H is the dry weight of the specimen after enzymatic treatment and W_I is the initial dry weight of the specimen.

3.4 *Thermal Analysis Techniques*

Determination of thermal properties of packaging material is widely employed in polymer industry, and it consists of series of analytical techniques. The properties of polymer such as energy uptake, differential temperature, dielectric constant, mechanical modulus, and evolved gases are measured as a measure of temperature. Since biodegradable and conventional packaging films exhibit the same physical and chemical nature, the well-established relationship between heat resistance and melting temperature as well as between glass transition and melting temperature in conventional plastics can be transferred to biopolymer also (Sencadas et al. 2004). The determination of thermal properties of biopolymer allows comparing the suitability of packaging material with conventional plastics. Existing biopolymer is inferior compared to conventional polymer in terms of heat resistance. Efforts are going on to improve the heat resistance of biopolymer by addition of various filler materials like nanoparticle (Rhim 2007). Thermal analysis consists of three different techniques.

3.4.1 Differential Scanning Calorimetry

When a material is heated or cooled, heat flow in or out of the system; also every material possesses a characteristic heat capacity. DSC measures the heat flow rate into the sample with reference to a sample of known heat capacity (ASTM D3418 2015). Different types of DSC are available according to the need of the study. The following are the standard methods adopted for testing of thermal properties of packaging materials:

- ASTM D3418 – 15 – Standard Test Method for Transition Temperatures and Enthalpies of Fusion and Crystallization of Polymers by Differential Scanning Calorimetry
- ASTM D6604 – 00 (2013) – Standard Practice for Glass Transition Temperatures of Hydrocarbon Resins by Differential Scanning Calorimetry
- ASTM E1356 – 08 (2014) – Standard Test Method for Assignment of the Glass Transition Temperatures by Differential Scanning Calorimetry
- ASTM E2602 – 09 (2015) – Standard Test Methods for the Assignment of the Glass Transition Temperature by Modulated Temperature Differential Scanning Calorimetry
- ISO 11357-2 Plastics – Differential Scanning Calorimetry (DSC) – Part 2: Determination of Glass Transition Temperature
- ISO 11357-3 Plastics – Differential Scanning Calorimetry (DSC) – Part 3: Determination of Temperature and Enthalpy of Melting and Crystallization

3.4.2 Thermogravimetric Analysis

TGA is generally done to study the thermal stability of the material. Thermogravimetric curve that is obtained from the test is an effective tool for monitoring the thermal degradation of polymers under inert and oxidative environment (Barneto et al. 2009). The method involves the measure of weight loss of samples while it was progressively heated. The temperature is increased at constant rate or through a series of steps, and the weight loss versus temperature curves are obtained. Negligible change in weight loss and slope in curves indicate the thermal stability of the material in desired temperature range (Singh et al. 2017).

3.4.3 Thermomechanical Analysis

In thermomechanical analysis the change in dimension of the packaging material under stress is measured while temperature of the sample is kept at desired temperature. In this method, the behavior of sample with change in stress, temperature, or both can be studied. The force exerted can be compression, tension, or torsion, and the temperature is maintained with the help of a furnace (Karagiannidis et al. 2008). The sample deforms because of thermal expansion and stress reorientation. The equipment (Fig. 4) consist of:

1. Furnace – maintain the required temperature
2. Force generator – apply required amount of force to the material
3. Probe– transfer the force from generator to sample
4. Length detector – measure the deformation of the sample
5. Thermocouple – measure the temperature of the sample

3.5 *Chemical Testing of Packaging Material*

3.5.1 Testing of Migration from Packaging Material

The packaging material was prepared by the addition of chemical compounds into various polymeric materials. These chemical compounds are added to improve the functional properties of the packaging material. During the storage and

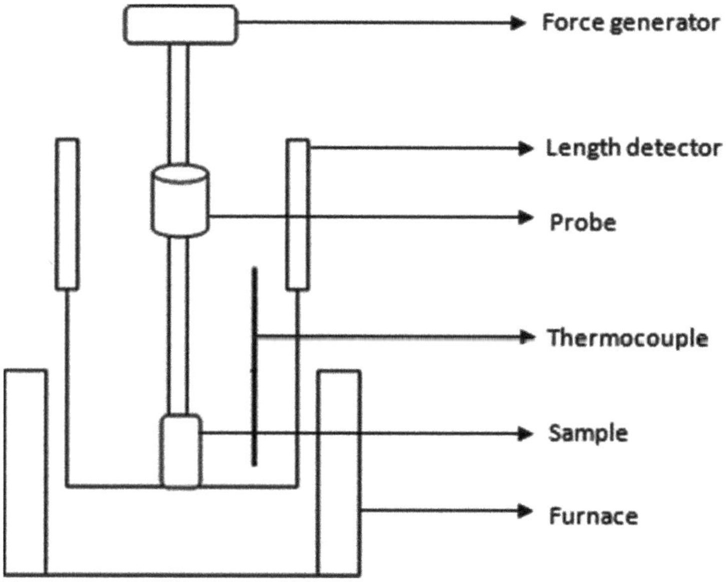

Fig 4 Schematic representation of thermomechanical analyzer

transportation of product, some of these chemicals may migrate from the packaging material to food products. If the migrated compounds in the food products exceeds specified limits serious safety and quality issues arises. The migration of chemical compound occurs through diffusion which depends on several parameters including nature of food, amount of chemical compounds in film, contact time, and environmental condition (Castle 2007). The quantification of migrants in the food material is important to ensure safety to the consumers. The migration can be studied by:

1. Determination using stimulants
2. Chromatographic methods
3. Mathematical modeling

In determination of migration using food stimulants, migration of compounds into various simulating liquids is studied instead of placing complex food product. The stimulant should be selected in such a way that it should resemble the food product in terms of interaction with good contact material. It will provide easy analysis of the migrants in the liquids. The stimulants should be carefully selected to obtain a reliable result. The stimulants such as water, ethanol, acetic acid, and oils are packed in the package to be tested and incubated at required temperature and time. The amount of migrant present in the stimulant is calculated by separating it from stimulant using various techniques. Various stimulants should be identified according to the nature of packaging material and food products. This limits the wide application of this technique in practical cases (Bhunia et al. 2013). Chromatographic technique involves the use of sophisticate equipments including GCMC and LCMS. The technique allows very low detection limit as low as ppb levels of migrants. But these techniques cannot be employed for determination of unknown compounds in the migrants. Both methods require sophisticated instrument and more time to determine migration. Mathematical modeling can be employed to avoid these disadvantages (Castle 2007).

4 Conclusion

Several standards are available for determining the property and performance of packaging material. The selection of particular technique or analyzing the material is crucial, because it affect the result. In order to compare properties of different packaging materials, the test material should be subjected to the same set of standard procedures. Various simulation tests are available to study the behavior of packaging material under various environmental conditions. The integrity of packaging material can be studied by keeping under accelerated environments, which will avoid extended and tedious studies. Each packaging material requires different sets of standard to analyze the property, so the standard needs to be extended for various types of packaging material.

References

ASTM D1653-13 (2013) Standard test methods for water vapor transmission of organic coating films. ASTM International, West Conshohocken www.astm.org
ASTM D2457-13 (2013) Standard test method for specular gloss of plastic films and solid plastics. ASTM International, West Conshohocken www.astm.org
ASTM D3078-02 (2013) Standard test method for determination of leaks in flexible packaging by bubble emission. ASTM International, West Conshohocken 2013. www.astm.org
ASTM D3418-15 (2015) Standard test method for transition temperatures and enthalpies of fusion and crystallization of polymers by differential scanning calorimetry. ASTM International, West Conshohocken www.astm.org
ASTM D3985-05(2010)e1 (2010) Standard test method for oxygen gas transmission rate through plastic film and sheeting using a coulometric sensor. ASTM International, West Conshohocken www.astm.org
ASTM D4003-98 (2015) Standard test methods for programmable horizontal impact test for shipping containers and systems. ASTM International, West Conshohocken 2015. www.astm.org
ASTM D523-14 (2014) Standard test method for specular gloss. ASTM International, West Conshohocken www.astm.org
ASTM D5277-92 (2015) Standard test method for performing programmed horizontal impacts using an inclined impact tester. ASTM International, West Conshohocken 2015. www.astm.org
ASTM D6604-00 (2017) Standard practice for glass transition temperatures of hydrocarbon resins by differential scanning calorimetry. ASTM International, West Conshohocken 2017. www.astm.org
ASTM D6868-17 (2017) Standard specification for labeling of end items that incorporate plastics and polymers as coatings or additives with paper and other substrates designed to be aerobically composted in municipal or industrial facilities. ASTM International, West Conshohocken www.astm.org
ASTM D880-92 (2015) Standard test method for impact testing for shipping containers and systems. ASTM International, West Conshohocken 2015. www.astm.org
ASTM E1356-08 (2014) Standard test method for assignment of the glass transition temperatures by differential scanning calorimetry. ASTM International, West Conshohocken 2014. www.astm.org
ASTM E2602-09 (2015) Standard test methods for the assignment of the glass transition temperature by modulated temperature differential scanning calorimetry. ASTM International, West Conshohocken 2015. www.astm.org
ASTM E398-13 (2013) Standard test method for water vapor transmission rate of sheet materials using dynamic relative humidity measurement. ASTM International, West Conshohocken www.astm.org
ASTM E96/E96M-16 (2016) Standard test methods for water vapor transmission of materials. ASTM International, West Conshohocken www.astm.org
ASTM F1249-13 (2013) Standard test method for water vapor transmission rate through plastic film and sheeting using a modulated infrared sensor. ASTM International, West Conshohocken www.astm.org
ASTM F1307-14 (2014) Standard test method for oxygen transmission rate through dry packages using a coulometric sensor. ASTM International, West Conshohocken www.astm.org
ASTM F1886/F1886M-16 (2016) Standard test method for determining integrity of seals for flexible packaging by visual inspection. ASTM International, West Conshohocken www.astm.org
ASTM F1921/F1921M-12e1 (2012) Standard test methods for hot seal strength (hot tack) of thermoplastic polymers and blends comprising the sealing surfaces of flexible webs. ASTM International, West Conshohocken www.astm.org

ASTM F1927-14 (2014) Standard test method for determination of oxygen gas transmission rate, permeability and permeance at controlled relative humidity through barrier materials using a coulometric detector. ASTM International, West Conshohocken www.astm.org
ASTM F2096-11 (2011) Standard test method for detecting gross leaks in packaging by internal pressurization (bubble test). ASTM International, West Conshohocken www.astm.org
ASTM F2338-09 (2013) Standard test method for nondestructive detection of leaks in packages by vacuum decay method. ASTM International, West Conshohocken 2013. www.astm.org
ASTM F2391-05 (2016) Standard test method for measuring package and seal integrity using helium as the tracer gas. ASTM International, West Conshohocken 2016. www.astm.org
ASTM F3039-15 (2015) Standard test method for detecting leaks in nonporous packaging or flexible barrier materials by dye penetration. ASTM International, West Conshohocken www.astm.org
ASTM F3169-16 (2016) Standard test method for leak detection in blister packaging by vacuum deflection method by laser measurement. ASTM International, West Conshohocken www.astm.org
ASTM F372-99 (2003) Standard test method for water vapor transmission rate of flexible barrier materials using an infrared detection technique (withdrawn 2009). ASTM International, West Conshohocken 1999. www.astm.org
ASTM F88/F88M-15 (2015) Standard test method for seal strength of flexible barrier materials. ASTM International, West Conshohocken www.astm.org
ASTM G21-15 (2015) Standard practice for determining resistance of synthetic polymeric materials to fungi. ASTM International, West Conshohocken www.astm.org
Arvanitoyannis IS, Bosnea L (2004) Migration of substances from food packaging materials to foods. Crit Rev Food Sci Nutr 44(2):63–76
Auras RA, Singh SP, Singh JJ (2005) Evaluation of oriented poly (lactide) polymers vs. existing PET and oriented PS for fresh food service containers. Packag Technol Sci 18(4):207–216
Barneto AG, Ariza J, Martín JE, Jiménez L (2009) Use of autocatalytic kinetics to obtain composition of lignocellulosic materials. Bioresour Technol 100:3963–3973
Bhunia K, Sablani SS, Tang J, Rasco B (2013) Migration of chemical compounds from packaging polymers during microwave, conventional heat treatment, and storage. Compr Rev Food Sci Food Saf 12(5):523–545
Castle L (2007) Chemical migration into food: an overview. In: Barnes KA, Sinclair CR, Watson DH (eds) Chemical migration and food contact materials. CRC Press, Boca Raton, pp 1–13
Frank B (2014) Corrugated box compression-a literature survey. Packag Technol Sci 27 (2):105–128
Guohua Z, Ya L, Cuilan F, Min Z, Caiqiong Z, Zongdao C (2006) Water resistance, mechanical properties and biodegradability of methylatedcornstarch/poly (vinyl alcohol) blend film. Polym Degrad Stab 91(4):703–711
Horvath L, Min B, Kim YT (2017) Testing of mechanical properties for plastic packaging materials. In: Singh P, Wani AA, Langowski HC (eds) Food packaging materials: Testing & Quality Assurance, 1st edn. CEC Press, New York, pp 103–122
ISO 2248:1985 (2017) International organization for standardization. https://www.iso.org/standard/7062.html. Accessed 10 Dec 2017
ISTA (2017) Test procedures. Retrieved from https://ista.org/test_procedures.php
Karagiannidis PG, Stergiou AC, Karayannidis GP (2008) Study of crystallinity and thermomechanical analysis of annealed poly (ethylene terephthalate) films. Eur Polym J 44 (5):1475–1486
Kaur I, Bhalla TC, Deepika N, Gautam N (2009) Study of the biodegradation behavior of soy protein-grafted polyethylene by the soil burial method. J Appl Polym Sci 111(5):2460–2467
Kim YT, Min B, Kim KW (2014) General characteristics of packaging materials for food system. In: Han JH (ed) Innovations in food packaging, 2nd edn. Academic press, pp 13–35
Lee B, Pometto AL, Fratzke A, Bailey TB (1991) Biodegradation of degradable plastic polyethylene by Phanerochaete and Streptomyces species. Appl Environ Microbiol 57(3):678–685

Marsh K, Bugusu B (2007) Food packaging—roles, materials, and environmental issues. J Food Sci 72(3)
Müller R J (2005) Biodegradability of polymers: regulations and methods for testing. Biopolymers Online
Natarajan S, Govindarajan M, Kumar B (2014) Fundamentals of packaging technology. PHI Learning Pvt. Ltd, New Delhi
Paine FA, Paine HY (2012) A handbook of food packaging. Springer Science & Business Media, New York
Puechner P, Mueller WR, Bardtke D (1995) Assessing the biodegradation potential of polymers in screening-and long-term test systems. J Environ Polym Degrad 3(3):133–143
Rhim JW (2007) Potential use of biopolymer-based nanocomposite films in food packaging applications. Food Sci Biotechnol 16(5):691–709
Robertson GL (2016) Food packaging: principles and practice. CRC Press, New York
Sencadas V, Lanceros-Mendez S, Mano JF (2004) Characterization of poled and non-poled β -PVDF films using thermal analysis techniques. Thermochim Acta 424(1):201–207
Sharma K, Kumar V, Kaith BS, Kumar V, Som S, Kalia S, Swart HC (2014) A study of the biodegradation behaviour of poly (methacrylic acid/aniline)-grafted gum ghatti by a soil burial method. RSC Adv 4(49):25637–25649
Shin PK, Kim MH, Kim JM (1997) Biodegradability of degradable plastics exposed to anaerobic digested sludge and simulated landfill conditions. J Polym Environ 5(1):33–39
Singh B, Sharma N (2008) Mechanistic implications of plastic degradation. Polym Degrad Stab 93:561–584
Singh P, Wani AA, Langowski H-C (2017) Food packaging materials: Testing & Quality Assurance. CRC Press, Boca Raton
Siracusa V, Rocculi P, Romani S, Dalla Rosa M (2008) Biodegradable polymers for food packaging: a review. Trends Food Sci Technol 19(12):634–643
Thakore IM, Desai S, Sarawade BD, Devi S (2001) Studies on biodegradability, morphology and thermo-mechanical properties of LDPE/modified starch blends. Eur Polym J 37(1):151–160
Tosin M, Weber M, Siotto M, Lott C, Degli Innocenti F (2012) Laboratory test methods to determine the degradation of plastics in marine environmental conditions. Front Microbiol 3
Twede D, Harte B (2003) Logistical packaging for food marketing systems. In: Coles R, McDowell D, Kirwan MJ (eds) Food packaging technology Vol. 5. CRC Press, New York, pp 98–99
Yang HS, Yoon JS, Kim MN (2005) Dependence of biodegradability of plastics in compost on the shape of specimens. Polym Degrad Stab 87(1):131–135
Zhang ZH, Han Z, Zeng XA, Xiong XY, Liu YJ (2015) Enhancing mechanical properties of chitosan films via modification with vanillin. Int J Biol Macromol 81:638–643

Functionality and Properties of Bio-based Materials

Pintu Pandit, Gayatri T. Nadathur, Saptarshi Maiti, and Baburaj Regubalan

Abstract This chapter reviews the impact of recent developments in bio-based sustainable materials with enhanced functionality and its properties related to moisture permeability, porosity and tunable gas permeability characteristics on storing and packing wet and dry foods and fresh produce. Bio-based polymers, plastics, biodegradable plastics and composites are gaining interest as reasonable substitutes for non-renewable petrochemical-based products. Natural fibres such as jute, hemp, flax, banana, wheat straw, etc. are significant sources for making biodegradable composites having commercial importance as food packaging materials. Combining plant-based fibrous materials and biopolymers/biomass-derived polymers gives environmentally friendly and biodegradable biocomposites with sufficient flexibility and mechanical strength comparable to petroleum-based polymers. Improved mechanical resistance, thermal insulation and enhanced physico-chemical properties which are key to the barrier and permeability features in bio-based packaging materials are achieved. Protein-based materials, which demonstrate good barrier properties, being impermeable to oxygen (in the absence of moisture) and aromatic compounds, have also been investigated as potential food packaging materials. This chapter presents a review of the literature available on such processes, techniques and methods applied to exploit these sustainable bio-based materials.

Keywords Bio-based polymers · Biodegradable · Physico-chemical properties · Food packaging

---

P. Pandit (✉) · G. T. Nadathur · S. Maiti
Department of Fibres and Textile Processing Technology, Institute of Chemical Technology, University Under Section -3 of UGC Act, Mumbai, India

B. Regubalan
Department of Food Engineering and Technology, Institute of Chemical Technology, University Under Section -3 of UGC Act, Mumbai, India

S. Ahmed (ed.), *Bio-based Materials for Food Packaging*,
https://doi.org/10.1007/978-981-13-1909-9_4

1 Introduction

A green and sustainable economy requires that a revolution occurs in the utilization of raw materials; it needs to substitute the overwhelming consumption of fossil fuel-based materials with processes and products obtained from plant-based and other renewable resources. Solar and wind power could likely drive the change in the energy supply sector, while the change will be driven, within the manufacturing industry, by bio-based materials and their composites in different applications such as food packaging, textile fibrous materials, geotextile, plastics industry, etc. (Storz and Vorlop 2013).

Bio-based materials have gained attractiveness in the last decades due to both ecological and economic concerns. The increased burden of pollution and waste has motivated the scientific and industrial communities to look at bio-based substitutes for fossil fuel-based materials. Bio-based materials can be synthesized mainly by the production of polymers directly from naturally occurring compounds or the production of bio-based monomers and their subsequent (bio)chemical polymerization, through processes such as fermentation. Whatever the pathway for production of bio-based materials, the main problem is to control its composition and reproducibility. The regulation of the size of the biopolymers appears even more complicated. In any case, the major hurdle is in making the (micro) organism perform the targeted biosynthesis consistently and repeatedly. To overcome this problem, one can be tempted to make more controlled chemical polymerization with the bio-based monomers. However, the polymerization of bio-based monomers often demands further developmental steps, as in the case of a polylactic acid and of polybutylene succinate. The production of partially bio-based materials (Soronas) or the production of bio-monomers identical or enhanced compared to currently used petroleum-based molecular units has been demonstrated (Storz and Vorlop 2013; Razzaq et al. 2016).

A simple method for the simultaneous extraction of protein and β-glucan and polymers from barley was employed to prepare biodegradable films. Cereal grains are sustainable sources of polymers that could be used to make environmentally friendly bio-based materials. The procedure was customized in that the extraction was completed in alkaline medium instead of normal pH water. Likewise, the good macroscopic performances of the film and the assessed chemical stability could suggest its first-line applications as a newer, cheaper and biodegradable plastic material for food packaging (Razzaq et al. 2016). A lot of research has taken place on the methods to convert these sustainable options into films, fibres, textiles, papers, packaging or any other form which has a prospective application in the field of bio-based materials.

With the increased importance placed on self-service marketing, the role of bio-based packaging materials is becoming quite significant. Bio-based packaging materials serve numerous purposes like physical coverage, information transmission, marketing, convenience, barrier protection and security. Currently, several bio-based materials like polymers, paper (brick carton, cardboard), foam, film, etc. exist as options to synthetic plastic which is the most common packaging material but problematic and costly with respect to disposal. The features common with all plastics are that they are light, strong, cheap to manufacture and easy to mould. Paper gets preference in food and textile packaging. The cardboard variety of paper used in the form of boxes, sheets, etc. has served as a container for pizzas, bread and pies, popcorn, coffee, chocolates and other takeout order from the fast-food restaurants (Johanson and Vahlne 1977; Sullins et al. 2017; Yan et al. 2014). This chapter presents a critical review of the current literature available on such processes and techniques that modify the functionality and characteristics of bio-based materials to exploit and utilize them to their fullest.

2 Classification and Methods of Bio-based Materials

2.1 Classification of Bio-based Materials

Bioplastics are not simply a uniform single class of polymers but a diverse class of materials that can differ significantly from one another. The term bioplastics denotes materials that are bio-based, biodegradable or both (refer Fig. 1).

The use of bio-based materials, i.e. fibres, paper, plastic coatings and solid articles made from sugar, corn, starch and other renewable raw materials, has multiplied in recent times. Biopolymers find versatile applications in drug delivery systems, surgical implant devices, scaffolds for tissue engineering and packaging, food containers, agriculture film, waste bags and packaging material. Biopolymers often have a firm and definite internal structure, though this may not be the main feature (e.g. lignocellulose).

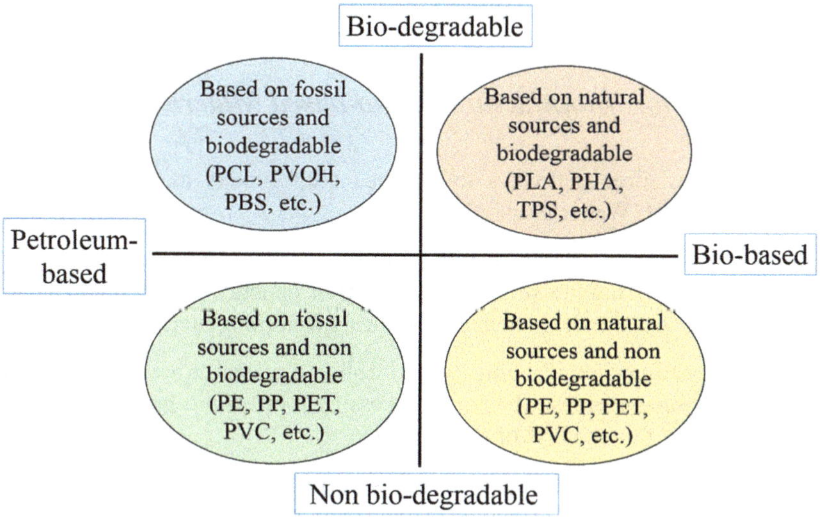

Fig. 1 Classification of bioplastics based on source and degradability

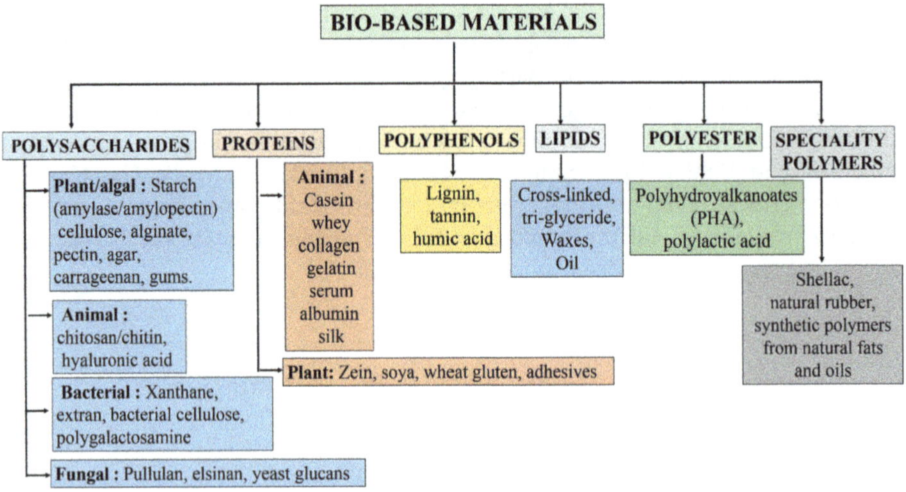

Fig. 2 Classification of bio-based materials

Among the biopolymers, polyethylene furanoate is observed to possess better mechanical properties. Cross-linking agents (e.g. glutaraldehyde, polycarboxylic acids) enhance the mechanical properties and aqueous stability of the biomaterials but may cause unfavourable modification to the functionality of the biopolymers or cause cytotoxicity (Türe et al. 2012; Petersen et al. 1999). All kinds of natural fibres like flax, bamboo and natural wool can be bonded together with synthetic and bio-based adhesives to form useful fibre-reinforced biocomposite composites (refer Fig. 2).

2.2 *Methods for Manufacture of Bio-based Materials*

Overall, there are three pathways to produce biopolymers as per the literature (Petersen et al. 1999; Weber et al. 2002).

(a) Directly extracted or separated from biomass
(b) Generated by microorganisms (fermentation) or genetically modified bacteria
(c) Manufactured by classical chemical synthesis from biomaterials

The basic repeating chemical units of the bio-based materials are almost similar to those of conventional plastics. The main processing route of bio-based materials can also be possible by this method of technology:

(a) Packaging film: extruded film, blown films, cast film
(b) Thermoformed sheets: trays, cups
(c) Injection moulding: drinking beakers, bottles, trays cutely, cups, plates, etc.

(d) Fibres and non-woven: diapers, feminine hygiene products, agricultural products, certain medical plastics, clothing
(e) Extrusion coating: laminated paper or films

3 Properties of Bio-based Materials

General attributes for enhanced properties and functionality of bio-based materials must be efficient, effective, clean and recyclable. The bio-based materials provide tremendous encouragement to green manufacturing by successfully enclosing and shielding products which are key properties for packaging. Also, the packaging system is intended to use materials and energy with minimal waste throughout the product life cycle. The properties that need to be reckoned with respect to food bio-based materials may include gas and water vapour permeability, sealing capability, thermoforming properties, transparency, antifogging capacity, printability, availability, UV resistance, water resistance, and resistance to acid, alkali, grease, etc. Cost is also one of the important properties for bio-based materials. Thus, bio-based materials in relation to end applications in food and other industries require favourable mechanical, thermal and gas barrier, water vapour permeability features and disposability of the materials. A product that contains 100% annually renewable raw materials may or may not be biodegradable/compostable. It is conditional on the molecular structure of the material itself, and if biodegradable, the products would pass when tested for biodegradability and compostability according to ASTM D6400 or ASTM D6868.

3.1 Gas Barrier, Moisture Resistance and Water Vapour Transmittance Properties

Many bio-based food packaging materials need the presence of specific atmospheric conditions to maintain the freshness and quality of food during storage. The packaging materials need to have certain gas barriers to retain a constant gas composition within the package. The gas mixture inside most of the package contains carbon dioxide, oxygen and nitrogen. Bio-based materials by using mineral oil-based polymer provide gas barrier property. Gas barrier properties of PA6 (polyamide) or EVOH (ethylene vinyl alcohol) which are vulnerable to moisture, when combined with LDPE, create a very restrictive water vapour barrier preventing the moisture from the foodstuffs interfering with the properties of PA6 or EVOH. Plasma technology was used for the deposition of glass like SiOx coatings on bio-based materials out of natural polymer and modified clay to enhance water vapour and gas properties (Fischer et al. 2000). Bio-based materials are mostly hydrophilic, and the gas barrier properties are influenced by the humidity conditions, such that the gas permeability of hydrophobic bio-based materials may increase drastically when humidity increases. Gas barriers based on PLA and PHA are likely to be independent of humidity (Petersen and Nielsen 2000).

Packaging for numerous food types requires materials that are impervious to moisture, which is a major challenge due to the contrary hydrophilic nature of most of the bio-based polymers. Inventive chemical modifications and treatments are being researched to solve this problem so that future bio-based materials must also be able to duplicate the water vapour barriers of the conventional materials. In many of the bio-based food packaging applications, water vapour barrier, along with gas barrier properties, is required.

3.2 Physical Blending and Chemical Processing

Current research in the properties of bio-based composite materials involves the development of technologies to innovate chemical and physical modifications for optimal composite properties related to air, water permeability, water resistance/hydrophobicity, surface properties of adhesion and frictional resistance, mechanical strength, fracture resistance, fire retardance, sound barrier, thermal insulation and optical and conductive properties.

Composites of bio-based materials find their major use as structural components in textiles, packaging and disposables, which may be extended to greater strength bearing the use in construction and transportation sectors, for obtaining weight reduction to aid fuel efficiency and reduce cost, without compromising on load-bearing ability. This may be achieved if properties, processing and hence performance are improved by suitable chemical and physical modifications. In the addition of bio-based reinforcing fibres in composites, this implies control of reinforcement orientation/alignment and concentration, the addition of nanofibres and nanoparticles to optimize mechanical, thermal features and increase durability.

3.3 Improvement in Mechanical Properties

Most of the bio-based materials function in a similar manner to conventional polymers. Bio-based polymers similar to polystyrene-like polymers relatively stiff at moderate service temperature or polyethylene-like polymers relatively flexible at moderate service temperature are available. The mechanical properties in terms of modulus and stiffness are quite close to those of conventional polymers. Bacterial cellulose polymer could be used in materials requiring noteworthy mechanical properties (Iguchi et al. 2000).

In terms of use of plant fibre/lignocellulosic fibre-reinforced composites/plastics as engineering materials, the important structural properties of tensile strength relative to the density of the material lie in the upper left region of the Ashby's

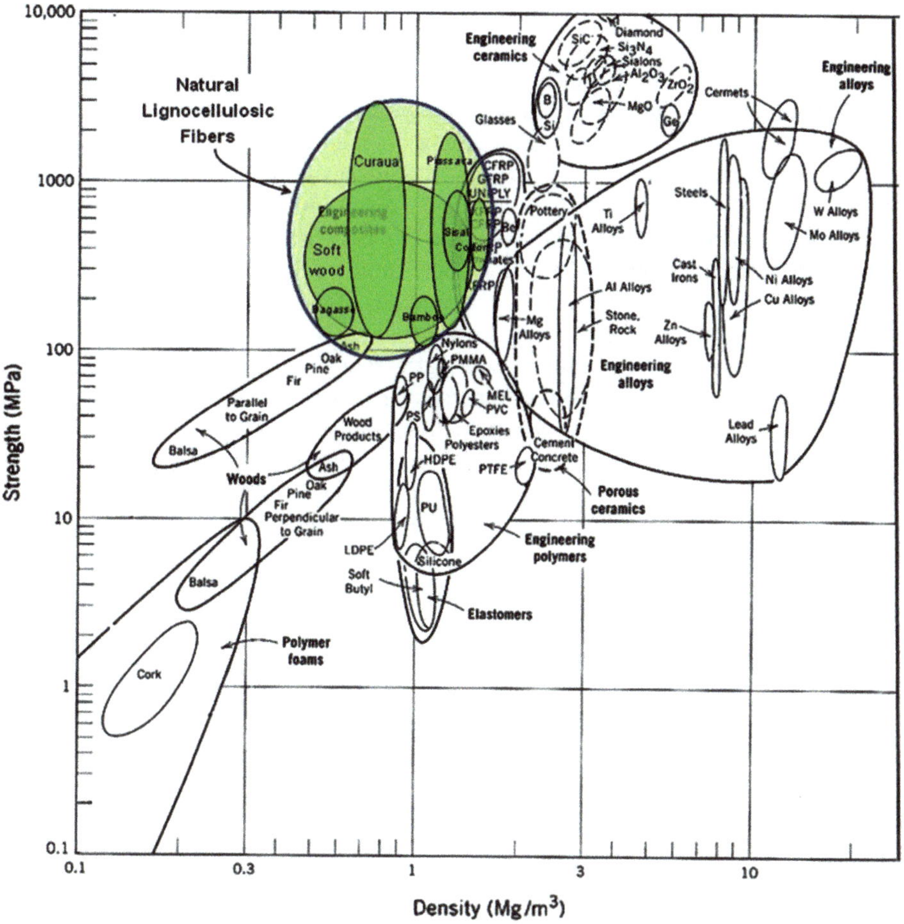

Fig. 3 Lignocellulosic fibres in Ashby's σ vs. ρ chart. (Monteiro et al. 2011)

material property chart (Fig. 3) (Monteiro et al. 2011). Their highest strength values are comparable to metallic alloys and ceramics and are favourable choices suiting guidelines of minimum weight design ($\sigma^{2/3}/\rho$, indexing the performance of a material related to bending stresses) and efficient mode of loading ($\sigma^{1/2}/\rho$). The existence of structural defects and dimensional heterogeneity in the natural source of plant fibres are a disadvantage that may be overcome by genetic selection and bioprocessing, for final use as high-performance thin fibres in composites. It was observed that plant fibres function as superlative reinforcement for polymers when the orientation of the fibres is unidirectional to the loading direction and the fibres are long. Tensile moduli and strengths of up to 40 GPa and 450 MPa, respectively, were reported for unidirectional plant fibre composites containing between 40 and 60 weight % flax, hemp, jute or ramie fibres (Fortea-Verdejo et al. 2017).

The tensile strength of glass fibres is at least twice as high as the tensile strength of plant fibres. Glass fibres are however inherently heavier than plant fibres (the density of glass fibres is 2.5 g cm^{-3} versus ca. 1.5 g cm^{-3} for plant fibres). Despite this, some randomly oriented plant fibre-reinforced polymer composites have very similar mechanical properties compared to randomly oriented GFRP, especially at low w_f (fibre loading fraction). At $w_f > 50$ wt-%, the tensile strengths of plant fibre-reinforced polymers showed lower values for tensile modulus and strength compared to PLLA (the bio-based polymer with the best mechanical properties). This may be due to the random orientation of plant fibres within the polymers, causing greater plasticity and hence lower tensile properties of the resulting composites. This situation is exacerbated when very short fibres are used, which may also be generated during the processing, resulting in less effective stress transfer from the matrix to the fibres. Constructive fibre reinforcement is achieved when the length of the fibre exceeds the critical fibre length (which is affected by the fibre-matrix combination and method of manufacturing), the fibre tensile strength at the critical length and the fibre diameter. Plant fibres substitute as cheaper filler possessing competitive stiffness, strength and fatigue properties compared to conventional engineering (polymer, glass or carbon) fibres, replacing some portion of the costlier polymers, leading to reduced costs while increasing the renewable fraction of the end composites.

The fabrication of bio-based anisotropic hierarchical fibres, combining small amounts of functionalized recombinant spider silk proteins with the widely available structural constituent, cellulose nanofibrils (CNFs), resulted in, for the first time, highly desirable mechanical performance with a stiffness of ~55 GPa, strength at break of ~1015 MPa and toughness of ~55 M m^{-3} (Mittal et al. 2017). The addition of small amounts of silk fusion proteins to cellulose nanofibrils (CNF) produced materials with advanced functionalities, which cannot be derived from wood-based CNF alone. The silk/CNF fibre surfaces also exhibited bioactivity allowing chemical-free immobilization of antibodies and adhesion of fibroblasts promoting their growth and spread, compatible with tissue culture, ascribed solely to the proteinaceous component of the silk.

Poly (L-lactide) (PLLA), a bio-based polymer, based on L-lactic acid, exhibits high tensile strength and Young's modulus, while its impact strength and heat deformation temperature (T_d) are notably low. To improve upon the properties of PLLA, a stereocomplex-type PLA (sc-PLA), obtained in a mixture (50/50) of preformed PLLA and PDLA, was developed (Ikada et al. 1987). The improved thermal resistance of sc-PLA is the reason why sc-PLA-based woven and knitted clothes can be treated to dyeing and hot-pressing and a knit of sc-PLA withstands laundering above 160 °C. In contrast, PLLA fibres are heat-sensitive, and processing above 120 °C destroys the fibre structure (Fan et al. 2004). When polymer films made by solution casting were studied by Tsuji and Ikada, the mechanical strength of sc-PLA was observed to be almost 60% higher than that of PLLA (Tsuji and Ikada 1999). The melt spinning of an oriented fibre of sc-PLA gave a product with very specific tensile strength and Young's modulus of 400 MPa and 5 GPa, respectively, in comparison to ordinary melt-spun fibre of PLLA which exhibited a greater range in tensile properties (Takasaki et al. 2003; Fukushima and Kimura 2006). Polymer films of sc-PLA may be used as heat-resistant insulators that are also optically transparent. The incombustible nature of PLA, having an oxygen index of 28, could be used in casing material for various electrical devices and automobile parts.

A polymer blend between PLA and natural rubber, which is a totally bio-based polymer, modulates the brittleness of PLA (strain at break ~5%), with NR as a toughening agent, to obtain a modulus of 300%. Ordinarily, thermoplastic elastomers are prepared from ductile polymers like polyethylene and nylon, rather than brittle plastic like PLA. Bio-based thermoplastic elastomers (TPE) made of natural rubber and poly(lactic acid), with a blend ratio of 60% of natural rubber and 40% of poly(lactic acid), were prepared by melt blending in an internal mixer (Tanrattanakul and Bunkaew 2014). The addition of plasticizers, such as tributyl acetyl citrate (TBAC), tributyl citrate (TBC), glycerol triacetate (GTA) and triethyl-2-acetyl citrate (TEAC), would increase the ductility of PLA by increasing the strain at break, on interaction with the continuous phase of plastic PLA, the solubility parameter of plasticizers being similar to PLA rather than NR. All plasticizers, except GTA, increased the tensile toughness of the TPE as indicated by the area under each stress-strain curve.

Blends with PE and starch displayed lower elongation values, tensile strength and tear strength but also that the mechanical strength parameter was dependent on the amount of plasticizer compared to synthetic polymers, LDPE and HDPE. The compression tests done on the cup forms indicated favourable results. PLA has values (around 280 N) similar to those of PS (polystyrene), while starch composites gave high values greater than 1000 N. In terms of water vapour permeability, increased starch content increased WVP, while PLA and PHB and cellophane showed lower values, favourable to prevent loss of moisture from fresh produce (Petersen et al. 2001).

3.4 Thermal Insulation Property

Rigid polyurethane (PU) and polyisocyanurate (PIR) foam materials are largely used as heat-insulating materials in the refrigeration industry and civil engineering due to its low thermal conductivity (λ), and other PU compositions are applied in coatings and adhesives. Renewable materials in the form of polyol, a hydroxyl derivative of vegetable and other natural oils, were incorporated into the PU polymer matrix, through condensation with isocyanates to give the PU product. Polyol sources which do not coincide with food sources, such as rapeseed oil, and nonagricultural mixtures, such as those of abietic acid and oleic acid, a by-product of cellulose production may be used, after transesterification with polyfunctional amine-based alcohol, like triethanolamine. Thermal insulation products with competitive fire retardance substituting halogenated fire retardants with nanosized expanded graphite and IF, a commercial product, based on mineral and glass fibres with inclusion of graphite particles, and co-introducing isocyanurate structures into PU matrix were approaches used for sustainable solutions (Kirpluks et al. 2016). The thermal conductivity of 22.0 mW/m·K, considered in the industry to be a standard for thermal insulation materials, is in the range of vegetable polyol-derived PU/PIR. Expanded graphite and EG in combination with polyphosphate-based conventional fire retardants, however, increase in thermal conductivity from the industry required standards, in comparison with IF alone. Increased isocyanurate content to an index of 200 increased the mechanical strength unlike the swelling and plasticization of PU foam microstructure introduced by EG. The use of adhesives was not required in the coating of IF on PU sandwich panels, achieving thermal insulation with decreased flammability. On flame exposure, they did not ignite but formed protective char layer on the surface with reduced smoke production. Isocyanurate cross-linking also favours fire retardance. Thus bio-based polyols, in combination with nanoparticles, may be used to develop sustainable flame retardants, in polyurethane-based materials, while substituting for petroleum-derived chemicals.

3.5 Improvement in Air Insulation Constructions

The use of bio-based insulation materials has expanded recently, typically within breathable wall constructions, despite the need to compensate for their poor hygrothermal characteristic, (exposure to moisture and heat lowering their mechanical properties), by chemical treatments and blending with a synthetic polymer. An air-permeable wall construction enables the insulation material to influence the indoor air quality. Volatile organic compounds (VOCs) with low range boiling points between 60 and 280 °C contaminate indoor air in buildings, and when they build up to sufficiently high concentrations, they can cause building-related illnesses. Bio-based insulation materials can afford satisfactory thermal conductivity at a smaller environmental footprint than conventional insulation materials. There is an increased chance for VOC emissions, when biomaterials with intrinsic carbon-based chemical structure are used. The purposeful use of bio-based insulation materials, including hemp-lime mixes, wood fibre board, cellulose flakes and sheep's wool, to reduce the total volatile organic compound (TVOC) emissions in components of toluene, and formaldehyde separately, was studied (Maskell et al. 2015).

Individual VOCs were largely unsaturated hydrocarbons, aldehydes, acetic acid and alcohols. Wood fibre sample emitted high concentrations of acetic acid and furfural. Hemp-lime mixes emitted but-1-ol. Wood-based materials, without adhesive resins, emitted formaldehyde due to oxidative, thermal, enzymatic and microbial degradation of the wood-based material (Mittal et al. 2017). Wood fibre-based insulation exhibited emission rates giving rise to concentrations in air of up to 607 μg m^{-3} TVOC and 13 μg m^{-3} formaldehyde. The VOC quenching by wool due to the reaction of amine groups of lysine and arginine and the amido groups of asparagine and glutamine with organic compounds such as formaldehyde is the reason for air purification, increasing indoor air quality (Baumann et al. 2000).

3.6 Sound Insulation Barrier Property

Noise reduction and noise control are essential considerations in modern urban construction, for fulfilling health and safety guidelines and reducing the impact on surroundings. The alternative, sustainable materials to replace asbestos-based and synthetic fibre-based sound-absorbing materials with the improved application is an area of widespread research, since the late 1990s. Natural fibres with lightweight, porous internal structure, containing a network of air cavities, in their hollow cell lumens, allow attenuation of sound energy by multiple mechanisms compared to the solid glass fibre. Sound absorption coefficient increases with a decrease in bamboo fibre diameter, as the number of fibres increases for the same cross-sectional area (Koizumi et al. 2002). When sound waves are incident on the natural fibre, the transmission into the fibre lumen with multiple pathways would cause loss of energy, along with conversion to heat, due to friction. The vibration of air in the lumen and bulk leads to fibre vibration. These pathways explain the excellent sound-absorbing coefficients of the fibre structure. Natural fibres are processed in a one-step cold/hot compacting with resinous fibre bonding agents to produce low-density and hot-pressed higher-density, one layered structured composites. With bamboo and kapok fibre, sound absorption coefficient increases with density, but an optimal sound-absorbing fibre mass density exists. Also, the higher the sound frequency, the shorter the wavelength, and the longer the propagation of sound waves in the network structure of the composites, causing greatest dissipation, producing best sound absorption performance at high frequencies. The different kinds of composites, which have been studied for sound absorption capability, are ramie, flax and jute composites with epoxy glue, corn particle board with wood glue, kapok fibre blended with polypropylene, hemp concrete, coconut fibre composite, tea-leaf fibres with polyurethane blend, rice straw-wood particle composite with formaldehyde resin, etc. (Zhu et al. 2013).

Activated carbon fibre composites, with the ACF non-woven structure blended along with fibres, allow greater air capacity to absorb sound. Addition of inorganic particles like precipitated calcium carbonate increased the stiffness of the composite while adding to the sound absorption. This was also seen in kenaf rubber composites containing calcium carbonate. Rigid, lightweight sound-absorbing materials can be generated from natural fibre sources, though their other properties such as fire retardancy may have to be improved. Multilayered structures such as sandwich structure and honeycomb structure have also been used to improve sound and vibration damping compared to synthetic fibre sandwich panels. Natural cork bundle as core covered with carbon fibre top sheets in sandwich form has exemplary sound absorption. Balsa, pinewood and synthetic foam as core with natural fibre composites as surface layers in the sandwich have low shear modulus cores matching structural damping values. Nano-fillers in composite mixes are another solution to achieve target properties of sound absorption with shading, insulation, moisture barrier and mechanical strength (Zhu et al. 2013).

3.7 Optical and Conductive Property

Biochar is produced from the thermal decomposition of biomass in the absence of atmospheric oxygen wherein fast pyrolysis process produces a solid residue (char), liquid condensate and gas products, while slow pyrolysis process produces carbonization to biochar/charcoal. They have been used as substitute electrode material as anode and cathode in microbial fuel cells, with comparable performance output, to activated carbon and graphitic granules (Huggins et al. 2014). Resistivity measurements showed that the biochar material, derived from high-temperature treatment of waste wood at 1000 °C, was highly conductive with a resistivity of $3.1 \pm 1\ \Omega\cdot cm^{-1}$. Increased treatment temperature produces a graphitic structure, with volatilization of dross organic matter. Biochar particle-filled (8%, 10%, 12%) PVA films possess suitable electrical conductivity and piezoresistive response and would be likely materials for piezoresistive sensors. The voltage output of the biochar/PVA sensor was highest for the highest loading of biochar, and the piezoresistive response did not vary significantly with film thickness (0.4–0.5 mm). They also did not degrade in response with multiple pressing, i.e. they had good recovery capability, essential for a pressure sensitive sensor (Nan and De-Vallance 2014). The use of textile fibres and fabrics coated with conductive nanoparticles such as polypyrrole has been another pathway to bio-based conductive materials.

Biologically derived optically transparent materials can be transformed into optoelectronic materials (Nan and De-Vallance 2014). DNA films can be utilized to function as a combination of hole-transporting and electron-blocking layer in optical devices. The DNA composition and derived physical properties can be manipulated by sequence selection, chemical functionalization and physical manipulation, to obtain required films to produce multilayer white polymer organic light-emitting diode (LED). Further research in charge transport in biopolymers would extend applications to medical devices, sensors and flexible biodegradable electronics for timed usage in implants and healing.

4 Bio-based Materials for Functionality Packaging

Plastic has become the most-used packing material in food and textile for consumer durables. However, plastic is a huge problem precisely because of its durability, and it poses a huge threat to our ecosystem. Discarded plastic bags have been linked to major disasters. Reduce, reuse and recycle of plastics are not sufficient to minimize the environmental impact and health threats. There exist many numerous research findings that the chemical additives in plastics, such as bisphenol-A (BPA) and phthalates (which keep plastics soft and pliable), are causing harmful effects on human health. The processing of plastics not only releases a lot of heat but a huge amount of carbon dioxide (Muneer 2013). Paper as food packaging also has its limitations. Corrugated containerboard is used to ship and transport packages ranging from electronics to fragile glassware to perishable goods; paperboard packages food, medicine and toiletries for handy storage and display. Paper bags give customers a sustainable choice to carry their purchases home. Brick carton's complex composition makes it difficult to decompose. Glass is considered to be pure and safe material for packaging. But if the glass is used to package products, one should be ready for its brittleness and high processing cost. Large amounts of energy are expended in the production of traditional packagings like plastics, corrugated boxes and plastic bags. Frequently, the energy is derived from burning fossil fuels that release large volumes of greenhouse gases like carbon dioxide and methane into the atmosphere, while disposed conventional packaging gets dumped in landfills or oceans causing soil, water and plant pollution. Hence in order to avoid all these harmful effects, we need to resort to sustainable means of packaging.

Biodegradable polymers and molecules which rely on feedstock sources—from annually grown crops like wheat (wheat gluten), soy (soy proteins), corn and potatoes (starch), cassava and its waste, green coconut shell extract, sugarcane and hemp—give a foundation for sustainable and eco-efficient products which can substitute and supersede plastics and composites made mainly from the petroleum-based feedstock. Bio-based materials maximize the use of recycled and renewable materials and low-impact production processes, thereby conserving raw materials. Conserving raw materials, in turn, preserves natural resources. Also, bio-based materials minimize exposure to potentially toxic and hazardous chemicals. Bio-based materials introducing eco-friendly and biodegradable product (cloth, paper and jute bags) are generally used in all shops for packaging and carrying items (refer Fig. 4).

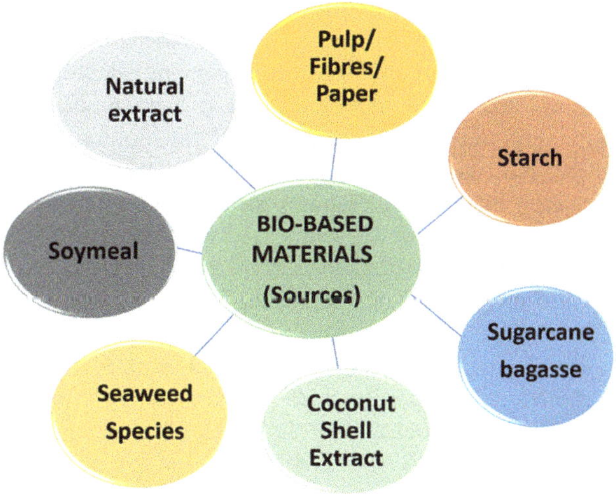

Fig. 4 Natural sources of bio-based materials

4.1 Bio-based Materials from Starch

Different glycerol contents were used for plasticizing the starch films. Cassava starch has been comprehensively used to make biodegradable films which feature non-toxic, isotropic, odourless, bio-degradable and colourless characteristics. Tree cassava films possessed better water solubility and water vapour permeability relative to those of cassava films irrespective of glycerol content (Souza et al. 2012). Scientists presented a study in which the effect of glycerol content and its blending method on barrier and tensile properties of biodegradable films from cassava starch were investigated. The influence of nano-clay particles and glycerol on tensile, barrier properties and glass transition temperature of biodegradable films produced from cassava starch was studied. The results established that films made from plasticized cassava starch reinforced with clay nanoparticles were appraised favourably as bio-based packaging material (Souza et al. 2012).

A review focused on properties of starch and sugar palm fibres and their products in green composites. The review also revealed the prospects of sugar palm fibres and biopolymer for industrial applications such as automotive, packaging, bioenergy and others (Sanyang et al. 2016).

Studies on the means and methods to prepare and characterize some polyvinyl alcohol and starch polymeric blends have been done. Polyvinyl alcohol (PVA) is positioned as the few biodegradable synthetic polymers that are easily and often modified owing to their properties. PVA dissolves easily in water as it contains multiple polar alcohol groups and may form hydrogen bonds with water. Starch, functioning as a storage polysaccharide in plants, is a frequently used biopolymer. It constitutes both linear and branched polysaccharides, like amylose and amylopectin. Starch is one of the most widely available and inexpensive polysaccharide sources which have the unique characteristic of "biodegradability", and it can be easily decomposed in water. It was seen that as content of the starch in PVA/starch blends increased, the melt processing characteristics reduce. It was observed that there was an increase in the melt viscosity of polyvinyl alcohol/starch blends with the increase of starch content which indicated a hard processability during mixing. Films containing starch showed lower UV transmittance compared with PVA/glycerol sample indicating that polyvinyl alcohol/starch films are a good defence against ultraviolet light. The moisture permeability of tested samples increased proportionally with the addition of starch. The results derived from experiments showed that the polymeric material made from polyvinyl alcohol and starch had favourable physical-mechanical characteristics, being suitable for packaging industry (Tănase et al. 2015).

The degree to which chemical structure of high-amylose maize starch films and its functional properties were influenced by temperature and time during the manufacturing process and the level of plasticizer content was examined. The films were prepared from a slurry of high-amylose maize starch and water (2.2 or 2.6 g/100 g water) that was heated in an autoclave and mixed thoroughly. The influence of variable film-forming conditions on its cohesiveness was compared. The strongest film was the non-plasticized. But glycerol was required to obtain films that were cohesive. Tensile strength was most affected by glycerol content, one of the processing factors. Longer heating times and hotter processing temperatures made films that were not contiguous but fragmented. Such stressful processing conditions caused starch degradation and brittle film. The temperature was more significant than time in influencing the progress of starch degradation. High molecular weight amylopectin and amylose were more likely to undergo thermal degradation. The aggregation of starch molecules in a definite microstructure through different stages of film formation to a final macrostructure was unfavourably affected by degradation. However, there were no similar changes in the mechanical properties with longer processing time or increasing heating temperature (Koch et al. 2010).

4.2 *Bio-based Materials from Coconut Shell Extract*

The multifunctional effect can be obtained from green coconut shell extract (CSE) applied on jute (lingo-cellulosic), cotton (cellulosic) and wool (proteinous) which can be a good low-cost bio-based packaging material. Application of CSE at different pH on these materials enhanced the thermal stability of the jute, cotton and wool fabric. Flame retardancy in the CSE-treated jute, cotton and wool fabric might be ascribed to the presence of inorganic metal salts, phosphate compounds, phenolic groups, etc. which promoted the production of foaming char which acted as a protective barrier and decreased volatilization and the formation of combustible gases. On jute and cotton fabric, CSE application was carried out in acidic, neutral and alkaline pH where as in wool it was carried out only in acidic pH 4.5 (as is extracted). Moreover, the inherent strength of jute, cotton and wool fabric was preserved. Treated jute and cotton fabric exhibited superior antibacterial property and very good ultraviolet protection properties. Moreover, treated jute, cotton and wool fabric showed inherent natural colouration, and it was found to be more intense at higher concentration. Therefore, CSE acts as multifunctional bio-based sources for jute, cotton and wool materials which can be used for various sustainable packaging materials. The raw material of the green coconut shell is widely available as waste product, and this kind of application leads to good low-cost value addition for bio-based packaging materials (Teli et al. 2017a; Teli and Pandit 2017a, 2018). Similarly, other natural sources such as banana pseudostem sap, *Sterculia foetida* fruit shell extract, *Delonix regia* stem shell waste, peanut husk, herbal-synthesized particle, etc. have been used by our research group on natural fibres as bio-based packaging materials (Teli and Pandit 2017b, c; Teli et al. 2017b; Basak et al. 2016; Pandey et al. 2018).

Fig. 5 Different products made from sugarcane bagasse

4.3 Bio-based Materials from Sugarcane Bagasse

Sugarcane bagasse cellulose was subjected to a homogenous chemical modification of esterification using phthalic anhydride reagent providing the acidic groups for esterifying the hydroxyl groups. The successful transformation to sugarcane bagasse phthalate films (considered a bioplastic) was achieved as proven by the absence of cracks or collapse in films obtained on drying after solution casting using DMF solvent. The sugarcane bagasse phthalate films displayed a uniform and consistent three-dimensional structure of their component polymeric chains. Increasing esterification however reduced the tensile strength of sugarcane bagasse phthalate films, with increasing weight percentage gain. A satisfactory breakthrough would be to obtain increased film flexibility while retaining strength in a bioplastic equivalent or better than in a natural banana leaf (refer Fig. 5) (Chen and Shi 2015).

4.4 Bio-based Materials from Variety of Seaweed Species

An in-depth review of a variety of seaweed species and seaweed products was done with a focus on the use of bioresource in the pharmaceutical industry and mostly applied in the medical field. Marine algal seaweed species are a bioresource, which has not been utilized to its maximum potential. Seaweeds are notable for the natural polysaccharides that can be extracted from them which are widely used in the area of biotechnology, microbiology, food technology and medicine. Carrageenan, agar and alginate are some of these polysaccharides. Since they are renewable biomass resources and are polymers of carbon chains composed of sugars, bioplastics may be prepared from them (refer Fig. 6). Additionally, seaweed and products derived from it have been expansively employed as a crucial ingredient in numerous pharmaceutical preparations like tablets, as rate retardant in sustained and extended release dosage forms and as bioplastic in the packaging industry. Bioplastics from

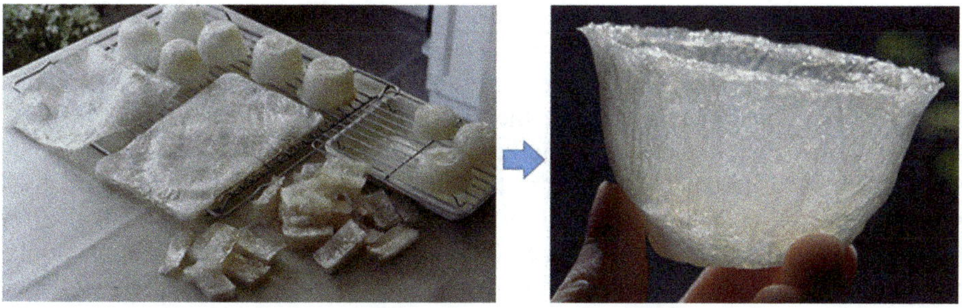

Fig. 6 Bioplastics from seaweed species

seaweed might be costly, but they have gained attention of late, because of their advantages over other biological sources (Gade et al. 2013).

Inexpensive and semi-refined kappa carrageenan, that is used as a seaweed polysaccharide in the preparation for strong biodegradable films, exhibited improved performance over conventional films. The formulation required no gelling salt, and the only other required additions to the film are glycerol and water. The films could be reprocessed and recycled if required (Meena et al. 2011).

4.5 *Bio-based Materials from Soymeal*

An attempt was made to use low-priced soymeal (SM) in place of costlier purified soy protein isolate and soy protein concentrate, as the raw material for manufacturing biodegradable and bio-based sustainable film. This required study of some of the modified initial processing steps, such as fermentation of soybean meal, the starting material, to dismantle some of the carbohydrates components while retaining the value of plasticizing ingredients (protein and pectin). Simultaneous saccharification and fermentation (SSF), yeast fermentation (YF) and natural fermentation (NF) were some of the multiple modes of fermentation process involved. Such fermentation degraded soymeal that was then plasticized with glycerol and then further blended with poly (butylene adipate-co-terephthalate) (PBAT); optimization of the proper mixing of plasticized soymeal with poly (butylene adipate-co-terephthalate) was done to better the performance of the resulting composite. Simultaneous fermentation and saccharification were found to be of benefit in function. The disadvantageous performance of proteins was overcome by their blending with other biomaterials in a cheap, efficient manner. The cost-efficient films so developed could have applications where biodegradable films are usable such as in consumer bags, packaging films, silage wraps and agricultural mulch films (Mekonnen et al. 2016).

Soybean-based fibre is a regenerated protein fibre, which may be blended with PVA, PE, nylon, cellulose, PP, PLA, PHA, etc. It is nonallergic, biodegradable and micro-biocidal. Soybean fibre itself is made from soybean cake after oiling by a bioengineering process. This involves, first, the distillation and refinement of spherical protein from soybean cake; the spatial structure of the spherical protein is then changed under the action of biological enzymes and auxiliary chemical treatment; after the liquid material is cooked, the 0.9–3.0 dtex fibre is generated by wet spinning, stabilized by acetalizing and lastly cut into short staples after curling and thermoforming. Fabrics produced from this have a good handle, draping property, moisture absorption and air permeability along with colour fastness, acid/alkali and light resistance, suitable for premium clothing. It may be blended with cotton, wool, polyester, etc. Chemical modification of soybean fibre during processing of the soybean protein meal could involve various routes to cross-linking, internal plasticization and compatibilization with hydrophobicity enhancers and/or blend with suitable resins to obtain favourable properties (Kumar et al. 2009; Guerrero et al. 2014; Vieira et al. 2011; Su et al. 2010).

4.6 *Bio-based Materials from Pulp/Paper/Fibre*

The production of paper was carried out wherein the main raw material for paper production that is pulp was obtained from the banana plant. This pulp could be blended with pulp obtained from bamboo which is less abundantly accessible to obtain oil-proof paper. The motivation behind the choice of banana as a raw material was that in banana plantations, after the fruits are harvested, the trunks or stems are discarded. Billion tons of stem and leaves are thrown away annually. In this project, pulp was produced from the banana waste stem by means of soda process and kraft process; also analysis of both processes based on their energy and raw material consumption yield was done, and a most efficient process was delivered (refer to Fig. 7) (Marella et al. n.d.).

The development of coatings and films, based on proteins, included suitable application in board or paper products, as well as prospective industrial potential. The signatory amide-based chemical structure of proteins gives rise to various properties but most significantly a high intermolecular binding potential such that protein-based films exhibit much improved mechanical properties than polysaccharide- and fat-based films. Two main types of generating protein coatings could be used: a wet process based on dispersion or solubilization of proteins, subsequently taken for drying, and a dry extrusion process based on the thermoplastic properties of proteins. Other processes using chemical transformation and cross-linking of proteins were also frequently utilized for modulating mechanical, barrier, surface and biodegradable properties (Coltelli et al. 2015).

Fig. 7 Bio-based materials from pulp/paper/fibres

4.7 Bio-based Materials from Sodium Alginate and Chitosan/Chitin

L-Guluronic acid and D-mannuronic acid are the two component monomer units which form the polymeric macromolecule of sodium alginate. Alginate fibres are used in commercial wound dressings as they resorb wound exudate while being a barrier to external infective agents sustaining a favourable microenvironment for tissue regrowth. Calcium alginate is prepared by reacting aqueous sodium alginate with the added aqueous calcium chloride. Chitin and chitosan may be considered as cellulose-like polymers in that the C-2 hydroxyl group of the ß-glucose monomer component is substituted with an acetamido group in chitin, while it is substituted with an amino group in case of chitosan. Based on the understanding of this chemical structure, removal of the acetyl group in the substituent produces the chitosan. Chitosan, having large-volume industrial utility, has a chemical structure composed of glucosamine units linked by 1–4 glycosidic bonds to N-acetyl glucosamine units. Being biocompatible, biodegradable and non-toxic gave it huge applicability in different industrial products such as those from food, biomedical, cosmetics, agricultural, textiles, pharmaceutical industries (Raafat and Sahl 2009).

5 Functionality of Bio-based Materials from Biopolymers

The key qualities of healthcare products include antiviral, bacteriostatic, non-toxic, fungistatic, high absorbent, nonallergic, breathable, haemostatic, biocompatible and ability to incorporate medications while also possessing reasonably good mechanical properties. Rigid packaging, flexible packaging, textiles, agriculture and horticulture (soil retention sheeting, agriculture film), non-woven fabrics for agriculture, filtration, hygiene and protective clothing are some such products. The production of biopolymers is increasing continuously from 1.5 million tons in 2012 to likely reach 6.7 million metric tons by 2018 (Johanson and Vahlne 1977). A few important biopolymers are discussed.

5.1 Polylactic Acid (PLA) and Bio-polyethylene

PLA, discovered in 1845, is the only melt-processable fibre from annually renewable natural resources such as cornstarch, tapioca or sugarcane. The process of its manufacture involves the extraction of starch from the raw natural plant source. The starch is degraded to its monomer dextrose and taken for fermentation without prior purification. The fermentation products are the D- and L-isomers of lactic acid, which is taken as the base monomer for preparation of mesolactide and final PLA polymer production. PLA is subsequently modified, as required, and drawn into fibres, cast as films or moulded into plastic products like bottles. The favourable properties of the bioplastic, PLA fibres are strong and durable, and their fabric has good drape and UV resistance. The limiting oxygen index (25) is higher than that of PET. It has low water uptake (0.4–0.6%) but greater than that of PET and PP (Douka et al. 2017; Choudhury 2018). Bio-based polyethylene has identical physical, chemical and mechanical properties to petrochemical polyethylene. The biological method of production involves an initial fermentation of sugarcane or other plant starches to bioethanol, which is distilled at high temperature over a solid catalyst to reduce it to ethylene monomer. This is then polymerized by microbes or a green polymerization to polyethylene (Choudhury 2018).

5.2 Poly(Alkylenedicarboxylate) Polyesters and Bacterial Polyesters

Common dicarboxylic and diol monomers found in poly(alkylenedicarboxylate) polyesters include succinic acid, adipic acid, ethylene glycol and 1,4 butanediol. They are used along with polyurethanes in the making of coatings, adhesives, flexible packaging, agricultural films and compostable bags, with other bio-based polymers to improve properties (Douka et al. 2017). Polyhydroxyalkanoates are biopolymers, which can also be produced biochemically by genetically modified microorganisms or plants. Polyhydroxyalkanoates can also be generated from glycerol, algae or aromatic sources likes plant lignins and tannins. Polyhydroxybutyrate copolymer named "Biopol" was developed by Zeneca Bioproducts (Choudhury 2018; Blackburn 2009).

6 Conclusions

Bio-based materials are significant because they decrease the ecological footprint at all the stages of the product's life cycle. Our exponentially increasing population and the global spread of economic development are both putting a demanding squeeze on the world's resources. The food, textile and plastic packaging industry have gained visibility in the forefront of the sustainability agenda, not because it is a major cause of environmental problems but because consumers are directly affected by their usage. Bio-based products became more attractive, and recycling of bio-based materials could be more sustainable than composting as a terminating procedure. Process development to tailor the physico-chemical properties of bio-based materials to their specific end application is an important tool, such as cross-linking matrix to increase mechanical strength. It is worth noticing that even if at the current stage of knowledge these materials look like a tiny compromise between sustainability and economic efficiency, the development of genetic tools allows hope in further full bio-based versions of these materials. Recent breakthroughs also concern the development of pure bio-based products identical to the petroleum-based materials.

References

Basak S, Samanta KK, Chattopadhyay SK et al (2016) Green fire retardant finishing and combined dyeing of proteinous wool fabric. Color Technol 132:135–143. https://doi.org/10.1111/cote.12200

Baumann MGD, Lorenz LF, Batterman SA, Guo-Zheng Z (2000) Aldehyde emissions from particleboard and medium density fiberboard products. For Prod J 50:75

Blackburn R (2009) Sustainable textiles: life cycle and environmental impact. Elsevier, Woodhead Publishing, and CRC Press, Cambridge, UK

Chen M-J, Shi Q-S (2015) Transforming sugarcane bagasse into bioplastics via homogeneous modification with phthalic anhydride in ionic liquid. ACS Sustain Chem Eng 3:2510–2515

Choudhury AKR (2018) Biopolymers in textile industry. In: Padinjakkara A, Thankappan A, Souza FG, Thomas S (eds) Biopolymers and biomaterials. Apple Academic Press, Toronto

Coltelli M-B, Wild F, Bugnicourt E et al (2015) State of the art in the development and properties of protein-based films and coatings and their applicability to cellulose based products: an extensive review. Coatings 6:1

Douka A, Vouyiouka S, Papaspyridi L-M, Papaspyrides CD (2017) A review on enzymatic polymerization to produce polycondensation polymers: the case of aliphatic polyesters, polyamides and polyesteramides. Prog Polym Sci 79:1–25

Fan Y, Nishida H, Shirai Y et al (2004) Thermal degradation behaviour of poly (lactic acid) stereocomplex. Polym Degrad Stab 86:197–208
Fischer S, Vlieger de J Kock T, Gilberts J, et al (2000) Green composites—the materials of the future—a combination of natural polymers and inorganic particles. In: Proceedings of the Food Biopack Conference, p 29
Fortea-Verdejo M, Bumbaris E, Burgstaller C, Bismarck A, Lee KY (2017) Plant fibre-reinforced polymers: where do we stand in terms of tensile properties? Int Mater Rev 62(8):441–464
Fukushima K, Kimura Y (2006) Stereocomplexed polylactides (Neo-PLA) as high-performance bio-based polymers: their formation, properties, and application. Polym Int 55:626–642
Gade R, Tulasi MS, Bhai VA (2013) Seaweeds: a novel biomaterial. Int J Pharm Pharm Sci 5:975–1491
Guerrero P, Leceta I, Peñalba M, De La Caba K (2014) Optical and mechanical properties of thin films based on proteins. Mater Lett 124:286–288
Huggins T, Wang H, Kearns J et al (2014) Biochar as a sustainable electrode material for electricity production in microbial fuel cells. Bioresour Technol 157:114–119
Iguchi M, Yamanaka S, Budhiono A (2000) Bacterial cellulose—a masterpiece of nature's arts. J Mater Sci 35:261–270
Ikada Y, Jamshidi K, Tsuji H, Hyon SH (1987) Stereocomplex formation between enantiomeric poly (lactides). Macromolecules 20:904–906
Johanson J, Vahlne J-E (1977) The internationalization process of the firm-a model of knowledge development and increasing foreign market commitments. J Int Bus Stud 8:23–32
Kirpluks M, Cabulis U, Avots A (2016) Flammability of bio-based rigid polyurethane foam as sustainable thermal insulation material. In insulation materials in context of sustainability. InTech
Koch K, Gillgren T, Stading M, Andersson R (2010) Mechanical and structural properties of solution-cast high-amylose maize starch films. Int J Biol Macromol 46:13–19
Koizumi T, Tsujiuchi N, Adachi A (2002) The development of sound absorbing materials using natural bamboo fibers. WIT Transactions on The Built Environment 59
Kumar R, Wang L, Zhang L (2009) Structure and mechanical properties of soy protein materials plasticized by thiodiglycol. J Appl Polym Sci 111:970–977
Marella JBR, Madireddy S Maripi AN (n.d.) Production of pulp from banana pseudo stem for grease proof paper. Table Content Top Page no 61
Maskell D, da Silva CF, Mower K, et al (2015) Properties of bio-based insulation materials and their potential impact on indoor air quality
Meena R, Lehnen R, Schmitt U, Saake B (2011) Effect of oat spelt and beech xylan on the gelling properties of kappa-carrageenan hydrogels. Carbohydr Polym 85:529–540
Mekonnen T, Misra M, Mohanty AK (2016) Fermented soymeals and their reactive blends with poly (butylene adipate-co-terephthalate) in engineering biodegradable cast films for sustainable packaging. ACS Sustain Chem Eng 4:782–793
Mittal N, Jansson R, Widhe M et al (2017) Ultrastrong and bioactive nanostructured bio-based composites. ACS Nano 11:5148–5159
Monteiro SN, Lopes FPD, Barbosa AP et al (2011) Natural lignocellulosic fibers as engineering materials—an overview. Metall Mater Trans A 42:2963
Muneer F (2013) Evaluation of the sustainability of hemp fiber reinforced wheat gluten plastics
Nan N, De-Vallance D (2014) Bio-based carbon/polyvinyl alcohol composite materials. In: Poster present. 2014 Bioelectron. Biosensing International Symposium. Morgantown, WV, April. p 28
Pandey R, Patel S, Pandit P, Nachimuthu S, Jose S (2018) Colouration of textiles using roasted peanut skin-an agro processing residue. J Clean Prod 172:1319–1326
Petersen K, Nielsen PV (2000) Potential biologically based food packaging: a Danish study. Food Biopack Conference Copenhagen (Denmark), 27–29 Aug 2000
Petersen K, Nielsen PV, Bertelsen G et al (1999) Potential of biobased materials for food packaging. Trends Food Sci Technol 10:52–68

Petersen K, Nielsen PV, Olsen MB (2001) Physical and mechanical properties of biobased materials starch, polylactate and polyhydroxybutyrate. Starch-Stärke 53:356–361

Raafat D, Sahl H (2009) Chitosan and its antimicrobial potential–a critical literature survey. Microb Biotechnol 2:186–201

Razzaq HAA, Pezzuto M, Santagata G et al (2016) Barley β-glucan-protein based bioplastic film with enhanced physicochemical properties for packaging. Food Hydrocoll 58:276–283

Sanyang ML, Sapuan SM, Jawaid M et al (2016) Recent developments in sugar palm (*Arenga pinnata*) based biocomposites and their potential industrial applications: a review. Renew Sust Energ Rev 54:533–549

Souza AC, Benze R, Ferrão ES et al (2012) Cassava starch biodegradable films: influence of glycerol and clay nanoparticles content on tensile and barrier properties and glass transition temperature. LWT Food Sci Technol 46:110–117

Storz H, Vorlop K-D (2013) Bio-based plastics: status, challenges and trends. Appl Agric For Res 63:321–332

Su J-F, Huang Z, Zhao Y-H et al (2010) Moisture sorption and water vapor permeability of soy protein isolate/poly (vinyl alcohol)/glycerol blend films. Ind Crop Prod 31:266–276

Sullins T, Pillay S, Komus A, Ning H (2017) Hemp fiber reinforced polypropylene composites: the effects of material treatments. Compos Part B Eng 114:15–22

Takasaki M, Ito H, Kikutani T (2003) Development of stereocomplex crystal of polylactide in high-speed melt spinning and subsequent drawing and annealing processes. J Macromol Sci Part B 42:403–420

Tănase EE, Popa ME, Râpă M, Popa O (2015) Preparation and characterization of biopolymer blends based on polyvinyl alcohol and starch. Rom Biotechnol Lett 20:10307

Tanrattanakul V, Bunkaew P (2014) Effect of different plasticizers on the properties of bio-based thermoplastic elastomer containing poly (lactic acid) and natural rubber. Express Polym Lett 8:387–396

Teli MD, Pandit P (2017a) Novel method of ecofriendly single bath dyeing and functional finishing of wool protein with coconut shell extract biomolecules. ACS Sustain Chem Eng 5 (9):8323–8333

Teli MD, Pandit P (2017b) Multifunctionalised silk using *Delonix regia* stem shell waste. Fibers Polym 18:1679–1690

Teli MD, Pandit P (2017c) A novel natural source *Sterculia foetida* fruit shell waste as colorant and ultraviolet protection for inen. J Nat Fibers 15(3):337–343

Teli MD, Pandit P (2018) Development of thermally stable and hygienic colored cotton fabric made by treatment with natural coconut shell extract. J Ind Text, 1528083717725113 48(1):87–118

Teli MD, Pandit P, Basak S (2017a) Coconut shell extract imparting multifunction properties to ligno-cellulosic material. J Ind Text, 1528083716686937 47(6):1261–1290

Teli MD, Sahoo MR, Pandit P (2017b) Antibacterial and UV-protective cotton fabric made by herbal synthesized silver nanoparticles, 04, 1310–1321

Tsuji H, Ikada Y (1999) Stereocomplex formation between enantiomeric poly (lactic acid) s. XI. Mechanical properties and morphology of solution-cast films. Polymer (Guildf) 40:6699–6708

Türe H, Gällstedt M, Hedenqvist MS (2012) Antimicrobial compression-moulded wheat gluten films containing potassium sorbate. Food Res Int 45:109–115

Vieira MGA, da Silva MA, dos Santos LO, Beppu MM (2011) Natural-based plasticizers and biopolymer films: a review. Eur Polym J 47:254–263

Weber CJ, Haugaard V, Festersen R, Bertelsen G (2002) Production and applications of biobased packaging materials for the food industry. Food Addit Contam 19:172–177

Yan L, Chouw N, Jayaraman K (2014) Flax fibre and its composites–a review. Compos Part B Eng 56:296–317

Zhu X, Kim B-J, Wang Q, Wu Q (2013) Recent advances in the sound insulation properties of bio-based materials. Bioresources 9:1764–1786

Potential Bio-Based Edible Films, Foams, and Hydrogels for Food Packaging

Baburaj Regubalan, Pintu Pandit, Saptarshi Maiti, Gayatri T. Nadathur, and Aranya Mallick

Abstract To improve the usage of sustainable materials and to reduce environmental pollution, bio-based materials are being encouraged for use in food packaging. This review introduces the recent developments in various potential bio-based materials that are being used for the preparation of edible films, foams, and hydrogels in food packaging applications. It also summarizes hydrocolloids studied to develop biodegradable packaging films and improvement on their functional properties like water vapor permeability and superior gas barrier functions. Recently, researchers are interested in the modification of hydrocolloids by novel techniques to improve their functional properties in order to replace the synthetic polymers. The biodegradable foams have been replaced by EPS (expanded polystyrene) using various cheap biopolymers like cassava, cellulose fibers, and sunflower proteins. The recent development in the formulation of hydrogels using bio-based materials has also been discussed. Application of these recent techniques to create edible films, foams, and hydrogels helps in the development of potential novel biodegradable food packaging applications.

Keywords Edible films · Foams · Hydrogels · Hydrocolloids · Biodegradable food packaging

B. Regubalan (✉)
Department of Food Engineering and Technology, Institute of Chemical Technology, University Under Section -3 of UGC Act, Mumbai, India

P. Pandit (✉) · S. Maiti · G. T. Nadathur · A. Mallick
Department of Fibres and Textile Processing Technology, Institute of Chemical Technology, University Under Section -3 of UGC Act, Mumbai, India

S. Ahmed (ed.), *Bio-based Materials for Food Packaging*,
https://doi.org/10.1007/978-981-13-1909-9_5

1 Potential Bio-Based Edible Films for Food Packaging

1.1 Introduction

Edible film is a thin layer of material that can be used to wrap or cover various foods to prolong their shelf life. It can be eaten together with the food. It improves mechanical properties, sensory, convenience, and barriers to prevent the loss of moisture and selective movement of gases like oxygen, carbon dioxide, and ethylene (Janjarasskul and Krochta 2010). The biopolymer available from various sources has been shown in Fig. 1. Bio-based polymer components are in the main, extracted from renewable resources, for example, polysaccharides (cellulose, starch, etc.), proteins (casein, gluten, etc.), and lipids. Similarly, other biopolymers are synthesized from renewable bio-derived monomers. Even genetically modified microbes also synthesize biopolymer like polyhydroxybutyrates.

The classification of edible films prepared from various components is shown in Fig. 2. Edible films are prepared by various components which are divided into three categories: hydrocolloids (such as polysaccharides and proteins), lipids (such as fatty acids and waxes), and composites (Park et al. 2002). Edible packaging has potential applications in the food industry to enhance the food product's quality and shelf life by the addition of food additives such as flavor and aroma compounds, antimicrobial agents, antioxidants, pigments, and vitamins.

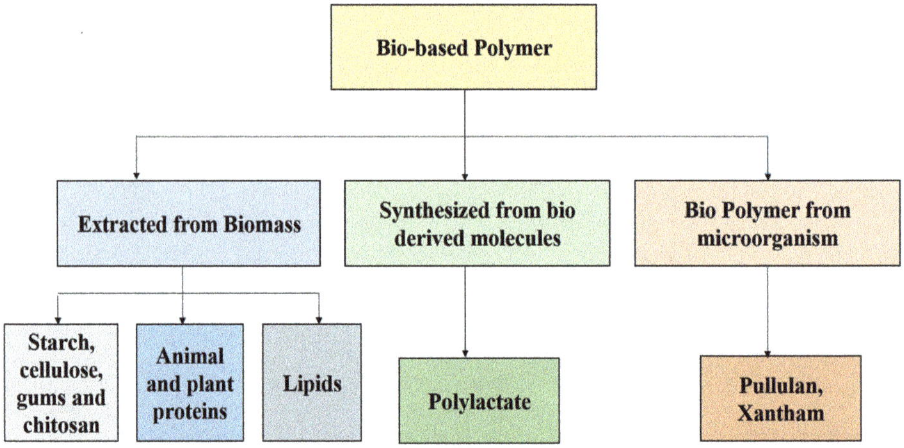

Fig. 1 Biopolymers available from different sources

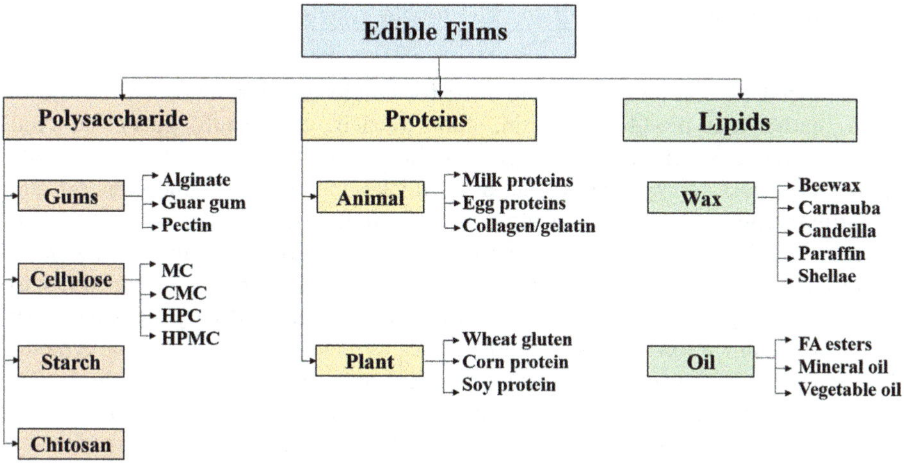

Fig. 2 Classification of edible film based on the nature of their components

1.2 Preparation of Edible Films

Typically, edible films are manufactured by dispersing film materials in a solvent such as water and alcohol. Generally, plasticizers, colors or flavors, and antimicrobial agents can be added followed by adjustment of pH of the film solution. Few polymeric film solutions require heating to facilitate dispersion. The film solution is cast and dried at a desired temperature to obtain stand-alone thin films. Several methods such as brushing, dipping, and spraying can be performed on food application.

1.3 Barrier Properties

Barrier properties of edible films are important parameters which include water vapor permeability (WVP), gas permeabilities, volatile permeability, and solute permeability. The effectiveness of edible film as desired is determined by investigating the water vapor permeability and oxygen permeability. The oxygen and carbon dioxide permeabilities are very important parameters that determine the rate of oxidation and respiration of foods. Edible films made up of polar polymer tend to crack due to high sensitivity to change in humidity. Lipid materials show very low water vapor permeability and also act as excellent moisture barriers. The disadvantage of lipid materials is that they are less effective gas barriers (Krochta 1992). Lipid materials do not form a film without film-forming agents such as polysaccharides and proteins.

1.4 Polysaccharide

The main polysaccharides which are used for edible film formation are cellulose, starch (native and modified), pectin, gums (acacia, tragacanth, guar), seaweed extracts (alginates, carrageenan, agar), pullulan, and chitosan. Polysaccharide imparts crispness, compactness, hardness, thickening quality, adhesiveness, and gel-forming ability to a variety of films. In general, polysaccharides are hydrophilic in nature which result in poor water vapor, gas barrier properties, and low mechanical properties. These drawbacks restricted the industrial applications. In order to overcome these problems, combination of polysaccharides and fatty acids has been produced. Modification of polysaccharides such as methylcellulose or carboxymethylcellulose has been used to enhance the barrier properties of the edible films. The synthesis of films formulated with chitosan and lipids by micro-fluidization process resulted in reduced water vapor permeability (Bonilla et al. 2012).

1.5 Protein

Edible films can be made from proteins of both plant and animal origin sources. In order to form an extended structure of the films, the protein must be denatured by heat, acid, or solvent. Eventually, the resulting interactive force produces a cohesive protein film matrix. These interactions are stabilized through uniform distribution of hydrogen bonding, van der Waals forces, covalent bonding, and disulfide bridges (Mchugh et al. 1994). The diverse range of amino acid functional groups in proteins can be altered chemically or enzymatically to improve the stability of the edible film. Thus cross-linked protein films often possess stable and longer shelf life than polysaccharide-based films. Casein, corn zein, gelatin, mung bean protein, peanut proteins, soy protein, wheat gluten, and whey protein are different types of proteins that have been used for edible film formulations (Bourtoom 2008).

1.6 Lipid

The hydrophobic film-forming barrier materials are waxes, lacquers, fatty acids and alcohols, acetylated glycerides, cocoa-based compounds, and their derivatives. Lipid-based edible films are thicker and more brittle due to low mechanical strength, but it provides higher moisture barrier than hydrocolloid-based edible films. Lipid-based edible films are susceptible to lipid oxidation and rancidity which eventually change the appearance and taste of foods. Therefore, the pure lipid has to mix with polysaccharides and proteins to provide a composite film with both barrier and mechanical properties (Kamper and Fennema 1985). Various structure formations are feasible to increase the mechanical strength of the composite film which includes the monolayer, multilayers, solid emulsion, and solid dispersion.

1.7 Plasticizer

Plasticizers are low molecular weight, nonvolatile substances with a high polarity that do not chemically bind to the polymeric chain and decrease intermolecular attractions between adjacent polymeric chains. Plasticizers improve the film elasticity, flexibility, stretchability, and toughness and reduce the brittleness and prevent cracking during handling and storage. Polysaccharide-based edible films require plasticizer, and it varies between 10% and 60% (w/w) of the hydrocolloid. Polyols, glycerol, sorbitol, polyethylene glycol, oligosaccharides, and water are the most commonly used plasticizers. The addition of plasticizer imparts free volume and molecular mobility which reduces film barrier properties (Sothornvit and Krochta 2005).

1.8 Other Additives

There is a wide range of food additives that have been used in edible films to achieve active packaging which includes antioxidants and antimicrobial compounds. Essential oils are the major antimicrobial compounds used in the formulation of edible film. Incorporation of nutraceutical inedible films is an approach to fortify the food. Other possible additives are vitamins, minerals, colorants, aromas, and prebiotics.

1.9 Applications

Active packaging is one of the most promising applications of edible films in foods. Instead of directly adding the active compounds or functional additives like antimicrobial compounds and antioxidant into the food, the sustainable release of these compounds is achieved by the active edible film. It can even act as a hurdle against spoilage of food. Various applications of edible films and their composition in food packaging are shown in Table 1.

1.10 Conclusion

The present chapter considers the recent literature concerning the classification of edible films and their application on food packaging. The edible films have been

Table 1 Composition and application of edible films in food packaging

Composition of films	Applications	References
Chitosan/thyme EO	Edible film on cured ham	Ruiz et al. (2015)
Argentine anchovy protein/sorbic/benzoic acids	Edible film for meat products	Rocha et al. (2014)
Chitosan/basil/thymus EOs	Edible film for pork	Bonilla et al. (2014)
Agar/green tea extract/probiotic strains	Edible fish for fish	De et al. (2014)
Gelatin/laurel leaf EO	Wrapping of trout	Alparslan et al. (2014)
Strawberry puree/carvacrol/methyl cinnamate	Shelf life extension of strawberry	Paeretto et al. (2014)
Apple, carrot, and hibiscus-based edible films	Shelf life improvement of green leafs	Zhu et al. (2014)
Cassava starch/mango pulp/yerba mate extract	Active films for palm oil	Reis et al. (2015)
Agar/nanocrystalline cellulose/savory essential oils	Active packaging	Atef et al. (2015)
k-Carrageenan/nanoclay/*Zataria multiflora Boiss* essential oil	Antimicrobial packaging	Shojaee et al. (2014)
Alginate/polycaprolactone/essential oils	Bioactive packaging	Salmieri et al. (2006)
Alginate/essential oils	Edible film on ham slices	Oussalah et al. (2007)
Starch/alginate	Edible films on beef patties	Wu et al. (2001)
Nanocrystalline cellulose/alginate	Biodegradable food packaging	Huq et al. (2012)
Carrageenan/grape fruit seed extract	Active packaging	Kanmani et al. (2014)
Agar/nanoclay	Biodegradable food packaging	Rhim et al. (2011)
Sodium caseinate/glycerol/maize germ oil bodies	Biodegradable food packaging	Matsakidou et al. (2013)
Starch/glycerol/xanthan gum/clove/cinnamon	Edible wrap on shrimps	Meenatchisundaram et al. (2016)
Whey protein/cinnamon/cumin	Preservation of red meat	Dohhi et al. (2014)
Sunflower protein/clove EO	Wrapping of fish patties	Salgado et al. (2013)
Methyl cellulose/clove/oregano EOs	Wrapping of sliced bread	Otoni et al. (2014)

prepared with a wide range of properties that could help food industry to reduce many problems encountered with the extension of shelf life of foods. Extensive research is still required to enhance the biopolymer-based film properties comparable to the synthetic films. Indeed, the improved biopolymer-based films also possess the potential applications.

2 Potential Bio-Based Foams for Food Packaging

2.1 Introduction

Food packaging is very important in case of food processing industries as it not only preserves the quality of the food and increases its span of consumption, but also it protects the food products from biological, physical, or chemical damages (Dallyn and Shorten 1988). In the beginning of the twentieth century, most of the consumer products were produced from plant-, animal-, or microbial-derived resources. However, after the advent of oil drilling, petroleum-derived chemicals came into existence. Now, recent developments are bringing back naturally derived resources to the forefront, once again making them the raw material source for the production of industrial products. In recent times, research scientists have successfully introduced developments and technologies that reduce cost and tailor the performance of bio-based products to the application (Weber et al. 2002). Nowadays, the interest in green manufacturing and environmentally friendly products is focusing on renewable resources derived from agriculture or forest output as alternative foodstocks. The advent of bio-based products like foams, edible films, hydrogels, etc. caused the growth of food packaging industries which was further boosted by the inception of new markets and industries. As said earlier, food packaging materials should be of such quality that they ensure retaining of quality for a long period of time and safe against all environmental adversities. It is a highly competitive area with huge demands for performance and economy. Utilization of petroleum derivatives, metals, glass, paper, and board was the most effective solution in this context due to their efficient performance at a very economical price. However, except boards and papers, which are biodegradable in nature, all others are made out of nonrenewable resources. A need of the food packaging materials made out of renewable resources such as polysaccharides extracted from natural resources became the far better choice in order to reach sustainability.

The agricultural policy of the European Union encourages the use of these agricultural crops as alternative sources of biopolymers like polysaccharides, casein, gluten, etc.; along with that, renewable bio-monomer-based polymers made by classical polymerization processes are also used, e.g., polylactate. In fact, these packaging materials based on alternative biological resources are at the research and development level for more than a decade and only recently have biomaterial-based food packaging materials from renewable resources that have been made commercially available. They have wide applications in food industries. Noticeably, bio-based and biodegradability are not exchangeable terms. The bio-based materials may have biodegradability as one of their properties, whereas biodegradable materials need not be bio-based.

Plastics have a wide application in preparing building material, packaging hygienic products, etc. (Popa et al. 2009). However, carelessness and negligence in dumping the plastic wastes caused havoc environmental damage, and with the course of time, it has increased exponentially in geographical spread and in quantity, with increased consumerism. Conventional plastic products made out of petroleum are non-biodegradable, thus considered as ecologically damaging waste. These polymers are very much persistent causing a significant environmental problem, intensified in urban places. Replacing the plastics with biodegradable foams can be a good solution. These new biopolymer-based products are biodegradable in nature, and their material cycle is similar to other natural products. The correspondence of properties between the food materials being packaged, and the bio-polymer foam based packaging materials, having potentially biodegradable, recyclable and disposable qualities, also attracts the sincere attention of consumers and manufacturing decision makers (Bucci et al. 2005; Lopez-Rubio et al. 2004).

2.2 *Production of Biodegradable Foams*

The biodegradable foams can be produced in three different ways as follows, baking or compression, extrusion, and microwave heating. The biodegradable foams are usually manufactured by the thermoforming process where copiously starch-based water mixtures are baked in closed molds. The renewable resources from agro-industrial residues like asparagus peel, cassava starch, cassava bagasse, coconut husk, kraft fiber, malt bagasse, sugarcane bagasse, and sunflower meal proteins have been explored for possible utilization of foams.

2.2.1 Baking

The process of production of biodegradable foams starts with a mixture of different compositions of raw materials followed by proper homogenization. A known quantity of homogenous solution is poured into a Teflon mold, and thermo-pressing is applied using the hydraulic press. The Teflon mold geometry totally depends on the requirement of the final foam's size and thickness. The compression mold is equipped with the programmable electric heating system to achieve the reproducible result. The evaporation of water at high temperature acts as a blowing agent to form a foam inside such mold. The temperature, hydraulic pressure, and holding time are crucial for the formation of foam. The viscosity of the initial solution before baking is the very important factor which also decides the end product quality. For example, the addition of starch and fibers leads to increase in viscosity which ultimately affects the expansion of the foam. In order to overcome limitations like poor mechanical properties, hydrophilicity, and low process yield, there is a need of various plasticizers and additives. Water and glycerol are the most preferred plasticizers, along with other additives like magnesium stearate. The magnesium stearate prevents the sticking of foams to the mold, and guar gum is used to avoid solid separation. Tween 80 and guar gum are also used for the foam production.

2.2.2 Extrusion

In general, the low-density foams have been conventionally manufactured from polystyrene, polypropylene, polyethylene, unplasticized PVC, and polylactic acids. Recently starch-based biodegradable foams have gained more importance for the production of biodegradable foam. Extrusion is one of the techniques to form foams in which water acts as both plasticizer and blowing agent in starch-based biodegradable foams. The expansion of steam occurs, and it exits through the die at high-pressure forming voids. The foam structure is majorly influenced by material compositions like starch-type, moisture content, and nucleating agent and extrusion conditions like screw speed, die diameter, and barrel temperature. The barrel temperature and screw speed usually fall in the range of 120–170 °C and 70–400 rpm, respectively, for the starch-based foams. The formation of the number of cells in the foams and expansion ratio increases with the increase in the barrel temperature and screw speed. The decrease in die size increases the expansion ratio and decreases the foam density.

$$\text{Expansion ratio/Expansion index (EI)} = (D/d)^2$$

D = Diameter of expanded foam after cooling
d = Diameter of die nozzle

The expansion of foam is significantly affected by moisture content. Typically, 15–18% water content is required to dispense the largest expansion to the foams. The low moisture content in feed gives low-density foam with highly expanded structure. Typically, talc is used as an inorganic nucleating agent and helps in the formation of uniform structure in the extruded foams. The type of starch especially the amylose content used in the feed composition plays a major role in the cell structure, density, compressibility, and resilience. The amylose content lesser than 25% resulted in brittle foams, while 70% amylose provides foam resembling the commercially available expanded polystyrene foams with low compressibility (Soykeabkaew et al. 2015).

2.2.3 Microwave Heating

The microwave is generally used to heat or cook the food but it can also be used to form foams. It is one of the cheap techniques to achieve baking through microwaves. The microwave treatment converts the moisture into steam bubbles and forms a cellular structure. The starch gets gelatinized and the steam is lost from the matrix which eventually forms a solid swollen foam. The microwave heating also depends on the moisture content of the initial starch composition. Generally, 10–13% moisture is recommended for microwave heat treatment to form the highly expandable foam. The types of starch, especially the percentage of amylose and amylopectin, also play a very important role in the foam formation because amylopectin-rich starch can trap more water content that could possess good swelling and trap bubbles (Glenn and Orts 2001).

2.2.4 Other Methods

The foams can be prepared by freeze-drying process. The freezing of water present in the starch-based composition resulted in ice crystals and separation of solvent and solute. Typically, ice crystals are formed by nucleation and growth of crystals. These crystals are sublimated to form cellular structures at primary drying. Thus the cellular structure of the foams is highly influenced by the size and distribution of crystals during freezing. The secondary drying is carried out to remove the residual moisture content of the foam (Svagan et al. 2008).

2.3 Application

2.3.1 Polyurethane Foam

Polymer industries have advanced a lot since its inception. Polyurethane, which is made out of organic units linked together by urethane links, is a very effective product to prepare foams, fibers, elastomers, etc. Polyurethane shows excellent performance, stability, and durability in preparing protective materials. Polyurethanes are susceptible to microbial attack, and thus antimicrobial agents may be required for better stability.

Antimicrobial products can either be incorporated throughout the product by mixing with polymer, or it can be placed superficially by coating onto the product. The methods are not very durable in delivering long-lived antimicrobial function. Therefore natural products and/or stabilizers are required to be added into the formulation of polyurethane to sustain bacteriostatic properties.

It has been found in researches that tannin is very effective in preventing bacterial growth due to the presence of polymeric phenolic compounds. Its molecular weights generally range from 500 to 3000. Wood tannin can be effectively included into polyurethane foam formulation which enhances its resistance against microorganism and at the same time makes it more biodegradable (Popa et al. 2009).

Polystyrene possesses insulating properties, which favors its use as popular material for food packaging that is also cheap in cost. But substitutes for polystyrene are increasingly in demand, as regulatory compliance and consumer awareness are convincing manufacturers to opt for eco-friendly choices. Conventional polystyrene takes hundreds of years to decompose under natural conditions, though the process can be speeded up by employing chemicals, mechanical destruction, etc. But the forced degradation is not economical and would cause increased pollution in the environment.

Since polystyrene is not naturally biodegradable, attempts have been made to make it so, by incorporating cellulose and starch, which is a substrate for decomposing by microbes or by addition of light-sensitive polymers that usually degrades on exposure to sunlight, producing free radicals that also attack the polystyrene bonds. These attempts have not been completely satisfactory in achieving biodegradation of polystyrene. Recently researchers have developed a solution-based system which can embed micro-sized hydrophilic resins all over the styrene prior to its polymerization and prepare a polystyrene-like material. This product swells in the water and also disintegrates to a biodegradable powder form. The scientists also claim that by modifying the ingredients ratio, the rate of disintegration may be controlled. Another important feature of theses materials is that they cost much less than conventional materials. It could be readily adopted by cost-conscious manufacturers who also need to be compliant with environmental regulations (Shen et al. 2010).

Another type of a new polystyrene foam for food packaging has been produced that undergoes accelerated degrading without any performance loss while in use. Canadian firm Cascades claims that Bioxo™ oxo-degradable polystyrene foam containers decompose within 3 years and perform effectively when used as a food packaging material. After usage, the foam can be degraded into a fine powder. Further, it can be utilized by the bacteria and other microorganisms. The decomposition process releases carbon dioxide into the atmosphere, and the solution adds to the problems of greenhouse gas emissions, even though it may not use up landfill space for its disposal. Bioxo has been approved as a food contact material by both Canadian and US regulators.

2.3.2 Green Cell Foam

Green Cell® Foam (GCF) is a commercially available biodegradable foam packaging material in various ranges of lamination and constructions. It is produced from a cornstarch-based blend. Thermal resistance (R-value) and cushioning ability are the fundamental properties required to substitute the use of synthetic foams. Since GCF is biodegradable and starch-based, there existed doubts that the susceptibility to moisture would unfavorably influence its physical properties.

Typically, ice melt test was used to determine the R-value, and ASTM method D1596 was followed to measure the cushioning properties. The 2-inch GCF had G-values matching that of synthetic foam materials. The R-values at higher humidities were determined to be less than that of synthetic materials (Arif et al. 2007). These studies gave a good basis for the acceptable use of GCF material in packaging.

2.3.3 Cassava Starch Foam

Many research studies have been conducted to partially replace the cassava starch with the addition of 5–10% of malt bagasse; this resulted in the highest production yield with proper distribution throughout the foam. The trays formed by utilizing malt bagasse were well shaped without any cracks and have good appearance and adequate expansion. The trays which were produced with 15% and 20% (w/w) malt bagasse showed fewer production yields (Mello and Mali 2014).

Another study involved the addition of cassava starch with the leaf sheath of betel nut palm, coconut husk, or kraft fibers, sunflower meal protein, and gluten protein forming a tray. The final formulation of 10% kraft fiber and 10% gluten protein composite trays showed a maximum flexural strength, better compressive strength, and water and oil absorptions at 25 °C with a relative humidity of 75% for 1 week. In both the studies, the density of the final trays was higher than the commercial EPS trays due to the incorporation of large amount of fibers (Kaisangsri et al. 2014). The commercial EPS trays can be replaced by these biodegradable trays for the packaging of perishable foods. The fiber and protein contents are optimized in cassava starch-based biodegradable trays by utilizing sunflower proteins and cellulose fibers. The addition of 10% protein isolate and 20% fiber showed the highest resistance and less water absorption capacity. It also exhibits composite and homogenous microstructure (Salgado et al. 2008).

2.3.4 Nano-cellulose Composite Foams

The microcrystalline cellulose (MCC) nanocomposite foams were formed by incorporating the surface-modified montmorillonite and Tween 80 along with cellulose. Thus thermal and mechanical properties and water vapor permeability of the nanocomposite foam were significantly improved than the pure cellulose foam. The dispersion of surface-modified montmorillonite was properly distributed and turned into a uniform and small cell size resulting in better properties. Thus, nanocomposite foam trays can be used to pack the low-moisture foods (Ahmadzadeh et al. 2015).

2.4 Conclusions

Food industry is one of the producers of maximum packaging waste, and this has driven the need for biodegradable food packages that have grown in importance due to reduced damage to the environment. The current applicability of bio-based foams for food packaging does not require completely ideal oxygen and/or water barrier properties, but it should satisfy the minimum criteria; this has promoted the commercial exploitation of different types of bio-based packaging materials. It is evident that bio-based packaging materials offer a variety of substrates to be modeled to the requirements of food packaging industries. To strictly determine the performance of the functionality of bio-based packaging forms, the testing with certain storage procedures is essential before such packaging foams can be freely marketed as suitable substitutes for conventional packaging materials.

3 Potential Bio-Based Hydrogel for Food Packaging

3.1 Introduction

Hydrogels are insoluble, three-dimensional (3D), water-swollen with hydrophilic polymeric network and cross-linked structures formulated by both natural and synthetic polymeric materials. Hydrogels have the ability to hold a large amount of water and biological fluids without dissolving in water. The hydrogen bonding and ionic interaction of the cross-linked structures make hydrogel insoluble in water. The interaction of water and polymeric chain networks happens through capillary, osmotic, and hydration forces which eventually expands the chain networks. The internal transport, diffusion characteristics, and mechanical strength are the inherent properties which determine the equilibrium state of the hydrogel. It can be made by both naturally occurring and synthetic polymers. Synthetic polymers like poly (hydroxyalkyl methacrylates), polyacrylate, polyacrylamide, and polymethacrylamide and its derivatives poly (*N*-vinyl-2-pyrolidone) are usually preferred. Synthetic hydrogels have a long service life, high capacity of water absorption, and high gel strength than hydrogel prepared from natural polymers (Laftah et al. 2011).

3.2 Technical Features of Hydrogel

The ideal hydrogel functional characteristics are mentioned below:

- Highest water absorption capacity
- Highest absorbency under load (AUL)
- Cheaper prize
- Appropriate absorption rate depends on the application
- Highest durability and stability
- Nontoxic, odorless, and colorless
- Highest biodegradability
- Good rewetting capability
- Photostability

It is impossible to obtain all ideal features of the hydrogel; despite this synthetic hydrogels are well known to achieve the maximum level of these features. The conventional hydrogels should have high absorption rate and lower residual monomer (Ahmed 2015).

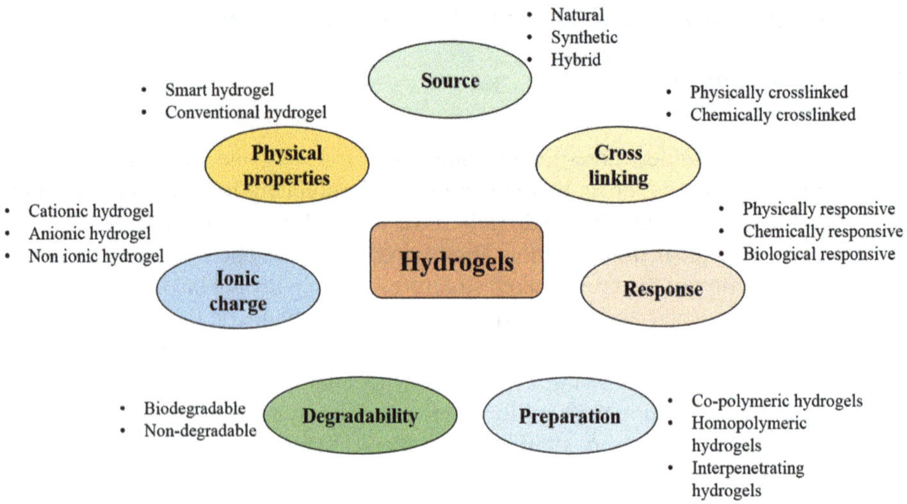

Fig. 3 Classification of hydrogels based on different properties

3.3 Classification of Hydrogels

Hydrogels can be classified on the basis of the method of preparation, ionic charges, nature of swelling, physical properties, and rate of biodegradation. The classification of hydrogels is shown in Fig. 3. Based on physical structure and chemical composition, hydrogels can be classified as follows: amorphous, semicrystalline (noncrystalline), and crystalline. The physical process such as crystallization, chain aggregation, hydrophobic association, and hydrogen bonding can be used to prepare physical gels. These hydrogels are reversible because the cross-linking process is physical and that may change their configuration. Meanwhile, chemical hydrogels are permanent and irreversible. Chemical hydrogels are prepared effectively by the chemical covalent cross-linking (Varaprasad et al. 2017). Since our major focus is on hydrogels in food packaging, the prominent hydrogels are briefly explained.

Based on sources, hydrogels are categorized into two main classes as artificial and natural. Further, it can be categorized into two main groups, polysaccharide hydrogel and protein hydrogel. Protein and polysaccharide pairs are capable of forming the most promising hydrogel from biopolymers, and it comes under poly-ion complex hydrogel in which ionic strength, biopolymer ratio, and pH are the major parameters for the formation of hydrogel (Turgeon et al. 2007). Generally, the natural-based hydrogels are prepared by a combination of some synthetic parts on the natural substrates, e.g., grafting of vinyl monomers on polysaccharides.

3.4 Polysaccharide-Based Natural Hydrogels

Polysaccharides show excellent properties such as water solubility, high swelling capacity, nontoxicity, easy to modify chemically, and a wide variety of chemical structures. The following polysaccharides can be used for hydrogel preparation which include alginate, agarose, carrageenan, chitosan, cellulose, gellan gum, gelatin, guar gum, locust beam gum, starch, pectin, xanthan gum, etc. (Cui et al. 2015). The hydrogel properties can be regulated and modulated by altering polysaccharides, temperature treatment, the nature of soluble additives, and the number of refreezing cycles.

3.5 Protein-Based Hydrogels

The hydrogel can be fabricated by a wide range of protein source which includes actin, collagen, elastin, fibrinogen, gelatin, keratin, and myosin. Peptides and proteins provide structural integrity to form a hydrogel and exhibit advantageous characteristics. Gelatin is a remarkable protein particularly for forming hydrogel packaging because it is relatively cheap and biodegradable. It can be modified on its amino, carboxyl, and hydroxyl group. Hydrogels prepared from gelatin are majorly affected by nature of raw material and fabrication process. The hydrogel formulated by cross-linking of oxidized pectin and gelatin with the addition of glycerol as plasticizer showed 324% of swelling equilibrium and improves the flexibility and handling ability (Gupta et al. 2014).

3.6 Application

Recently, the poly(DMA-co-SDM) hydrogels with a pH-responsive ionizable sulfonamide group were synthesized through free radical polymerization. This hydrogel can be used as freshness indicators by changing in the transparency response to metabolites released which ultimately determines the quality of the food products like meat and fish (Baek et al. 2017). The hydrogel added pastries showed a lower tendency to moisture redistribution and a higher increase in firmness when stored at ambient temperature (Manzocco et al. 2012). Agar-based hydrogel incorporated with antimicrobial compounds improves the shelf life of cheese through active packaging without affecting the sensory parameters (Incoronato et al. 2011).

Active packaging of molecular imprinted hydrogels (MIHs) has a great potential in the food industry. The ferulic acid-loaded MIHs protected the butter from oxidation and increased the shelf life during cold storage (Benito-Peña et al. 2016). The vegetables and fruits that are easily susceptible to early spoilage can be preserved by hydrogel-based packaging. The hydrogel-based PVP-CMC food package is used to increase the shelf life of table grapes up to 3 weeks at room temperature (Saha et al. 2015). The polyunsaturated lipids are encapsulated and protected by casein hydrogel microspheres. The hydrogel is fabricated by electrostatic complexation of low-methoxy pectin and caseinate at pH 4.5.The transglutaminase enzyme is used to improve the stability of the casein-rich hydrogel matrices. The hydrogel rich in casein could protect the polyunsaturated lipids in foods (Zhang et al. 2014). The stability of B-complex vitamins is improved by protective hydrogels. Folic acid and thiamine are well-known B vitamins that easily undergo oxidative degradation. At accelerated conditions of oxidative stress, catechin-based cross-linked hydrogels protect the B-complex vitamins (Spizzirri et al. 2013).

4 Conclusion

The hydrogels, which exhibit extraordinary characteristics, have been employed as vital tools for applications in almost all the fields. The hydrogels are made out of bio-based materials, and their applications on the food package have been discussed in this chapter. To fulfill the demands of food packaging, extensive research is ongoing, to provide novel hydrogels with improved features, for the expanding future needs of food packaging.

References

Ahmadzadeh S, Nasirpour A, Keramat J, Hamdami N, Behzad T, Desobry S (2015) Nanoporous cellulose nanocomposite foams as high insulated food packaging materials. Colloids Surf A Physicochem Eng Asp 468:201–210

Ahmed EM (2015) Hydrogel: preparation, characterization, and applications: a review. J Adv Res 6:105–121

Alparslan Y, Baygar T, Baygar T et al (2014) Effects of gelatin-based edible films enriched with laurel essential oil on the quality of rainbow trout (*Oncorhynchus mykiss*) fillets during refrigerated storage. Food Technol Biotechnol 52:325

Arif S, Burgess G, Narayan R, Harte B (2007) Evaluation of a biodegradable foam for protective packaging applications. Packag Technol Sci 20:413–419

Atef M, Rezaei M, Behrooz R (2015) Characterization of physical, mechanical, and antibacterial properties of agar-cellulose bionanocomposite films incorporated with savory essential oil. Food Hydrocoll 45:150–157

Baek S, Kim D, Jeon SL, Seo J (2017) Preparation and characterization of pH-responsive poly (N, N-dimethyl acrylamide-co-methacryloyl sulfadimethoxine) hydrogels for application as food freshness indicators. React Funct Polym 120:57–65

Benito-Peña E, González-Vallejo V, Rico-Yuste A et al (2016) Molecularly imprinted hydrogels as functional active packaging materials. Food Chem 190:487–494

Bonilla J, Atarés L, Vargas M, Chiralt A (2012) Effect of essential oils and homogenization conditions on properties of chitosan-based films. Food Hydrocoll 26:9–16

Bonilla J, Vargas M, Atarés L, Chiralt A (2014) Effect of chitosan essential oil films on the storage-keeping quality of pork meat products. Food Bioprocess Technol 7:2443–2450
Bourtoom T (2008) Factors affecting the properties of edible film prepared from mung bean proteins. Int Food Res J 15:167–180
Bucci DZ, Tavares LBB, Sell I (2005) PHB packaging for the storage of food products. Polym Test 24:564–571
Cui Z, Wang J, Zhang H, Jia H (2015) Investigation of high-intensity magnetic hydrogels in the application of membrane physical cleaning. Sep Purif Technol 154:301–308
Dallyn H, Shorten D (1988) Hygiene aspects of packaging in the food industry. Int Biodeterior 24:387–392
Dohhi E-BS (2014) Evaluation of the antimicrobial action of whey protein edible films incorporated with cinnamon, cumin and thyme against spoilage flora of fresh beef. Int J Agric Res 9:242–250
Glenn GM, Orts WJ (2001) Properties of starch-based foam formed by compression/explosion processing. Ind Crop Prod 13:135–143
Gupta B, Tummalapalli M, Deopura BL, Alam MS (2014) Preparation and characterization of in-situ crosslinked pectin–gelatin hydrogels. Carbohydr Polym 106:312–318
Huq T, Salmieri S, Khan A et al (2012) Nanocrystalline cellulose (NCC) reinforced alginate based biodegradable nanocomposite film. Carbohydr Polym 90:1757–1763
Incoronato AL, Conte A, Buonocore GG, Del Nobile MA (2011) Agar hydrogel with silver nanoparticles to prolong the shelf life of Fior di Latte cheese. J Dairy Sci 94:1697–1704
Janjarasskul T, Krochta JM (2010) Edible packaging materials. Annu Rev Food Sci Technol 1:415–448
Kaisangsri N, Kerdchoechuen O, Laohakunjit N, Matta FB (2014) Cassava starch-based biodegradable foam composited with plant fibers and proteins. J Compos Biodegrad Polym 2:71–79
Kamper SL, Fennema O (1985) Use of an edible film to maintain water vapor gradients in foods. J Food Sci 50:382–384
Kanmani P, Rhim J-W (2014) Development and characterization of carrageenan/grapefruit seed extract composite films for active packaging. Int J Biol Macromol 68:258–266
Krochta JM (1992) Control of mass transfer in food with edible coatings and films. In: Adv food eng, pp 517–538
de Lacey AML, López-Caballero ME, Montero P (2014) Agar films containing green tea extract and probiotic bacteria for extending fish shelf-life. LWT-Food Sci Technol 55:559–564
Laftah WA, Hashim S, Ibrahim AN (2011) Polymer hydrogels: a review. Polym-Plast Technol Eng 50:1475–1486
Lopez-Rubio A, Almenar E, Hernandez-Muñoz P et al (2004) Overview of active polymer-based packaging technologies for food applications. Food Rev Int 20:357–387
Manzocco L, Anese M, Calligaris S et al (2012) Use of monoglyceride hydrogel for the production of low fat short dough pastry. Food Chem 132:175–180
Matsakidou A, Biliaderis CG, Kiosseoglou V (2013) Preparation and characterization of composite sodium caseinate edible films incorporating naturally emulsified oil bodies. Food Hydrocoll 30:232–240
Mchugh TH, Aujard J, Krochta JM (1994) Plasticized whey protein edible films: water vapor permeability properties. J Food Sci 59:416–419
Meenatchisundaram S, Chandrasekar CM, Udayasoorian LP et al (2016) Effect of spice-incorporated starch edible film wrapping on shelf life of white shrimps stored at different temperatures. J Sci Food Agric 96:4268–4275
Mello LRPF, Mali S (2014) Use of malt bagasse to produce biodegradable baked foams made from cassava starch. Ind Crop Prod 55:187–193
Otoni CG, Pontes SFO, Medeiros EAA, Soares N de FF (2014) Edible films from methylcellulose and nanoemulsions of clove bud (*Syzygium aromaticum*) and oregano (*Origanum vulgare*) essential oils as shelf life extenders for sliced bread. J Agric Food Chem 62:5214–5219

Oussalah M, Caillet S, Salmieri S et al (2007) Antimicrobial effects of alginate-based films containing essential oils on Listeria monocytogenes and Salmonella typhimurium present in bologna and ham. J Food Prot 70:901–908
Park HJ, Rhim JW, Weller CL et al (2002) Films and coatings from proteins of limited availability. In: Protein-based film coatings, 12, pp 305–328
Peretto G, Du W-X, Avena-Bustillos RJ et al (2014) Increasing strawberry shelf-life with carvacrol and methyl cinnamate antimicrobial vapors released from edible films. Postharvest Biol Technol 89:11–18
Popa M, Mitelut A, Niculita P et al (2009) Biodegradable materials for food packaging applications. In: Int. work. environ. relantionsh. fram. EU policy, pp 16–17
Reis LCB, de Souza CO, da Silva JBA et al (2015) Active biocomposites of cassava starch: the effect of yerba mate extract and mango pulp as antioxidant additives on the properties and the stability of a packaged product. Food Bioprod Process 94:382–391
Rhim J-W (2011) Effect of clay contents on mechanical and water vapor barrier properties of agar-based nanocomposite films. Carbohydr Polym 86:691–699
da Rocha M, Loiko MR, Tondo EC, Prentice C (2014) Physical, mechanical and antimicrobial properties of Argentine anchovy (*Engraulis anchoita*) protein films incorporated with organic acids. Food Hydrocoll 37:213–220
Ruiz-Navajas Y, Viuda-Martos M, Barber X et al (2015) Effect of chitosan edible films added with Thymus moroderi and Thymus piperella essential oil on shelf-life of cooked cured ham. J Food Sci Technol 52:6493–6501
Saha N, Benlikaya R, Slobodian P, Saha P (2015) Breathable and polyol based hydrogel food packaging. J Biobased Mater Bioenergy 9:136–144
Salgado PR, Schmidt VC, Ortiz SEM et al (2008) Biodegradable foams based on cassava starch, sunflower proteins and cellulose fibers obtained by a baking process. J Food Eng 85:435–443
Salgado PR, López-Caballero ME, Gómez-Guillén MC et al (2013) Sunflower protein films incorporated with clove essential oil have potential application for the preservation of fish patties. Food Hydrocoll 33:74–84
Salmieri S, Lacroix M (2006) Physicochemical properties of alginate/polycaprolactone-based films containing essential oils. J Agric Food Chem 54:10205–10214
Shen L, Worrell E, Patel M (2010) Present and future development in plastics from biomass. Biofuels Bioprod Biorefin 4:25–40
Shojaee-Aliabadi S, Mohammadifar MA, Hosseini H et al (2014) Characterization of nanobiocomposite kappa-carrageenan film with Zataria multiflora essential oil and nanoclay. Int J Biol Macromol 69:282–289
Sothornvit R, Krochta JM (2005) Plasticizers in edible films and coatings. In: Innov. food packag. Elsevier, pp 403–433
Soykeabkaew N, Thanomsilp C, Suwantong O (2015) A review: starch-based composite foams. Compos Part A Appl Sci Manuf 78:246–263
Spizzirri UG, Cirillo G, Curcio M et al (2013) Stabilization of oxidable vitamins by flavonoid-based hydrogels. React Funct Polym 73:1030–1037
Svagan AJ, Samir MAS, Berglund LA (2008) Biomimetic foams of high mechanical performance based on nanostructured cell walls reinforced by native cellulose nanofibrils. Adv Mater 20:1263–1269
Turgeon SL, Schmitt C, Sanchez C (2007) Protein–polysaccharide complexes and coacervates. Curr Opin Colloid Interface Sci 12:166–178
Varaprasad K, Raghavendra GM, Jayaramudu T et al (2017) A mini review on hydrogels classification and recent developments in miscellaneous applications. Mater Sci Eng C 79:958–971
Weber CJ, Haugaard V, Festersen R, Bertelsen G (2002) Production and applications of biobased packaging materials for the food industry. Food Addit Contam 19:172–177
Wu Y, Weller CL, Hamouz F et al (2001) Moisture loss and lipid oxidation for precooked ground-beef patties packaged in edible starch-alginate-based composite films. J Food Sci 66:486–493

Zhang Z, Decker EA, McClements DJ (2014) Encapsulation, protection, and release of polyunsaturated lipids using biopolymer-based hydrogel particles. Food Res Int 64:520–526

Zhu L, Olsen C, McHugh T et al (2014) Apple, carrot, and hibiscus edible films containing the plant antimicrobials carvacrol and cinnamaldehyde inactivate Salmonella Newport on organic leafy greens in sealed plastic bags. J Food Sci 79:M61–M66

Nanotechnology and Edible Films for Food Packaging Applications

Paula J. P. Espitia and Caio G. Otoni

Abstract Nanotechnology has been exploited in the last decade as an innovative approach that allows obtaining nanoscaled organic and inorganic compounds featuring unique properties due to their size. When nanomaterials are incorporated in polymers intended to act as film-forming matrices for food packaging, they are known to improve their physical-mechanical performances by acting as fillers. Other properties (e.g., biological activity against foodborne pathogens and/or spoilage microorganisms) may also be developed and used as key features in biopolymer-based materials for food preservation. Concerning food-grade biopolymers, edible films may be produced and denote alternative packaging materials that are promising due to the possibility of taking advantage of film-forming polysaccharides and polypeptides that would be wasted to produce edible packaging featuring biodegradability and nutritional and sensory values. These edible materials may also have boosted active roles (antimicrobial and antioxidant, in particular) when specific nanoscaled additives are used. Therefore, this chapter reviews the most recent advances on the development of edible films incorporated with nanoparticles or nanoencapsulated compounds to be used as packaging materials for food preservation purposes. Moreover, physical-mechanical characterization techniques (such as mechanical resistance, barrier properties, and microstructure) and antimicrobial properties are addressed. Finally, future trends are discussed considering potential effects on consumers. This contribution presents the state of the art on edible nanocomposite films for food packaging applications.

Keywords Active packaging · Nanocomposite · Antimicrobial activity · Antioxidant activity · Nanoemulsion · Nanocapsule · Nanocellulose · Nanoclay · Nanostarch · Nanochitosan · Peptide

---

P. J. P. Espitia (✉)
Nutrition and Dietetics School, Universidad del Atlántico, Carrera 30 Número 8- 49. Puerto Colombia, Atlántico, Colombia

C. G. Otoni
National Nanotechnology Laboratory for Agribusiness, Embrapa Instrumentacão, São Carlos, Brazil

S. Ahmed (ed.), *Bio-based Materials for Food Packaging*,
https://doi.org/10.1007/978-981-13-1909-9_6

1 Introduction

Food packaging has always played a key role in food supply chain not only by unitizing items and attracting consumers with a marketing appeal but also by protecting the packaged product from external spoiling agents, leading to augmented food safety, quality, and shelf life. The environment denotes indeed an inexhaustible source of oxygen, moisture, light, heat, and microorganisms. When it comes to food, it is an overall consensus that these components are detrimental, requiring that packaging materials – the only obstacles between food and environment – act as efficient barriers to avoid food contamination and spoilage. In this sense, they should present suitable mechanical properties in order to withstand handling and shipping, as well as good thermal and barrier properties to prevent quality losses.

When neither glass nor metal is appropriate (e.g., due to cost, weight, or other technical limitations), the aforementioned requirements are often fulfilled by synthetic plastics. Their most common source (i.e., fossil fuels), however, is nonrenewable, and they are mostly nonbiodegradable, giving rise to concerns about raw material availability and waste accumulation as a result of extensive petroleum exploitation and disposal of petroleum-derived polymers in the environment (Otoni et al. 2017). Much attention has been given to polymers that are biodegradable and/or obtained from renewable sources so that to address these issues (Babu et al. 2013; Khan et al. 2017). Furthermore, various biopolymers (e.g., polysaccharides and polypeptides) are also edible, opening up a range of possibilities and potential applications, as extensively listed elsewhere (Otoni et al. 2017).

Nevertheless, the physical-mechanical properties of biopolymer-based packaging are generally inferior to those of conventional plastics, encouraging the development of strategies to overcome technical hurdles and make their practical application feasible. Among such strategies, nanotechnology emerges as the most promising approach of boosting packaging performance. It involves structures that range in size within the nanoscale and that are expected to feature unique characteristics in comparison to their micro- and macroscale analogues (Nitin et al. 2017).

Nanotechnology has been combined with edible films not only to improve their physical-mechanical properties but also to provide them with an additional protection against food contamination and spoilage. These nanocomposites are in line with the increasing trend of active packaging, i.e., packaging whose roles are no longer limited to physical protection but extended to antimicrobial and antioxidant actions, to mention a few (Otoni et al. 2016b). Food packaging, nanocomposites, and edible films have been widely reviewed elsewhere, though separately. Because the research activity on the intersection of these topics is increasing rapidly, as demonstrated by the significant change in the number of publications and citations during the past 10 years (Fig. 1), we identified a lack in the literature on nanotechnology and edible films for food packaging.

In this chapter, we review recent developments on nanostructured edible films for food packaging applications. Emphasis is given to the literature published during the last 5 years in order to hit the state of the art in the use of nanotechnology to provide

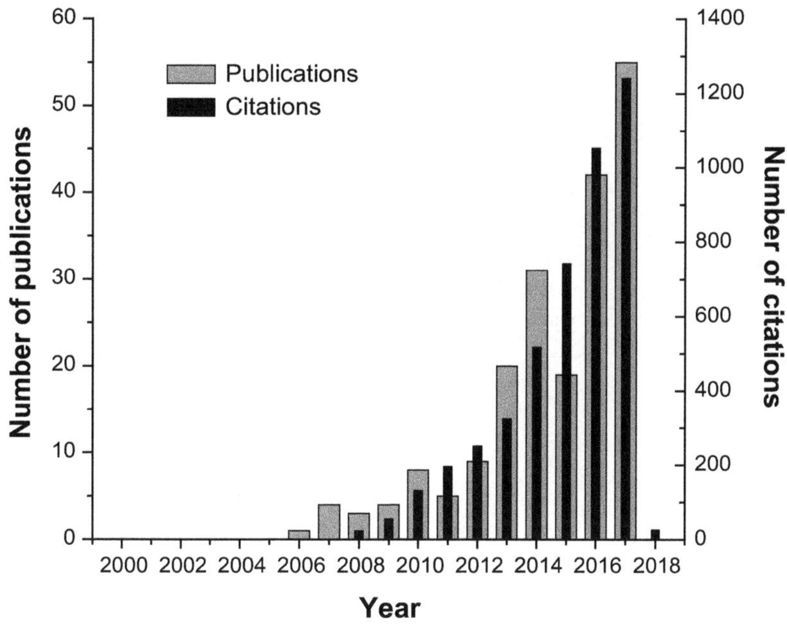

Fig. 1 Number of publications and citations on nanotechnology and edible films for food packaging applications [topic: nano* and (edible or eatable or digestible) and (film or sheet) and food and package*] retrieved from Web of Science™ Core Collection

edible films with (i) ameliorated physical-mechanical performances to pave the route to commercial applications and/or (ii) active roles (antimicrobial and antioxidant activities, in particular) toward active food packaging. Finally, future trends are presented and discussed.

2 Nanostructured Antimicrobial Edible Films

The application of antimicrobial compounds – including antimicrobial nanostructures – into food packaging has been widely studied and reviewed elsewhere (de Azeredo 2013; Sung et al. 2013). The major health concerns are related to the potential migration of active compounds from food contact materials toward food matrix. Concerning edible packaging materials, all of their components must be considered as part of the food product itself, provided that the whole system may be eaten without posing health threats (de Azeredo 2013). Two main mechanisms of adding antimicrobial compounds into food-grade polymer matrices have been exploited to produce antimicrobial edible films: nanoemulsions containing essential oils and nanoencapsulated antimicrobial peptides. Recent advances on both strategies are compiled in Table 1, and further information are discussed below.

Table 1 Antimicrobial edible films containing nanoemulsified, nanostructured, and nanoencapsulated active compounds developed during the past 5 years

Matrix	Nano-sized active compound	Tested microorganisms	References
Alginate	Lemongrass essential oil	*Escherichia coli*	Acevedo-Fani et al. (2015)
	Sage essential oil		
	Thyme essential oil		
Basil seed gum	*Zataria multiflora* essential oil	*Bacillus cereus*	Hashemi Gahruie et al. (2017)
		E. coli	
Casein	Nisin	*Listeria monocytogenes*	Boelter and Brandelli (2016)
Gelatin		*Staphylococcus aureus*	
		Clostridium perfringens	
Chitosan	Phage	*E. coli* O157:H7	Cui et al. (2017)
Hazelnut meal protein	Clove essential oil	*Bacillus subtilis*	Gul et al. (2018)
		E. coli	
		L. monocytogenes	
		Pseudomonas aeruginosa	
		S. aureus	
Hydroxypropyl methylcellulose (HPMC)	*Thymus daenensis* essential oil	*Acinetobacter baumannii*	Moghimi et al. (2017)
		B. subtilis	
		Candida albicans	
		Enterococcus faecalis	
		Enterococcus faecium	
		E. coli	
		Klebsiella pneumoniae	
		Salmonella typhi	
		Shigella dysenteriae	
		Shigella flexneri	
		S. aureus	
		Staphylococcus epidermidis	
HPMC	Nisin		Imran et al. (2012)
HPMC/starch	Nisin	*Listeria innocua*	Basch et al. (2013)
	Potassium sorbate	*Zygosaccharomyces bailii*	
Isolated soy protein	Carvacrol	–	Otoni et al. (2016a)
	Cinnamaldehyde		

(continued)

Table 1 (continued)

Matrix	Nano-sized active compound	Tested microorganisms	References
Methylcellulose	Clove bud essential oil	*Aspergillus niger*	Otoni et al. (2014b)
	Oregano essential oil	*Penicillium* sp.	
Mesquite seed gum	Palm fruit oil	–	Rodrigues et al. (2016)
Pectin	Cinnamon essential oil	–	Moura et al. (2014)
Pectin	Cinnamaldehyde	*E. coli*	Otoni et al. (2014a)
		L. monocytogenes	
		Salmonella enterica	
		S. aureus	
Pectin	Eugenol	*E. coli*	Sasaki et al. (2016)
		S. aureus	
Pullulan	Rosemary essential oil	*E. coli* O157:H7	Morsy et al. (2014)
	Oregano essential oil	*L. monocytogenes*	
		Salmonella typhimurium	
		S. aureus	
Tara gum	Chitosan	*E. coli*	Antoniou et al. (2015)
		S. aureus	

Nanoemulsions are colloidal, kinetically stable systems comprising at least three components: an oil phase, an aqueous phase, and surfactants. They can be used to encapsulate and deliver a wide range or hydrophilic and lipophilic compounds, including drugs for treating cancer treatment (Sasikumar and Kamalasanan 2017) and several other medical applications (Singh et al. 2017). In food science, nanoemulsions have been extensively exploited to encapsulate plant essential oils and deliver antimicrobial activity to increase food safety and shelf life (Pathakoti et al. 2017). They can also be used to incorporate other bioactive compounds, including fatty acids, carotenoids, antioxidants, phytosterols, and quinones (Salvia-Trujillo et al. 2017), and have the potential to be incorporated in edible films to produce active food packaging systems (Fu et al. 2016).

Moghimi et al. (2017) reported the antimicrobial activity of hydroxypropyl methylcellulose (HPMC)-based edible films incorporated with *Thymus daenensis* essential oil against a range of Gram-positive and Gram-negative bacteria as well as a fungus. The authors demonstrated that depending on the essential oil origin (i.e., wild or cultivated *T. daenensis*), the resulting films were more antimicrobially efficient against a specific class of microorganisms as a result of their different compositions.

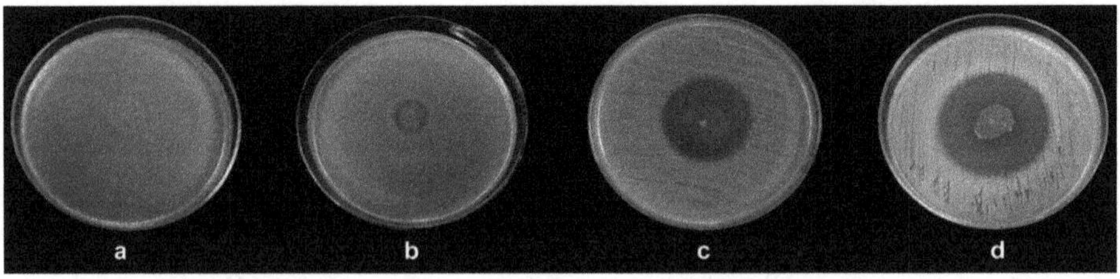

Fig. 2 Inhibition zones created by pectin/papaya puree-based edible films incorporated with cinnamaldehyde nanoemulsions comprising droplets of ca. 270 (**b**), 120 (**c**), and 40 nm (**d**) as well as control, i.e., nanoemulsion-free film (**a**). (Adapted from Otoni et al. (2014) with permission from Elsevier)

Otoni et al. (2014a) produced edible films based on low and high methylester pectins and papaya puree incorporated with cinnamaldehyde nanoemulsions of varying droplet sizes. Because of their increased surface areas and bioavailabilities, smaller droplets let to increased antimicrobial efficiencies against food spoilage and pathogenic bacteria. This is illustrated in Fig. 2, which demonstrates the greater inhibition zones created by smaller droplets for the same active compound concentration. In line with this observation, Otoni et al. (2014b) reported increased antimicrobial efficiency of MC-based edible films when these were added by nanoemulsions of clove bud and oregano essential oils in comparison with coarse-emulsified films. Similar outcomes have been reported for other systems (Gul et al. 2018).

Although essential oil nanoemulsions have been extensively applied to edible matrices in order to provide films with antimicrobial properties against foodborne pathogens, they often affect the physical-mechanical properties of the materials. *Thymus daenensis* essential oil nanoemulsion has been shown to plasticize hydroxypropyl methylcellulose-based films (Moghimi et al. 2017). The same has been observed for clove bud and oregano essential oils in a methylcellulose matrix (Otoni et al. 2014b); nanoemulsions of thyme, lemongrass, and sage essential oils in alginate edible films (Acevedo-Fani et al. 2015); palm fruit oil nanoemulsion in mesquite seed gum films (Rodrigues et al. 2016); and carvacrol and cinnamaldehyde nanoemulsions in isolated soy protein films (Otoni et al. 2016a). On the other hand, anti-plasticizing effects of nanoemulsions in edible films have also been demonstrated (Hashemi Gahruie et al. 2017; Otoni et al. 2014a; Sasaki et al. 2016). Rutin-loaded nanoemulsions, in particular, have been found to increase both tensile strength and elongation at break of gelatin-based films (Dammak et al. 2017). The authors attributed this behavior to the cross-linking of subjacent peptide molecules induced by rutin.

The barrier properties of edible films are also altered upon the incorporation of nanoemulsions. The barrier to moisture of pectin/papaya puree-based edible films has been increased by cinnamaldehyde nanoemulsions as a result of their hydrophobic nature (Otoni et al. 2014a). Similar outcomes have been reported for eugenol nanoemulsion in pectin films (Sasaki et al. 2016), oregano essential oil nanoemulsion in isolated soy protein films (Bilbao-Sáinz et al. 2010), sage essential oil nanoemulsion in alginate films (Acevedo-Fani et al. 2015), and palm fruit oil nanoemulsion in mesquite seed gum films (Rodrigues et al. 2016). Rutin-loaded nanoemulsions showed the opposite behavior in pigskin gelatin-based films (Dammak et al. 2017).

Protein fragments called peptides denote another important class of antimicrobial compounds used for food preservation. A range of naturally occurring – i.e., produced by microorganisms, amphibians, and mammals – and synthetic antimicrobial peptides have been combined with film-forming materials to produce antimicrobial food packaging, as comprehensively reviewed elsewhere (Perez Espitia et al. 2012). In addition to having an antimicrobial packaging system, this strategy is interesting because it avoids biological activity losses of the active compounds as a result of their deactivation when directly added into complex food systems (Imran et al. 2012). It allows controlling the diffusion of the antimicrobial compound to food surface from packaging materials, some of which may be food-grade or edible. Pintado et al. (2009) were able to control the growth of *L. monocytogenes* strains isolated from cheeses by using whey protein isolate-based edible films enriched with nisin, a bacteriocin – i.e., peptide synthesized at bacterial ribosome – with widely reported antimicrobial activity against many spoilage and pathogenic microorganisms (Ahmad et al. 2017; Barbosa et al. 2017). *L. monocytogenes* has also been inhibited by HPMC-based edible films added by nanoencapsulated nisin (Imran et al. 2012). Nisin has been combined with potassium sorbate and added into HPMC/starch blends to produce active edible films that were effective in inhibiting the growths of *Listeria innocua* and *Zygosaccharomyces bailii* (Basch et al. 2013). Boelter and Brandelli (2016) encapsulated nisin in liposomes, which were then combined with gelatin and casein to produce active edible films. These nanocomposites were effective in inhibiting the growths of *L. monocytogenes*, *S. aureus*, and *Clostridium perfringens*. These authors also added halloysite nanotubes, but were not able to achieve improved physical-mechanical properties. Finally, chitosan-based edible films have been produced and incorporated with zinc oxide nanoparticles (Bajpai et al. 2010). The produced films showed great antimicrobial activity against *E. coli*.

3 Edible Films Containing Nano-sized Fillers

Although edible films have numerous advantages over conventional plastics, their physical-mechanical properties are typically inferior than those of the latter and often limit their commercial applicability (Otoni et al. 2017). Concerning food packaging, the major drawbacks of edible films are their mechanical and water barrier properties. In order to fulfill the technical requirements for food packaging applications, edible films have been added by different nanostructured fillers. The recent developments on edible nanocomposite films are summarized in Table 2 and described in details in this section.

Table 2 Edible films containing nano-sized fillers developed during the past 5 years

Matrix	Nano-sized filler	Target properties	References
Banana flour	Montmorillonite (MMT)	Barrier to moisture	Orsuwan and Sothornvit (2017)
	Starch nanoparticles	Mechanical properties	
Chitosan and cornstarch	Cellulose nanofibrils	Barriers to light, oxygen, and moisture	Yu et al. (2017)
		Mechanical properties	
Gelatin	Amino-functionalized MMT	Barrier to light	Ge et al. (2017)
		Mechanical properties	
		Resistance to moisture	
Gelatin	Bacterial cellulose nanocrystals	Affinity to moisture	George and Siddaramaiah (2012)
		Barrier to moisture	
		Mechanical properties	
		Thermal properties	
Gelatin	Hydroxyapatite	Mechanical properties	Wu et al. (2017)
		Thermal properties	
Hydroxypropyl methylcellulose (HPMC)/guava puree	Chitosan nanoparticles	Affinity to moisture	Lorevice et al. (2012)
		Barrier to moisture	
		Mechanical properties	
		Thermal properties	
HPMC/papaya puree	Chitosan nanoparticles	Barrier to moisture	Lorevice et al. (2014)
		Mechanical properties	
		Thermal properties	
Mango puree	Cellulose nanofibrils	Barrier to moisture	Azeredo et al. (2009)
		Mechanical properties	
Pectin	$Mg(OH)_2$ nanoplates	Mechanical properties	Moreira et al. (2013)
		Thermal properties	
Pectin/banana puree	Chitosan nanoparticles	Barrier to moisture	Martelli et al. (2013)
		Mechanical properties	
Pectin/cocoa puree	Chitosan nanoparticles	Barrier to moisture	Melo et al. (2017)
		Mechanical properties	
Pectin	Cellulose nanocrystals	Barrier to moisture	Chaichi et al. (2017)
		Mechanical properties	

(continued)

Table 2 (continued)

Matrix	Nano-sized filler	Target properties	References
Pectin	Chitosan nanoparticles	Barrier to moisture	Lorevice et al. (2016)
		Mechanical properties	
Pigskin gelatin	MMT	Affinity to moisture	Alexandre et al. (2016)
		Mechanical properties	
Potato starch	Starch nanocrystals	Barrier to moisture	Sessini et al. (2016)
		Mechanical properties	
		Thermal properties	
Tara gum	Chitosan nanoparticles	Affinity to moisture	Antoniou et al. (2015)
		Mechanical properties	

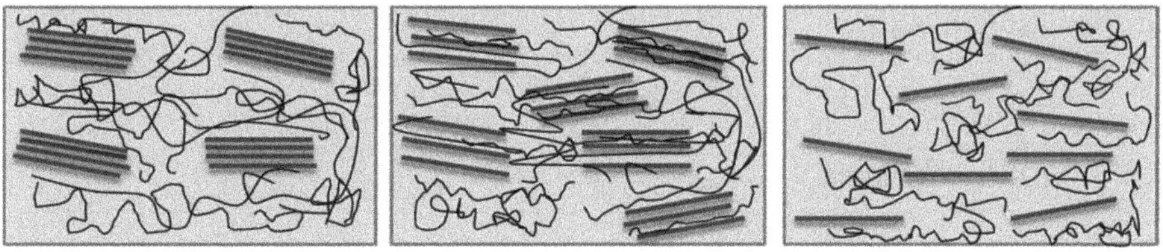

Fig. 3 Possible morphologies of nanoclays depending on their dispersion within polymer chains: aggregated (left), intercalated (center), and exfoliated (right). (Reproduced from Tan and Thomas (2016) with permission from Elsevier)

3.1 Nanoclays

Nanoclays are known to improve the mechanical resistance and barrier to moisture of polymeric films (Tan and Thomas 2016). These inorganic nanofillers are characterized by high aspect ratio and interfacial area. Nanoclays may be present in three main dispersion levels within polymer-based nanocomposites: aggregated, intercalated, and exfoliated (Fig. 3). Regardless of distribution within the nanocomposite matrix, clay tactoids may be aggregated (i.e., not delaminated), partially delaminated so polymer chains may be intercalated with them, and completely exfoliated into single-layered platelets (Tan and Thomas 2016). The last morphology is often the most desired since it has been shown to lead to the best thermal, barrier, and mechanical improvements even at low contents (Sinha Ray and Okamoto 2003), characteristics that are in line with technical requirements for efficient food packaging materials.

Montmorillonite is by far the most exploited nanoclay for reinforcement purposes in edible films. It has a negatively charged surface and a highly crystalline structure arising from a two-dimensional layer comprising a central octahedral sheet of aluminum oxide and magnesium oxide combined with two external silica tetrahedrons (Alexandre et al. 2016). Banana (*Musa sapientum* Linn) flour has been intercalated with MMT (Cloisite Na^+) for producing mechanically reinforced films (Orsuwan and Sothornvit 2017). The addition of such natural nanoclay also improved the water barrier properties of the edible films. The authors pointed out the higher hydrophilicity of MMT-Na^+ over organically modified nanoclays as an advantage because of the improved compatibility with food-grade biopolymers, which are mostly hydrophilic. The incorporation of MMT Cloisite Na^+ and Cloisite 30B has improved the tensile properties and barrier to moisture of glycerol-, sorbitol-, sucrose-, fructose-, and polypropylene glycol-plasticized red algae-based edible films (Jang et al. 2011). Similar outcome was observed by Alexandre et al. (2016), who reported pigskin gelatin-based edible films featuring enhanced mechanical properties as well as lower affinity to water when added by MMT. Ge et al. (2017) produced amino-functionalized MMT to serve as reinforcement fillers in gelatin-based edible nanocomposites cross-linked by dialdehyde xanthan gum. In addition to displaying suitable physical-mechanical properties and tunable degradability in soil, these edible films were added by free nisin and showed antimicrobial effectiveness against *S. aureus*.

3.2 Nanocelluloses

Cellulosic nanomaterials – i.e., cellulose nanocrystals (CNC) and cellulose nanofibrils (CNF), illustrated in Fig. 4 – have been extensively combined with film-forming polymers to produce films featuring improved mechanical (Azeredo et al. 2017) and barrier (Ferrer et al. 2017) properties for food packaging applications.

The remarkable mechanical reinforcement capacity of cellulose nanomaterials arises from their extremely high stiffness (Azeredo et al. 2017), whereas improvements on barrier properties upon the use of CNC and CNF are a result of the dense networks that are formed by nanofibrils as well as small, uniform particles (Ferrer et al. 2017). Examples involving edible films, in particular, are not numerous, but

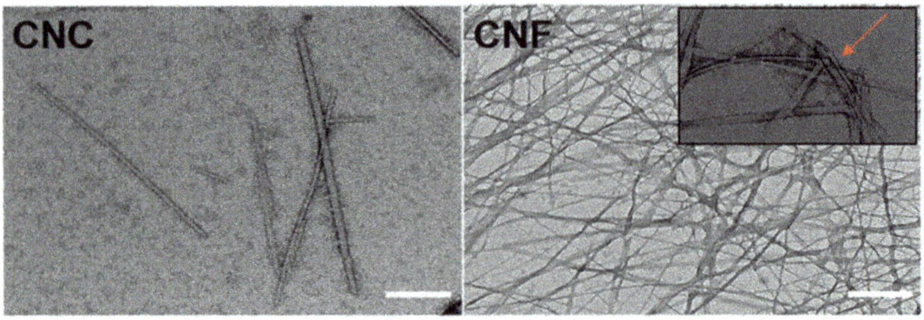

Fig. 4 Transmission electron microscopy images showing the different morphologies of cellulose nanocrystals (CNC; scale bar: 100 nm) and nanofibrils (CNF; scale bar: 200 nm). (Image reproduced from Yu et al. (2016))

considerable progress has been already made lately. Bilbao-Sainz et al. (2011) produced HPMC-based films reinforced with CNC. The authors were able to increase tensile strength and Young's modulus by 22% and 55%, respectively, whereas film was only slightly reduced – the films still allowed an 86–89% visible light transmission. Pectin has also been combined with CNC into edible nanocomposites featuring increased tensile strength and elongation at break as well as decreased water vapor permeability (Chaichi et al. 2017). The authors claimed that CNC is more promising than other strengthening strategies to develop renewable, strong, and completely biodegradable food packaging materials. CNC have been isolated through acid hydrolysis from bacterial cellulose produced by *Gluconacetobacter xylinus* and incorporated into a gelatin matrix to produce edible nanocomposites displaying improved thermal and mechanical properties, in addition to lessened affinity to moisture and water vapor permeability (George and Siddaramaiah 2012). CNF have also been exploited for the same purpose. These particles have reinforced edible films made up of cornstarch and chitosan (Yu et al. 2017) as well as mango puree (Azeredo et al. 2009).

3.3 Nanostarch

Starch is a naturally occurring, renewable, biodegradable, nontoxic, and biocompatible polymer that has been extensively investigated both as a matrix and as a ligand due to its outstanding film-forming properties (Li et al. 2015; Otoni et al. 2017). More recently, the scientific community has directed attention toward nanoparticles made up of starch (Le Corre et al. 2010), at first for drug-carrying purposes (Han et al. 2013) and then for reinforcing biopolymer-based films (Le Corre and Angellier-Coussy 2014; Zheng et al. 2009). Figure 5 present images of starch nanowhiskers or nanocrystals from different origins.

Regarding edible films, specifically, banana starch nanoparticles have been prepared by miniemulsion cross-linking and incorporated into banana flour-based edible films, which presented improved mechanical and water barrier properties upon the addition of such nanofiller (Orsuwan and Sothornvit 2017). Sessini et al.

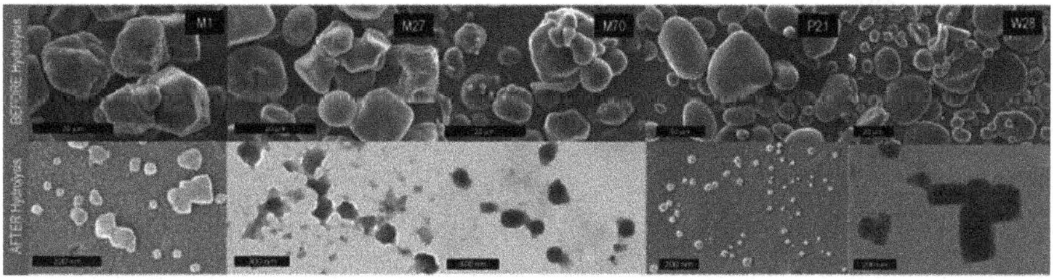

Fig. 5 Micrographs of native starch granules, i.e., before hydrolysis (top), and their corresponding nanocrystals, i.e., after hydrolysis (bottom). (Image reproduced from LeCorre et al. (2012) with permission from Elsevier)

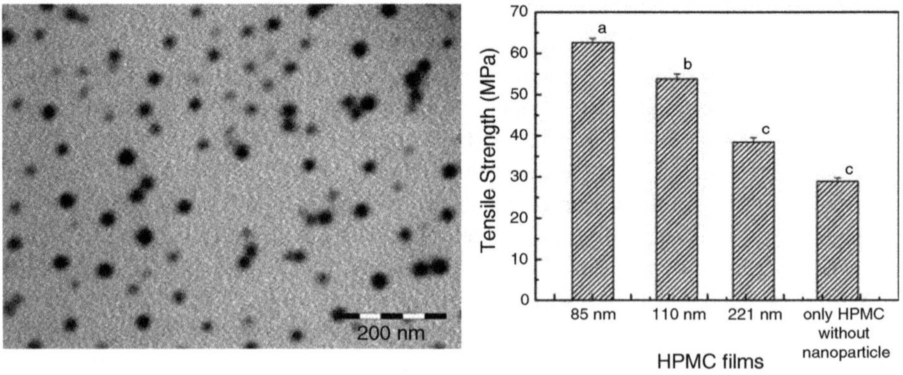

Fig. 6 Electron transmission microphotography of chitosan nanoparticles (CSNPs) synthesized through ionic gelation with sodium tripolyphosphate (left) and mechanical properties of hydroxypropyl methylcellulose (HPMC)-based edible films added by CSNPs of different diameters. (Images reproduced from de Moura et al. (2009) with permission from Elsevier)

(2016) extracted starch nanocrystals from waxy cornstarch granules through acid hydrolysis and used these particles to increase the mechanical attributes and thermal stability of glycerol-plasticized potato starch edible films. A similar approach was used by Kristo and Biliaderis (2007) to reinforce and control water adsorption of pullulan-based edible films upon the addition of starch nanocrystals.

3.4 Nanochitosan

Chitosan is a nontoxic, biocompatible, bioactive, biodegradable, and cationic polysaccharide obtained from chitin through chemical or enzymatic deacetylation reactions (de Moura et al. 2011; Ghormade et al. 2017; Muxika et al. 2017). The most common source of chitin – one of the most abundant naturally occurring polymers – is marine invertebrate shells, but in addition to crustaceans and mollusks, it can be extracted from the exoskeletons of insects and fungi (Ghormade et al. 2017; Muxika et al. 2017). Chitosan has been used to produce nanoparticles (Fig. 6) through the ionic gelation of its protonated amino groups with multivalent polyanions, among which sodium tripolyphosphate (STPP) and poly(methacrylic acid) are the most common due to their compatibility with biological systems (de Moura et al. 2008, 2009, 2011).

Antoniou et al. (2015) have combined tara gum as a film-forming matrix with bulk chitosan to produce edible blends as well as with chitosan nanoparticles to produce edible nanocomposites. The addition of nanostructured chitosan improved mechanical properties (tensile strength was increased in *ca.* 35 MPa) and affinity and barrier to moisture (water solubility and water vapor permeability were decreased by *ca.* 74% and 23%, respectively). Bulk chitosan did not show the same efficiency in

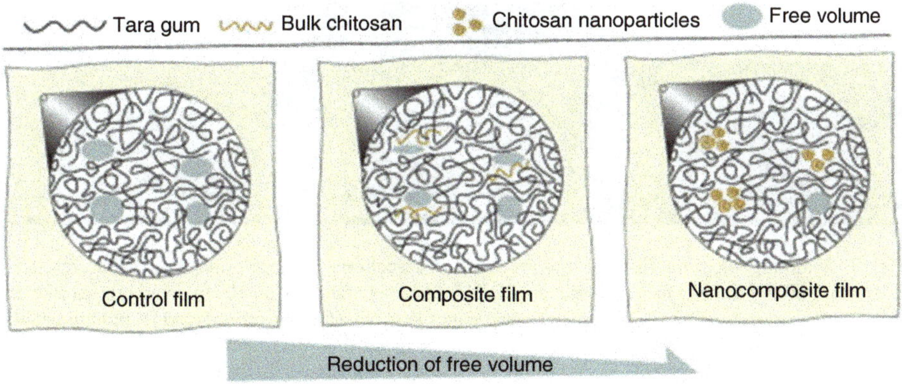

Fig. 7 Proposed model for the efficiency of bulk chitosan and chitosan nanoparticles in reducing the free volume among tara gum chains. (Illustration reproduced from Antoniou et al. (2015) with permission from Elsevier)

improving the physical-mechanical properties of tara gum-based edible films. The authors attributed this finding to the more compact structure of chitosan nanoparticles, which was more capable of reducing polymer free volume, therefore increasing the tortuosity of moisture diffusion pathway, as illustrated in Fig. 7.

de Moura et al. (2009) have successfully improved the mechanical and water barrier properties of HPMC-based edible films through the addition of chitosan/STPP nanoparticles. Interestingly, the authors observed that the mechanical reinforcement efficiency of these nanostructures depended upon their sizes, as shown in Fig. 6. They attributed this observation to the increased surface area of smaller nanoparticles, area which is available for interaction with the matrix. Similar outcome was obtained in film's barrier to moisture, as smaller chitosan nanoparticles were found to decrease water vapor permeability values. Similarly to Antoniou et al. (2015), the authors stated that smaller nanoparticles can occupy the empty among HPMC chains, thereby preventing water diffusion. Different behaviors were observed for chitosan nanoparticles synthesized with poly(methacrylic acid) (De Moura et al. 2008, 2011).

Chitosan nanoparticles have been also demonstrated to successfully reinforce and decrease the permeability of moisture through pectin- (Lorevice et al. 2016), pectin/cocoa puree- (Melo et al. 2017), pectin/banana puree- (Martelli et al. 2013), HPMC/guava puree- (Lorevice et al. 2012), and HPMC/papaya puree-based edible films (Lorevice et al. 2014).

3.5 *Other Food-Grade Nanostructured Fillers*

Wu et al. (2017) used up to 20% (wt.) of nanoscale hydroxyapatite – the major inorganic component of bones and teeth – to improve the mechanical and thermal

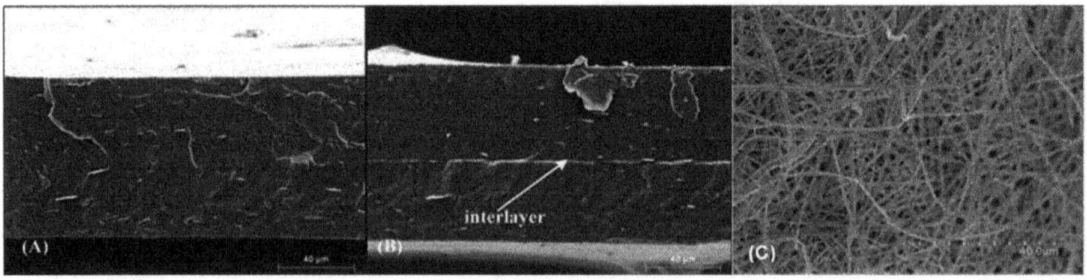

Fig. 8 Scanning electron micrographs of the cross sections of neat poly(hydroxybutyrate-*co*-valerate) (PHBV) films (**a**) and multilayer (PHBV) films separated by an electrospun zein interlayer (**b**) as well as of the surface of electrospun zein nanofibers (**c**). (Images reproduced from Fabra et al. (2014) with permission from Elsevier)

properties of bovine skin-derived type A gelatin films. The water barrier of the edible nanocomposite was also improved.

Moreira et al. (2013) prepared magnesium hydroxide nanoplates and incorporated them into a pectin matrix to produce edible bionanocomposites featuring improved mechanical and thermal properties, in addition to serving as a dietary matrix. The authors carried out migration assays and demonstrated that these films were capable of delivering $Mg(OH)_2$ upon contact with arugula leaves. Because release did not take place in the absence of direct contact, using the bionanocomposite as an edible film would increase the efficiency of magnesium supplementation.

Another concept of increasing the barrier of edible films is through the use of interlayers comprised of food-grade nanofibers, such as those produced by electrospinning, and made up of zein, whey protein isolate, pullulan, and zein/whey protein isolate and zein/pullulan blends (Fabra et al. 2013, 2014). These materials have been demonstrated to improve the barrier properties of biopolymer-based films to water, oxygen, and limonene. Such nanostructures (Fig. 8) have been applied between layers of polyhydroxyalkanoates but could be also applied to separate edible layers, giving rise to edible multilayer, nanostructured films.

4 Nanostructured Antioxidant Edible Films

The direct addition of nutritional and antioxidant compounds into food systems may result in activity loss because of their interactions with food components, so bioactive compounds have been incorporated into edible films to produce antioxidant active packaging (Ganiari et al. 2017). Sekhon (2010) pointed out a twofold advantage of incorporating bioactive nanoparticles into food packaging materials: provide traditionally passive packaging with active roles and allow the continuous release of antioxidants and other bioactive molecules from packaging toward food surface. In the case of antioxidant packaging, this is of particular interest because food oxidation takes place more intensively in the surface of food items. Different nanostructured

Table 3 Nanostructured antioxidant edible films developed during the past 5 years

Matrix	Active compound	Nano-sized delivery system	References
Chitosan	Quercetin	Poly(lactic acid) nanoparticles	Basu et al. (2017)
Chitosan hydrochloride	Epigallocatechin gallate	Zein nanocapsules	Liang et al. (2017)
Gelatin	Rutin	Nanoemulsions	Dammak et al. (2017)
Gelatin, gelatin-sodium caseinate, and gelatin-chitosan	α-Tocopherol, cinnamaldehyde, and garlic essential oil	Nanoemulsions	Pérez Córdoba and Sobral (2017)
Gelatin	Ginger essential oil	Nanoemulsions	Alexandre et al. (2016)
Hazelnut meal protein	Clove essential oil	Nanoemulsions	Gul et al. (2018)
Hydroxypropyl methylcellulose	Green tea extract	Poly(lactic acid) nanoparticles	Wrona et al. (2017)
Pectin	Peppermint essential oil	Halloysite nanotubes	Biddeci et al. (2016)
–	Vitamin C and vitamin E	Chitosan nanoparticles	Aresta et al. (2013)
Starch	Lycopene	Poly(-ε-caprolactone) nanocapsules	Assis et al. (2017)

antioxidant compounds have been combined with food-grade polymer matrices to produce edible films featuring antioxidant activity, as summarized in Table 3.

The antioxidant, anti-inflammatory, and nutraceutical activities of quercetin, a flavonoid naturally found in plants (e.g., onion, broccoli, and grapes), have been broadly demonstrated in different food systems (Lesjak et al. 2018). However, such bioflavonoid is unstable to changes in pH and chemical environment as well as exposure to light. To enhance its bioactivity and stability, Basu et al. (2017) have synthesized poly(lactic acid) nanoparticles to serve as quercetin carriers and embedded these particles in a chitosan matrix to produce antioxidant edible films. These films also presented antimicrobial activity against *S. aureus*.

Chitosan-based films have also been provided with antioxidant capability through the incorporation of epigallocatechin gallate-loaded nanocapsules, as indicated by their greater DPPH radical scavenging activity when compared to neat chitosan films (Liang et al. 2017). Liang et al. (2017) have claimed that such nanoencapsulation is promising because the direct incorporation of this green tea polyphenol into food systems might result in interactions with food components, therefore leading to prompt oxidative degradation. In addition, the hydrophilic character of this active compound would restrict its direct addition into oily foods.

Poly(lactic acid) nanoparticles have been synthesized and loaded with green tea extract and combined with hydroxypropyl methylcellulose to produce antioxidant edible films (Wrona et al. 2017). Particle size was manipulated in order to control the release of the antioxidant compound from the films. Similar nanoparticles have been used to encapsulate and release other antioxidant compounds – including aureusidin, quercetin, and *Albizia chinensis*-isolated quercitrin – in a controlled fashion (Kumari et al. 2010, 2011; Roussaki et al. 2014).

Aresta et al. (2013) produced chitosan nanoparticles cross-linked by sulfobutylether-β-cyclodextrin and loaded with vitamins C and E. In this study, in vitro tests demonstrated that the antioxidant compounds were slowly and continuously released from chitosan nanoparticles throughout 7 days. The authors developed such system aiming at incorporating it into edible food packaging for increasing the shelf lives of both nanoencapsulated vitamins and packaged food.

Nanoclays have also been used to encapsulate active compounds and allow their sustained release to target systems. Halloysite clay nanotubes, for instance, have been demonstrated to encapsulate a range of drugs, biocides, cosmetics, and other functional compounds (Lvov and Abdullayev 2013; Lvov et al. 2016a, b). Concerning edible films, a pectin matrix has been incorporated with halloysite nanotubes loaded with peppermint essential oil (Biddeci et al. 2016). The authors of this interesting study first functionalized the nanotube surface with cucurbit[6]uril to improve its affinity to the essential oil. Indeed, modified halloysite nanotubes adsorbed *ca.* ten times more essential oil than their unmodified counterparts. DPPH assays indicated a remarkable antioxidant activity of these edible films, which also displayed antimicrobial effect against *E. coli* and *S. aureus*.

5 Concluding Remarks and Future Trends

As demonstrated in this contribution, much activity has been identified lately in topics related to nanotechnological approaches to provide edible films with improved physical-mechanical performances as well as unique active roles, out of which antimicrobial and antioxidant stand out as the most targeted functions for food packaging purposes.

Although we have divided this chapter into three main topic (i.e., nano-sized materials featuring antimicrobial, antioxidant, and reinforcement capabilities), we have identified the trend of developing nanostructures that are able to provide edible films with two or more roles at the same time. For instance, chitosan nanoparticles have been classified as fillers, but some authors have reported their antimicrobial activity against food spoilage and pathogenic microorganisms (Pilon et al. 2015). Indeed, this property of bulk chitosan-containing films has been extensively reported (Ji et al. 2017), but some authors have demonstrated that nanostructuring chitosan is expected to result in loss of antimicrobial efficiency (Antoniou et al. 2015). Such dual-action approach has also been applied to essential oil nanoemulsions, which have shown both antimicrobial and antioxidant activities, in addition to modifying edible film properties (particularly mechanical and barrier to moisture).

Finally, as new nano-sized materials are developed, however, questions related to their toxicity and biocompatibility gain evidence, making studies that address these issues of utmost importance. While legislation regarding the use of nanostructured edible packaging materials is lacking, it is recommended that these systems are considered as food products, being subjected to the same regulatory scenario.

References

Acevedo-Fani A, Salvia-Trujillo L, Rojas-Graü MA, Martín-Belloso O (2015) Edible films from essential-oil-loaded nanoemulsions: physicochemical characterization and antimicrobial properties. Food Hydrocoll 47(Supplement C):168–177

Ahmad V, Khan MS, Jamal QMS, Alzohairy MA, Al Karaawi MA, Siddiqui MU (2017) Antimicrobial potential of bacteriocins: in therapy, agriculture and food preservation. Int J Antimicrob Agents 49(1):1–11

Alexandre EMC, Lourenço RV, Bittante AMQB, Moraes ICF, Sobral PJ d A (2016) Gelatin-based films reinforced with montmorillonite and activated with nanoemulsion of ginger essential oil for food packaging applications. Food Packag Shelf Life 10(Supplement C):87–96

Antoniou J, Liu F, Majeed H, Zhong F (2015) Characterization of tara gum edible films incorporated with bulk chitosan and chitosan nanoparticles: a comparative study. Food Hydrocoll 44 (Supplement C):309–319

Aresta A, Calvano CD, Trapani A, Cellamare S, Zambonin CG, De Giglio E (2013) Development and analytical characterization of vitamin(s)-loaded chitosan nanoparticles for potential food packaging applications. J Nanopart Res 15(4):1592

Assis RQ, Lopes SM, Costa TMH, Flôres SH, Rios A d O (2017) Active biodegradable cassava starch films incorporated lycopene nanocapsules. Ind Crop Prod 109(Supplement C):818–827

Azeredo HMC, Mattoso LHC, Wood D, Williams TG, Avena-Bustillos RJ, McHugh TH (2009) Nanocomposite edible films from mango puree reinforced with cellulose nanofibers. J Food Sci 74(5):N31–N35

Azeredo HMC, Rosa MF, Mattoso LHC (2017) Nanocellulose in bio-based food packaging applications. Ind Crop Prod 97(Supplement C):664–671

Babu RP, O'Connor K, Seeram R (2013) Current progress on bio-based polymers and their future trends. Prog Biomater 2(1):8

Bajpai SK, Chand N, Chaurasia V (2010) Investigation of water vapor permeability and antimicrobial property of zinc oxide nanoparticles-loaded chitosan-based edible film. J Appl Polym Sci 115(2):674–683

Barbosa AAT, Mantovani HC, Jain S (2017) Bacteriocins from lactic acid bacteria and their potential in the preservation of fruit products. Crit Rev Biotechnol 37(7):852–864

Basch CY, Jagus RJ, Flores SK (2013) Physical and antimicrobial properties of tapioca starch-HPMC edible films incorporated with Nisin and/or potassium sorbate. Food Bioprocess Technol 6(9):2419–2428

Basu A, Kundu S, Sana S, Halder A, Abdullah MF, Datta S, Mukherjee A (2017) Edible nano-bio-composite film cargo device for food packaging applications. Food Packag Shelf Life 11 (Supplement C):98–105

Biddeci G, Cavallaro G, Di Blasi F, Lazzara G, Massaro M, Milioto S, Parisi F, Riela S, Spinelli G (2016) Halloysite nanotubes loaded with peppermint essential oil as filler for functional biopolymer film. Carbohydr Polym 152(Supplement C):548–557

Bilbao-Sáinz C, Avena-Bustillos RJ, Wood DF, Williams TG, McHugh TH (2010) Nanoemulsions prepared by a low-energy emulsification method applied to edible films. J Agric Food Chem 58 (22):11932–11938

Bilbao-Sainz C, Bras J, Williams T, Sénechal T, Orts W (2011) HPMC reinforced with different cellulose nano-particles. Carbohydr Polym 86(4):1549–1557

Boelter JF, Brandelli A (2016) Innovative bionanocomposite films of edible proteins containing liposome-encapsulated nisin and halloysite nanoclay. Colloids Surf B: Biointerfaces 145(Supplement C):740–747

Chaichi M, Hashemi M, Badii F, Mohammadi A (2017) Preparation and characterization of a novel bionanocomposite edible film based on pectin and crystalline nanocellulose. Carbohydr Polym 157(Supplement C):167–175

Cui H, Yuan L, Lin L (2017) Novel chitosan film embedded with liposome-encapsulated phage for biocontrol of Escherichia coli O157:H7 in beef. Carbohydr Polym 177(Supplement C):156–164

Dammak I, de Carvalho RA, Trindade CSF, Lourenço RV, do Amaral Sobral PJ (2017) Properties of active gelatin films incorporated with rutin-loaded nanoemulsions. Int J Biol Macromol 98 (Supplement C):39–49

de Azeredo HMC (2013) Antimicrobial nanostructures in food packaging. Trends Food Sci Technol 30(1):56–69

De Moura MR, Avena-Bustillos RJ, McHugh TH, Krochta JM, Mattoso LHC (2008) Properties of novel hydroxypropyl methylcellulose films containing chitosan nanoparticles. J Food Sci 73(7): N31–N37

de Moura MR, Aouada FA, Avena-Bustillos RJ, McHugh TH, Krochta JM, Mattoso LHC (2009) Improved barrier and mechanical properties of novel hydroxypropyl methylcellulose edible films with chitosan/tripolyphosphate nanoparticles. J Food Eng 92(4):448–453

de Moura MR, Lorevice MV, Mattoso LHC, Zucolotto V (2011) Highly stable, edible cellulose films incorporating chitosan nanoparticles. J Food Sci 76(2):N25–N29

Fabra MJ, Lopez-Rubio A, Lagaron JM (2013) High barrier polyhydroxyalcanoate food packaging film by means of nanostructured electrospun interlayers of zein. Food Hydrocoll 32(1):106–114

Fabra MJ, López-Rubio A, Lagaron JM (2014) On the use of different hydrocolloids as electrospun adhesive interlayers to enhance the barrier properties of polyhydroxyalkanoates of interest in fully renewable food packaging concepts. Food Hydrocoll 39(Supplement C):77–84

Ferrer A, Pal L, Hubbe M (2017) Nanocellulose in packaging: advances in barrier layer technologies. Ind Crop Prod 95(Supplement C):574–582

Fu Y, Sarkar P, Bhunia AK, Yao Y (2016) Delivery systems of antimicrobial compounds to food. Trends Food Sci Technol 57(Part A):165–177

Ganiari S, Choulitoudi E, Oreopoulou V (2017) Edible and active films and coatings as carriers of natural antioxidants for lipid food. Trends Food Sci Technol 68(Supplement C):70–82

Ge L, Zhu M, Xu Y, Li X, Li D, Mu C (2017) Development of antimicrobial and controlled biodegradable gelatin-based edible films containing nisin and amino-functionalized montmorillonite. Food Bioprocess Technol 10(9):1727–1736

George J, Siddaramaiah (2012) High performance edible nanocomposite films containing bacterial cellulose nanocrystals. Carbohydr Polym 87(3):2031–2037

Ghormade V, Pathan EK, Deshpande MV (2017) Can fungi compete with marine sources for chitosan production? Int J Biol Macromol Part B 104:1415–1421

Gul O, Saricaoglu FT, Besir A, Atalar I, Yazici F (2018) Effect of ultrasound treatment on the properties of nano-emulsion films obtained from hazelnut meal protein and clove essential oil. Ultrason Sonochem 41(Supplement C):466–474

Han F, Gao CM, Liu MZ (2013) Fabrication and characterization of size-controlled starch-based nanoparticles as hydrophobic drug carriers. J Nanosci Nanotechnol 13(10):6996–7007

Hashemi Gahruie H, Ziaee E, Eskandari MH, Hosseini SMH (2017) Characterization of basil seed gum-based edible films incorporated with Zataria multiflora essential oil nanoemulsion. Carbohydr Polym 166(Supplement C):93–103

Imran M, Revol-Junelles A-M, René N, Jamshidian M, Akhtar MJ, Arab-Tehrany E, Jacquot M, Desobry S (2012) Microstructure and physico-chemical evaluation of nano-emulsion-based antimicrobial peptides embedded in bioactive packaging films. Food Hydrocoll 29(2):407–419

Jang SA, Shin YJ, Seo YB, Song KB (2011) Effects of various plasticizers and nanoclays on the mechanical properties of red algae film. J Food Sci 76(3):N30–N34

Ji N, Qin Y, Xi T, Xiong L, Sun Q (2017) Effect of chitosan on the antibacterial and physical properties of corn starch nanocomposite films. Starch Stärke 69(1–2):1600114 n/a

Khan B, Bilal Khan Niazi M, Samin G, Jahan Z (2017) Thermoplastic starch: a possible biodegradable food packaging material—a review. J Food Process Eng 40(3):e12447 n/a

Kristo E, Biliaderis CG (2007) Physical properties of starch nanocrystal-reinforced pullulan films. Carbohydr Polym 68(1):146–158

Kumari A, Yadav SK, Pakade YB, Singh B, Yadav SC (2010) Development of biodegradable nanoparticles for delivery of quercetin. Colloids Surf B: Biointerfaces 80(2):184–192

Kumari A, Yadav SK, Pakade YB, Kumar V, Singh B, Chaudhary A, Yadav SC (2011) Nanoencapsulation and characterization of *Albizia chinensis* isolated antioxidant quercitrin on PLA nanoparticles. Colloids Surf B: Biointerfaces 82(1):224–232

Le Corre D, Angellier-Coussy H (2014) Preparation and application of starch nanoparticles for nanocomposites: a review. React Funct Polym 85:97–120

Le Corre D, Bras J, Dufresne A (2010) Starch nanoparticles: a review. Biomacromolecules 11 (5):1139–1153

LeCorre D, Bras J, Dufresne A (2012) Influence of native starch's properties on starch nanocrystals thermal properties. Carbohydr Polym 87(1):658–666

Lesjak M, Beara I, Simin N, Pintać D, Majkić T, Bekvalac K, Orčić D, Mimica-Dukić N (2018) Antioxidant and anti-inflammatory activities of quercetin and its derivatives. J Funct Foods 40 (Supplement C):68–75

Li X, Qiu C, Ji N, Sun C, Xiong L, Sun Q (2015) Mechanical, barrier and morphological properties of starch nanocrystals-reinforced pea starch films. Carbohydr Polym 121(Supplement C):155–162

Liang J, Yan H, Zhang J, Dai W, Gao X, Zhou Y, Wan X, Puligundla P (2017) Preparation and characterization of antioxidant edible chitosan films incorporated with epigallocatechin gallate nanocapsules. Carbohydr Polym 171(Supplement C):300–306

Lorevice MV, Moura MR d, Aouada FA, Mattoso LHC (2012) Development of novel guava puree films containing chitosan nanoparticles. J Nanosci Nanotechnol 12(3):2711–2717

Lorevice MV, Moura MR d, Mattoso LHC (2014) Nanocompósito de polpa de mamão e nanopartículas de quitosana para aplicação em embalagens. Química Nova 37:931–936

Lorevice MV, Otoni CG, Moura MR d, Mattoso LHC (2016) Chitosan nanoparticles on the improvement of thermal, barrier, and mechanical properties of high- and low-methyl pectin films. Food Hydrocoll 52(Supplement C):732–740

Lvov Y, Abdullayev E (2013) Functional polymer–clay nanotube composites with sustained release of chemical agents. Prog Polym Sci 38(10):1690–1719

Lvov YM, DeVilliers MM, Fakhrullin RF (2016a) The application of halloysite tubule nanoclay in drug delivery. Expert Opin Drug Deliv 13(7):977–986

Lvov Y, Wang W, Zhang L, Fakhrullin R (2016b) Halloysite clay nanotubes for loading and sustained release of functional compounds. Adv Mater 28(6):1227–1250

Martelli MR, Barros TT, de Moura MR, Mattoso LHC, Assis OBG (2013) Effect of chitosan nanoparticles and pectin content on mechanical properties and water vapor permeability of banana puree films. J Food Sci 78(1):N98–N104

Melo PTS, Aouada FA, Moura MR d (2017) Fabricação de filmes bionanocompósitos à base de pectina e polpa de cacau com potencial uso como embalagem para alimentos. Química Nova 40:247–251

Moghimi R, Aliahmadi A, Rafati H (2017) Antibacterial hydroxypropyl methyl cellulose edible films containing nanoemulsions of Thymus daenensis essential oil for food packaging. Carbohydr Polym 175(Supplement C):241–248

Moreira FKV, De Camargo LA, Marconcini JM, Mattoso LHC (2013) Nutraceutically inspired pectin–mg(OH)2 nanocomposites for bioactive packaging applications. J Agric Food Chem 61 (29):7110–7119

Morsy MK, Khalaf HH, Sharoba AM, El-Tanahi HH, Cutter CN (2014) Incorporation of essential oils and nanoparticles in pullulan films to control foodborne pathogens on meat and poultry products. J Food Sci 79(4):M675–M684

Moura MRD, Aouada FA, Souza JR, Mattoso LHC (2014) Preparação de novos nanobiocompósitos comestíveis ativos contendo nanoemulsão de canela e pectina. Polímeros 24:486–490

Muxika A, Etxabide A, Uranga J, Guerrero P, de la Caba K (2017) Chitosan as a bioactive polymer: processing, properties and applications. Int J Biol Macromol 105(Part 2):1358–1368

Nitin K, Preetinder K, Surekha B (2017) Advances in bio-nanocomposite materials for food packaging: a review. Nutr Food Sci 47(4):591–606

Orsuwan A, Sothornvit R (2017) Development and characterization of banana flour film incorporated with montmorillonite and banana starch nanoparticles. Carbohydr Polym 174(Supplement C):235–242

Otoni CG, Pontes SFO, Medeiros EAA, Soares NDFF (2014a) Edible films from methylcellulose and nanoemulsions of clove bud (*Syzygium aromaticum*) and oregano (*Origanum vulgare*) essential oils as shelf life extenders for sliced bread. J Agric Food Chem 62(22):5214–5219

Otoni CG, Moura MR d, Aouada FA, Camilloto GP, Cruz RS, Lorevice MV, Soares N d FF, Mattoso LHC (2014b) Antimicrobial and physical-mechanical properties of pectin/papaya puree/cinnamaldehyde nanoemulsion edible composite films. Food Hydrocoll 41(Supplement C):188–194

Otoni CG, Avena-Bustillos RJ, Olsen CW, Bilbao-Sáinz C, McHugh TH (2016a) Mechanical and water barrier properties of isolated soy protein composite edible films as affected by carvacrol and cinnamaldehyde micro and nanoemulsions. Food Hydrocoll 57(Supplement C):72–79

Otoni CG, Espitia PJP, Avena-Bustillos RJ, McHugh TH (2016b) Trends in antimicrobial food packaging systems: emitting sachets and absorbent pads. Food Res Int 83(Supplement C):60–73

Otoni CG, Avena-Bustillos RJ, Azeredo HMC, Lorevice MV, Moura MR, Mattoso LHC, McHugh TH (2017) Recent advances on edible films based on fruits and vegetables—a review. Compr Rev Food Sci Food Saf 16(5):1151–1169

Pathakoti K, Manubolu M, Hwang H-M (2017) Nanostructures: current uses and future applications in food science. J Food Drug Anal 25(2):245–253

Pérez Córdoba LJ, Sobral PJA (2017) Physical and antioxidant properties of films based on gelatin, gelatin-chitosan or gelatin-sodium caseinate blends loaded with nanoemulsified active compounds. J Food Eng 213(Supplement C):47–53

Perez Espitia PJ, de Fátima Ferreira Soares N, dos Reis Coimbra JS, de Andrade NJ, Souza Cruz R, Alves Medeiros EA (2012) Bioactive peptides: synthesis, properties, and applications in the packaging and preservation of food. Compr Rev Food Sci Food Saf 11(2):187–204

Pilon L, Spricigo PC, Miranda M, de Moura MR, Assis OBG, Mattoso LHC, Ferreira MD (2015) Chitosan nanoparticle coatings reduce microbial growth on fresh-cut apples while not affecting quality attributes. Int J Food Sci Technol 50(2):440–448

Pintado CMBS, Ferreira MASS, Sousa I (2009) Properties of whey protein-based films containing organic acids and Nisin to control listeria monocytogenes. J Food Prot 72(9):1891–1896

Rodrigues DC, Cunha AP, Brito ES, Azeredo HMC, Gallão MI (2016) Mesquite seed gum and palm fruit oil emulsion edible films: influence of oil content and sonication. Food Hydrocoll 56 (Supplement C):227–235

Roussaki M, Gaitanarou A, Diamanti PC, Vouyiouka S, Papaspyrides C, Kefalas P, Detsi A (2014) Encapsulation of the natural antioxidant aureusidin in biodegradable PLA nanoparticles. Polym Degrad Stab 108(Supplement C):182–187

Salvia-Trujillo L, Soliva-Fortuny R, Rojas-Graü MA, McClements DJ, Martín-Belloso O (2017) Edible nanoemulsions as carriers of active ingredients: a review. Annu Rev Food Sci Technol 8 (1):439–466

Sasaki RS, Mattoso LHC, de Moura MR (2016) New edible bionanocomposite prepared by pectin and clove essential oil nanoemulsions. J Nanosci Nanotechnol 16(6):6540–6544

Sasikumar A, Kamalasanan K (2017) Nanomedicine for prostate cancer using nanoemulsion: a review. J Control Release 260(Supplement C):111–123
Sekhon BS (2010) Food nanotechnology – an overview. Nanotechnol Sci Appl 3:1–15
Sessini V, Arrieta MP, Kenny JM, Peponi L (2016) Processing of edible films based on nanoreinforced gelatinized starch. Polym Degrad Stab 132(Supplement C):157–168
Singh Y, Meher JG, Raval K, Khan FA, Chaurasia M, Jain NK, Chourasia MK (2017) Nanoemulsion: concepts, development and applications in drug delivery. J Control Release 252(Supplement C):28–49
Sinha Ray S, Okamoto M (2003) Polymer/layered silicate nanocomposites: a review from preparation to processing. Prog Polym Sci 28(11):1539–1641
Sung S-Y, Sin LT, Tee T-T, Bee S-T, Rahmat AR, Rahman WAWA, Tan A-C, Vikhraman M (2013) Antimicrobial agents for food packaging applications. Trends Food Sci Technol 33 (2):110–123
Tan B, Thomas NL (2016) A review of the water barrier properties of polymer/clay and polymer/ graphene nanocomposites. J Membr Sci 514(Supplement C):595–612
Wrona M, Cran MJ, Nerín C, Bigger SW (2017) Development and characterisation of HPMC films containing PLA nanoparticles loaded with green tea extract for food packaging applications. Carbohydr Polym 156(Supplement C):108–117
Wu X, Liu Y, Wang W, Han Y, Liu A (2017) Improved mechanical and thermal properties of gelatin films using a nano inorganic filler. J Food Process Eng 40(3):e12469 n/a
Yu SI, Min SK, Shin HS (2016) Nanocellulose size regulates microalgal flocculation and lipid metabolism. Sci Rep 6:35684
Yu Z, Alsammarraie FK, Nayigiziki FX, Wang W, Vardhanabhuti B, Mustapha A, Lin M (2017) Effect and mechanism of cellulose nanofibrils on the active functions of biopolymer-based nanocomposite films. Food Res Int 99(Part 1):166–172
Zheng H, Ai F, Chang PR, Huang J, Dufresne A (2009) Structure and properties of starch nanocrystal-reinforced soy protein plastics. Polym Compos 30(4):474–480

Bio-nanocomposites in Packaging: Business Model for Products' Commercialisation

Hezekiah Oladimeji and Shalini Singh

Abstract The business potential of nanotechnology is expected to alleviate global social challenges as this emerging technology is expected to serve as an important driving force in modernistic agriculture for sustainable food production. Nanotechnology has the prospect to transform the packaging system of the food industry through improved packaging methods for better food quality and customer satisfaction. However, the difficulty lies in commercialising these inventions to put them to use. Notwithstanding the potential risks associated with nanotechnology-based products, this chapter explores the essential elements for the successful commercialisation of engineered nanomaterials. It focuses on bridging the link between the innovation, development and markets in the commercialisation of food packaging products from bio-nanocomposites. Critical factors that could be considered in the management decision-making for the commercialisation of bio-nanocomposite applications were identified. It is expected that, within a global market, a thorough understanding and adoption of these identified factors for successful commercialisation of active packaging technologies will yield system-based solution.

Keywords Food packaging · Nanocomposite · Commercialisation

H. Oladimeji
Department of Business Administration, College of Social & Management Sciences, CALEB University, Lagos, Nigeria

Department of Operations and Quality Management, Durban University of Technology, Durban, South Africa

S. Singh (✉)
Department of Operations and Quality Management, Durban University of Technology, Durban, South Africa
e-mail: shalinis@dut.ac.za

S. Ahmed (ed.), *Bio-based Materials for Food Packaging*,
https://doi.org/10.1007/978-981-13-1909-9_7

1 Introduction

Nanotechnology is a peculiar facet creating much attention (Hosseini et al. 2011). It is an emerging field that is increasingly becoming a fact of life and business with major relevance in all human endeavours (Glenn 2006) including advancements in the incorporation of nanomaterials in other products (Mansoori and Soelaiman 2005) and in various industrial areas such as food safety and packaging (Soleimanpour et al. 2011), in global agricultural development (Rockefeller Foundation 2005) and in thermal and insulation materials in building industry (Gholami-Rostam et al. 2015). According to Aithal and Aithal (2016), the business potential of nanotechnology is expected to alleviate global social challenges. For example, nanotechnology is expected to serve as an important driving force in modernistic agriculture for sustainable food production (Alfadul et al. 2017).

Commercialisation of nanotechnology products is long and hard despite its commercial prospects (Mazzola 2003). Historically, the first nanotechnology-based product is perhaps carbon black and has been in commercial use for centuries (Romig et al. 2007). Nanotechnology, particularly engineered nanomaterials (ENMs), has expectantly impacted on the world's economy through the diversity of their applications (Nogueira et al. 2013). Lo et al. (2012) iterated that the commercialisation of nanotechnology is important as it ensures that this emerging technology achieves performance, reliability and economic requirements of the causal-effect assumption that the acceptance of consumers has a significant effect towards a successful commercialisation process. Nanotechnology is fast offering complete solutions to food matters from food production to its packaging (Singh et al. 2017). It has the prospect to transform the entire packaging system of food industry through improved packaging methods for better food quality and customer's satisfaction.

To this end, there are many nanostructures and their applications conceptualised, while their potential for commercialisation is yet to be explored (Garg et al. 2017; Bradley et al. 2011), a case being that the application of ENMs in the food industry is fast gaining attention with some promising prospects in the global market and economy (Honarvar et al. 2016). However, food packaging technologies have experienced limited breakthroughs despite the innovational potentials due to multiple commercialisation challenges (Werner et al. 2017). Hence, there is urgent need for an in-depth analysis of the commercialisation of nanotechnology. This serves as the basis for the current embrace of nanotechnology as the potential 'next great commercial opportunity' (Uldrich and Newberry 2003). Of particular interest, business models for emerging technologies are evolving, and the expansion of the commercialisation of these technologies is reshaping the existing business system (Inigo et al. 2017).

While there is a range of recent reports and reviews on applications of nanotechnology and ENM in the food industry (e.g. Sekhon 2010; Bradley et al. 2011; Chaudhry and Castle 2011; Sekhon 2014; Bumbudsanpharoke and Ko 2015; Rossi et al. 2017; Singh et al. 2017), this chapter focuses on recent developments

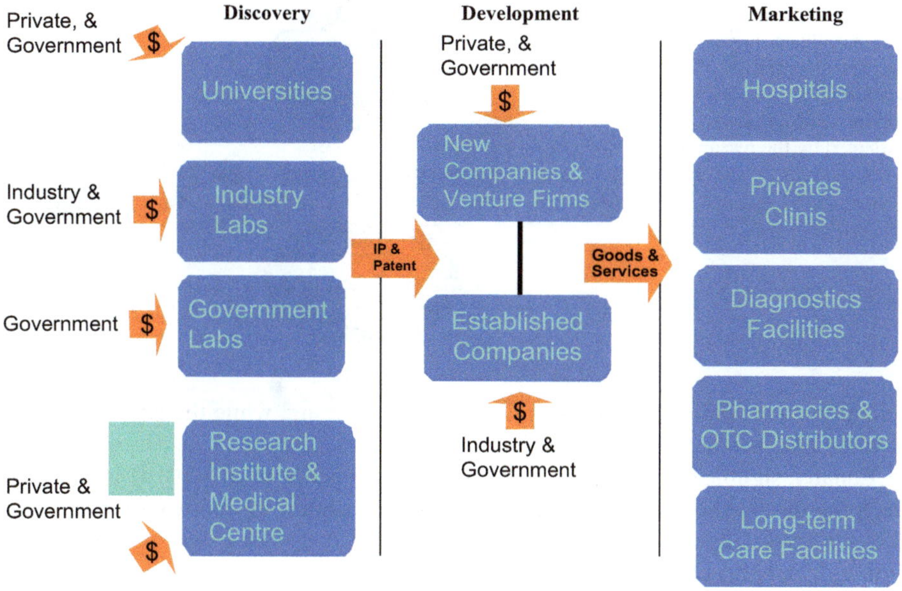

Fig. 1 Typical nanotechnology commercialisation stages. (Hobson 2009)

in business modelling for commercialisation of engineered bio-nanocomposites for food packaging. The commercialisation of nanotechnology has potential applications in many product sectors, food packaging and medical applications inclusive (Hobson 2009; Sekhon 2010; Othman 2014) as depicted in Fig. 1.

Several outputs of nanotechnologies are increasingly being used for commercial purposes (Andreev et al. 2009). According to Lopez (2013), commercialisation of ENM is an opportunity for scientific and economic growth and development through further job expansions and business opportunities.

Ziamou (2002) argues that there is a link between the innovation, development and markets in the commercialisation of nanotechnology products. He contends that through innovation interfaces, products from emerging technologies are marketed to discover the consumer's interest and interaction with a developed product in order to determine the product's functionality. Therefore, this systematic flow or linkage depicts the interface between innovation, development and markets can progress applications of scientific knowledge in our daily life, thus allowing the market to aid in successful direction of scientific research (Lo et al. 2012).

Meanwhile, Aithal and Aithal (2016) highlighted some elements such as innovative materials, market share, market potential and the prevailing economic situation as essential for initiating the successful commercialisation of ENMs. Thus, this chapter now turns to the business perspective and forms a link by bridging the gap between technological invention (innovation) and technological adoption as depicted in Fig. 2. The invention stage of innovation is the act of generating new

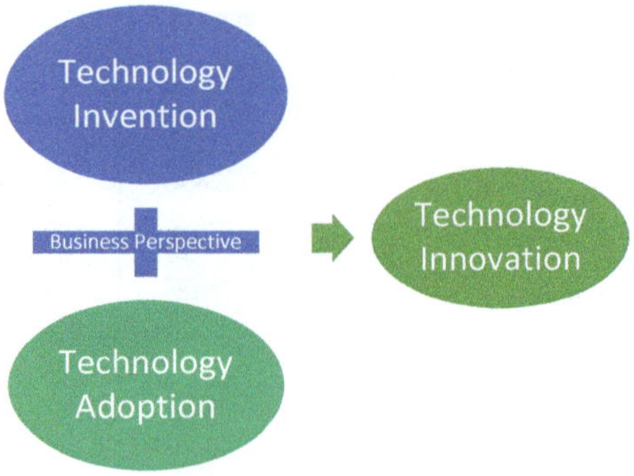

Fig. 2 A schematic diagram illustrating a business perspective of technology commercialisation. (Adapted by the researcher)

scientific and technical advances in materials engineering, while the adoption stage is where the new advances are put into practice.

Equally significant to business development, Shapira et al. (2016) argue that mapping the technology invention and commercialisation for innovation of ENMs is important to business decision-makers for the purpose of business investment and policy. In support of Shapira et al. (2016) and Aithal and Aithal (2016) confirm that several organisations are initiating developments in nanotechnology to take monetary advantage of business opportunities in nanotechnology developments through successful product commercialisation. To this end, the business viewpoint of the nanotechnological innovations and the critical factors that are to be considered and their order of preference or priority towards successful commercialisation of the engineered nanomaterials are of primary importance.

Hansen et al. (2016) confirm that the present consumer market is characterised by approximately 2500 food packaging products which are nanotechnology-enabled as they contain ENMs. Many innovations of ENMs are being adopted into consumer products as a result of their biocidal effect as they can prevent any form of survival of organisms that could be hazardous to human health in nano-enabled products (Mackevica et al. 2016). Similarly, for improvements in food safety and cost-price-efficiency, these materials are being incorporated into biopolymers, hence proffering active properties to packaging systems (Azerendo et al. 2011).

2 Drivers of Engineered Nanomaterials for Food Packaging

Bradley et al. (2011) enumerated the main merits offered by ENMs in food packaging for optimal applications and commercialisation. As presented in Fig. 3, 'Innovation and new product development' is the primary driver of nanomaterials in food packaging applications. Light-weighting involves the use of less packaging materials while maintaining the same technical performance against the traditional packaging materials. Greater protection and preservation of the food helps in the maintenance of food quality and increases the shelf-life without the mixing of

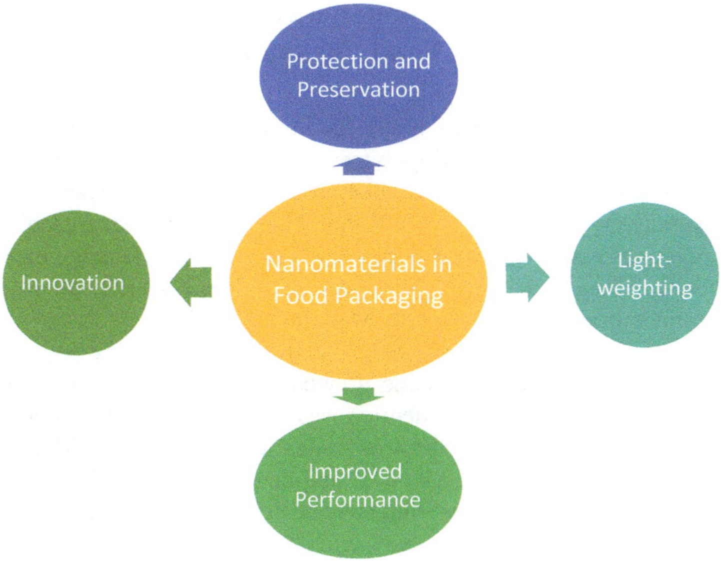

Fig. 3 Drivers for applications of nanomaterials in food packaging materials. (Adapted from Bradley et al. (2011))

Table 1 Benefits of the drivers for applications of nanomaterials in food packaging materials

Innovation	Light-weighting	Protection and preservation	Improved performance
Gives greater consumer choice and convenience	Lower material usage	Potential cheaper and more reliable food supply	Synthetic polymers are substituted by locally sourced bio-based materials
	Lower carbon footprint		
Supports social change and lifestyles	Lesser manufacturing handling	Better nutrition	
Opens new markets	Minimised transportation cost	Lesser food wastage	
Drives economic growth			
Creates employment			

preservatives into the packaging material which could introduce harmful chemicals into the body systems of potential consumers. The incorporation of nanomaterials and nanotechnology for improved production methods is the fourth driver of applications of nanomaterials in food packaging materials. The benefits of each of these drivers are as highlighted in Table 1.

These drivers are essential in the development and diffusion or commercialisation of the biocomposite applications for food packaging.

3 Current Applications of Biopolymers and Nanocomposites in Food Packaging

There has been increasing use of polymers in food packaging from renewable sources (Mensitieri et al. 2011). Biopolymers are derived from polymers with the prefix 'bio-' indicating biodegradable, which indicates the ability to be degraded into carbon dioxide (CO_2) and water (H_2O) (Othman 2014). The biodegradability property of biopolymers makes them preferable options for food packaging materials that are eco-friendly (Tang et al. 2012) by rapidly degrading (Jayaramudu et al. 2013) and totally mineralising in environment (Majeed et al. 2013).

Bio-nanocomposites have been established for food packaging applications as improvement of biopolymers has some drawbacks such as poor mechanical and barrier properties (Othman 2014). Attention was drawn to bio-nanocomposites with the addition of additives of nano-fillers in biodegradable polymers (Maurizio et al. 2005). Nanotubes, nanoparticles, nanorods and nano-fibrils are a few of the many types of nanosized fillers (<100 nm) that have been used for the enhancement of the performance of biopolymers as food packaging materials (Othman 2014).

Several studies (Rouhi et al. 2013; Kanmani and Rhim 2014; Rafieian et al. 2014) have been conducted to study the different properties of bio-nanocomposite materials produced from the incorporation of fillers into biopolymer. The various properties are identified in Fig. 4. A study of these properties is highly significant for the

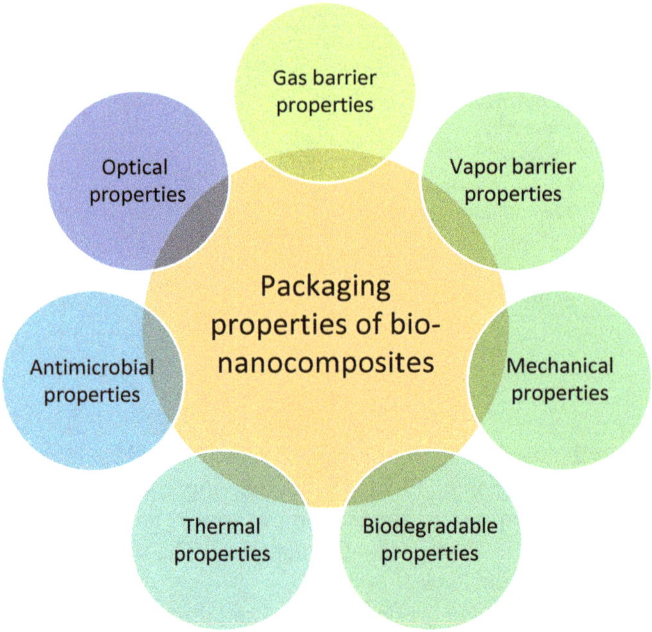

Fig. 4 Properties of bio-nanocomposites for food packaging applications

enhancement of food packaging quality, safety and commercialisation after the post-processing stages (Rhim et al. 2013).

4 Modelling-Based Commercialisation of Bio-nanocomposites in Food Packaging

Synonymously, technology commercialisation is often used with technology transfer and is referred to as the logical process of transforming scientific innovations, inventions and discoveries into products and services involving companies, organisations or even an entire state (Mohannak and Samtani 2014). As mentioned earlier, ENMs have increasingly been applied to a number of commercialised products especially those involving food packaging (Sonker et al. 2017). According to Reynolds (2007), one of every four food products in the market is packaged by the use of ENMs.

Hosseini et al. (2011) described a series of steps to achieve optimum market entry of nanotechnology materials. Figure 5 illustrates these as necessary and sufficient steps for effective commercialisation of innovations. This illustration contributed to the development of commercialisation theories of this chapter.

Nanotechnology has virtually adjourned as the new growth entrepreneurial innovator (Ludhiyani et al. 2011). Commercialisation of ENM has the potential to drive innovative and competitive industry to change by enabling key advances in many manufacturing sectors such as energy, automotive, aerospace, construction, electronics and biotechnology. Meanwhile, many challenges confronting nanotechnology research call for solutions from a multidisciplinary approach as Lu et al. (2009) opined. They further argued that statistical techniques used in the modelling process

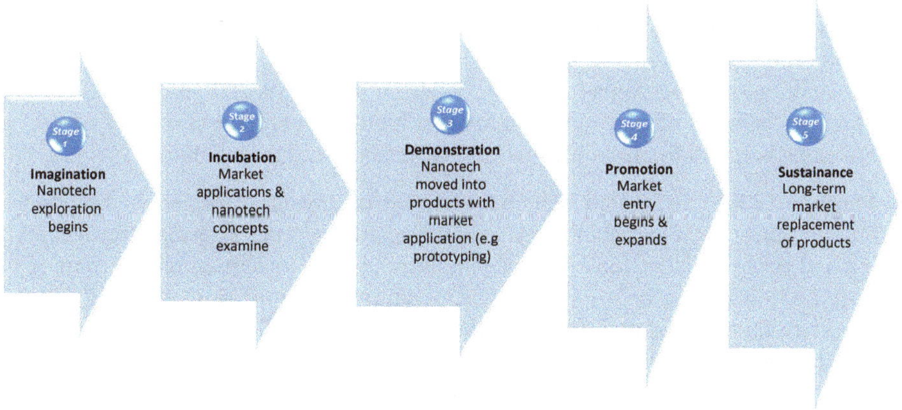

Fig. 5 Commercialisation stages for optimum market entry. (Adapted by the researcher)

have made sizable impacts in many other technology fields in the past. Statistics is expected to address these challenges and boost the development of nanotechnology.

With the fast development of nanotechnology and the introduction of mass production of nano-devices, statistics is expected to play an increasingly important role in both academic and industrial fields (Lu et al. 2009). Complex processes can be better understood with the aid of statistical tools and methods which are primarily concerned with the study of variability that all natural process typically exhibit (Wikle and Royle 2004). Meanwhile, soon after the early days of the Scientific Revolution, pragmatic scientists and industrialists, by mutual inventiveness, got the novel idea that the scientific method of observation and experimentation could be used to discover the cause-and-effect relations administering the manufacturing of high-quality consumer products; however, this is time-consuming and required much complex problem-solving and trouble-shooting (Bisgaard 1997). Bisgaard (1997) further suggested that statistical concepts and tools could help in making the process of scientific investigation more efficient and increasing the probability of discovering new ways to enhance productivity and the quality of products for the satisfaction of customers.

Statistics plays a key central role in business and industry. Statistical models are used for the following purposes: products' quality designs and processes (Box and Bisgaard 1987), forecasting the market behaviour of a new brand to prepare its launch and managing existing brands (Zhigljavsky 2011), among others. These roles of statistical applications are being controlled by a series of factors, which include technologies designed purposely to generate functional and applicable data with high potentials, more and better computing power, new statistical and graphical methods and methodology and useful statistical software for data analysis using modern statistical methods (Keller-McNulty et al. 2005). Statistics is seen to play a major part in making technological advances and improving processes of all kinds (Keller-McNulty et al. 2005).

The challenges highlighted below provide opportunities for statistical applications in nanotechnology (Lu et al. 2009):

- Statistical procedures help us to learn more about the formation processes of nanocomposites.
- Statistical modelling helps to deal with special data types and processes.
- Statistics helps to improve low-quality and high-defect processes.
- Statistics helps to improve low and unpredictable product reliability.

Mazzola (2003) suggested that there should be greater attention in academic institutions through interdisciplinary working teams to accelerate the rate of innovation and commercialisation of nanotechnology. As shown earlier, nanotechnology has been adjourned to be a convergence of several technologies operating at the interface which required that different skills be taught at all levels of an educational process (Romig et al. 2007). Many challenges confronting nanotechnology research call for solutions from a multidisciplinary approach. Statistical and Operational Research (S&OR) techniques play an important role in addressing these challenges and boosting the development of nanotechnology, as a major similarity between these disciplines is that both are multidisciplinary (Oladimeji and Singh 2013). Quality improvement activities are actually strongly based on the adoption of appropriate statistical models (Echeveste et al. 2007).

Critical assessment and evaluation of the product drivers are, however, important in the formulation, implementation and evaluation of effective innovation and efficient commercialisation policies.

5 Factors for Commercialisation of Engineered Bio-composites for Packaging

Several studies have been conducted proposing various critical factors that should be adopted for effective commercialisation process of emerging technologies. This section of the chapter reviews some of these critical factors. For technologists and business managers to implement successful technology commercialisation strategies, there is a need to identify the critical factors required. According to Luthra et al. (2015a, b), these critical factors are important enablers which are necessary for success. This section therefore seeks to broaden this empirical concept of critical factors and modelling for successful nanotechnology commercialisation. Several studies, as briefly outlined in Table 2, have been conducted on these concepts, critical factors and framework. The work of Kumar et al. (2015) on discourses of these concepts is being used as the departure point of this section.

Kumar et al. (2015) recognised the vital role of enablers or critical factors for the implementation of successful technology commercialisation by technologists and managers. Corroborating Luthra et al. (2015a, b), these enablers or critical factors are conceptualised as factors which are essential for the successful adoption of technology inventions. They identified and evaluated 5 critical dimensional factors and grouped them in 24 sub-factors.

Lee et al. (2013) identified four common categories and four relevant categorised critical factors of nanotechnology commercialisation:

- Field of nanotechnology – this classification and further prediction for commercialisation period were first carried out by Roco (2004). The four fields that were identified were nanoequipment, nanomaterials, nano-devices and nano-bios.
- Nanotechnology feature – the three critical factors are convergence between nanotechnology and other emerging technologies like information and communication technology (ICT) or biotechnology (BT), toxicity and energy and environmental relevance.
- R&D strategy of nanotechnology organisations – the four factors considered to ease commercialisation are high R&D intensity, explorative/exploitative optimal innovation strategies to develop new markets, government R&D support and research partnership with universities.
- Size and age of nanotechnology organisations.

Table 2 A brief overview of studies on technology commercialisation process

Author(s) and year	Model type used	Factors considered
Kumar and Jain (2003)	Descriptive statistics, correlation and regression analyses	Funds availability
		Support for technology supplier
		Product engineering to market needs
		Commitment of entrepreneur/company
		Technical/market staff training
		Product and packaging aesthetics
		Pricing, product positioning and launch
		Longer and low interest rate loan repayment
Bandarian (2007)	Delphi and fuzzy logic	Technology area (process evaluation, technical evaluation)
		Market area (economic evaluation, market evaluation, perception evaluation)
		Legal area (regulatory/policy evaluation)
Mahboudi and Ananthan (2010)	AHP	Cultural factors (information development, diffusion of scientific attitude)
		Global factors (training in international companies, employment of expatriates in technological field)
		Recipient organisation factor (investment in R&D, technology development strategic planning, managerial and organising skills' development in the organisation)
		Structural factors (standard creation, localisation of importing technology)
		Absorption and application factors (creating research and production relationship, absorption potential of importing technology, expertise training)
		Infrastructural factors (personnel infrastructure, informational and organisational infrastructure)
		Technological factors (commercialised technology price, technology's complicity and simplicity level, technology's development and improvement level)

(continued)

Table 2 (continued)

Author(s) and year	Model type used	Factors considered
Shen et al. (2010)	Delphi, AHP and patent co-citation approach (hybrid model)	Technological merit (advancement of technology, innovation of technology, key of technology, propriety of technology)
		Generics of technology (technological connections, technological extendibility)
		Business effect (potential return on investment, current market share effect, new market potential, market size potential, technology timing)
		Technology development potential (technical resources availability, equipment support, opportunity for technical success)
		Risk (technical risk, technical difficulties, commercial risk)
Amadi-Echendu and Rosetlola (2011)	Delphi	Finance-related issues
		Marketing strategy issues
		Intellectual property issues
		Commercialisation environment issues
		Technology management issues
		Innovation development issues
Borzouei et al. (2011)	Kaiser-Meyer-Olkin (KMO) and Bartlett's test and Kaiser criterion for factor analysis	Infrastructural factors (government investment interest, support for researchers, motivation for producers, etc.)
		Economic factors (budget provision, income security improvement, dynamic market existence, etc.)
		Cultural factors (creating consumer belief, nano-awareness creation, etc.)
		Informative factors (private sector investment in nanotechnology, media attention, social culture creation), financial factors (research budget)
		Provision, research organisations' finance system
Lo et al. (2012)	Importance-performance analysis	Compatibility
		Complexity
		Relative advantage

(continued)

Table 2 (continued)

Author(s) and year	Model type used	Factors considered
		Trialability
		Observability
		Brand
		Price
		Country-of-origin effects
		Quality
Lee et al. (2013)	Literature and logistic and Tobit regression models	Field of technology (nanoequipment, nanomaterials, nano-devices and nano-bios)
		Feature of technology (convergence between nanotechnology and other emerging technologies like information and communication technology or biotechnology, toxicity and energy and environmental relevance)
		R&D strategy (R&D intensity, explorative/exploitative R&D, government R&D support and research partnership with universities)
		Feature of organisation (size, age)
Lan and Sheng (2014)	Delphi and fuzzy AHP	Personal factors (gender and age, education, income, level of consumption, awareness, stakeholders' crucial opinions)
		Environment factors (cultural value, government environmental protection policy, government subsidy measures, economic circumstances)
Mohannak and Samtani (2014)	Delphi and mean, standard deviation and inter-rater agreement (IRA)	Technological readiness (technology stage of development, possible replicability, technological nature and sophistication, proof of application, potential to combine with other technologies, prototype availability, technical feasibility, uniqueness of technology, potential for further development)
		Economical and market factors (growth contribution, market needs, financial risk, market impact, time to market, level of competition, potential for attracting required resources)

(continued)

Table 2 (continued)

Author(s) and year	Model type used	Factors considered
		Social benefits (employment provision, knowledge spillover, social networks, environmental impact, brand creation, cost advantage to users)
		Legal and regulatory (exclusive patent, protection of IP rights, open innovation, new areas of application, need for complimentary technologies)
Hassanali and Zahra (2015)	Delphi and AHP	Environmental factors (government legislation, legal/regulatory, government support)
		Marketing factors (customer need, advert, pricing policy, competition strategy)
		Technology factors (R&D capabilities, market oriented, timely innovative)
		Management factors (management, management skill, management acceptance)
Kamolkittiwong and Phruksaphanrat (2015)	Delphi and AHP	External factors (suppliers, regulation, consumer/market, competitors, social/stakeholder)
		Internal/organisational factors (organisation strategy, top management support, reverse logistics, economic benefits)
Koç and Burhan (2015)	AHP	Environmental factors (land mass and security, transportation, climate, urbanisation rate)
		Cost of investment
		Sectoral factors (customer potential, market proximity, warehouse proximity, regional commercial activity, possible competitors)
		Human resource potential
		Regional potential (facilities available in the region, automobile insurance rate)
Kumar et al. (2015)	Delphi and AHP	Marketing benefits and forces (new areas' penetration, use of existing customers, end users' support, market demand, competition, timing judgement)

(continued)

Table 2 (continued)

Author(s) and year	Model type used	Factors considered
		Technical features (scientific changes, suppliers' technical strength, local suitability of technology, compatibility, functionality, reliability, trialability, observability)
		Regulatory concerns (government authorities, environmental concerns, international bodies)
		Management strategic issues (strategic implications, training and development, human resources, commitment)

Adapted by the Researcher

Though this is a most rewarding form of new product development, successful commercialisation is risky as there exists a considerable uncertainty in characterising the type of consumer needs that can be satisfied by a new technology (Ziamou 2002). Kim et al. (2010) identified three issues serving as potential barriers to successful commercialisation of nano-products, and they are presented below:

- The need for extensive research on basic discipline. Hobson (2009) and Kim et al. (2010) argued that many fields and subareas of nanotechnology require more critical scientific investigations to be launched into the consuming markets which require more interdisciplinary collaboration disciplines.
- The challenge of large investments against start-ups and micro, small and medium enterprises who operate on limited financial resources. According to Kim et al. (2010), this may continue to be a big hurdle in developing countries.
- Public acceptance of nanotechnology products due to issues like toxic effects and safety. This vague concern is however based on limited knowledge and lack of sufficient and accurate information on the subject of toxicity and safety.

A functional input-output model on technology commercialisation process was proposed by Luthra et al. 2015a, b and depicted in Fig. 6.

While the factors considered in Table 2 are relevant to new technologies, several are essentially critical for commercialisation of applications of ENMs. The critical factors were summarised and categorised into ten factors. Each of these factors was further classified into various corresponding sub-factors as depicted in Table 3. Meanwhile, even for the best managed business organisations, the effective transition of scientific and technological ideas from the research laboratory to the consumer through the operational and transformational processes of production, marketing and sales has over time proven to be an arduous task. Hence, there is essential and sufficient need to identify the critical factors and develop theories for effective and efficient commercialisation of products from ENMs' applications.

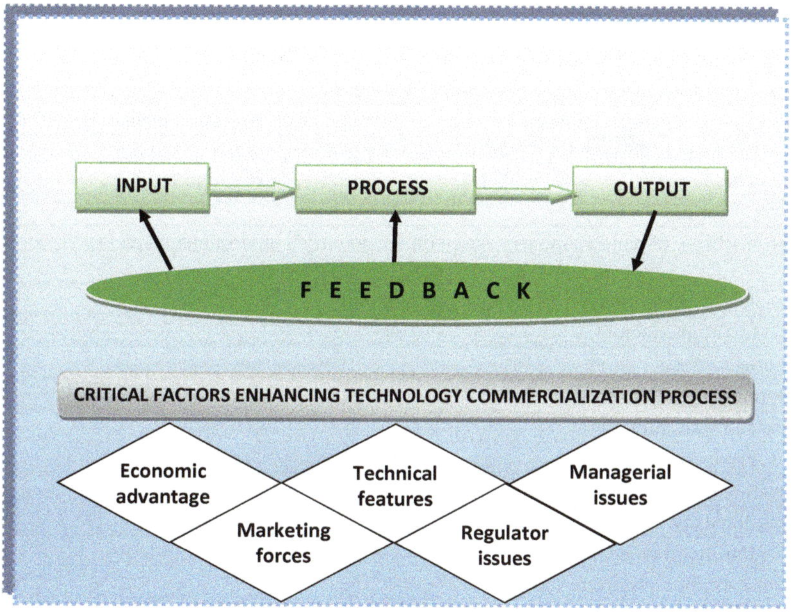

Fig. 6 Modified input-output model on technology commercialisation process. (Adapted by the Researcher from Luthra et al. 2015a, b)

These critical factors are referred to as key factors or facilitators necessary for any organisation to be successful. To this end, this chapter has been able to present main critical factors essential for effective and efficient commercialisation of ENMs' applications.

6 Business Model for NEMs' Application

Scholars of technology commercialisation and operational research experts (e.g. Hernandez and Bennison 2000; Kuo et al. 2002; Priya and Venkatesh 2012) proposed models that need to be considered for the effective commercialisation of ENMs. Two analytical techniques have been identified in extant literature to cover a large number and wide scope of organisational decision-making for quality control and commercialisation of technologically processed materials (Hernandez and Bennison 2000; Kuo et al. 2002; Priya and Venkatesh 2012). These are as illustrated in Fig. 7.

Conventional or analogue-based approaches only provide the algorithm for the commercialisation decision, but not the relationship between the critical factors for commercialisation decision-making. An example is the checklist methods. *Statistical approaches*, on the other hand, the Delphi method, are able to provide a step-by-step procedure for the commercialisation decision and likewise identify the

Table 3 Identified critical factors for commercialisation of ENMs' applications

Factor 1: Technical and technological features of the nano-product
Intersection of nanotechnology with biotechnology
Convergence of nanotechnology with information and communication technology (ICT)
Scientific changes through high level of technological support and technology management effectiveness
Technical abilities of suppliers
Local suitability of nanotechnology based on the country's environmental and socio-economic state
Technological functionality, trialability and reliability
Factor 2: Economic factors
Cost-effectiveness
Profit margins
Increase in sales turnover
Factor 3: Production factors
Production cost issues
Labour intensity
Factor 4: Informative factors
Publicity through press and media
Private-public partnership
Factor 5: Cultural factors
Researcher-investor constructive collaboration
Awareness and emphasis of national ministry of technology towards nanotechnology commercialisation
Acculturation with respect to commercialisation of nanotechnology
Factor 6: Social benefits
Enhancement of social infrastructures/network
Employment creation
Product branding and brand recognition
Cost advantages to customers
Factor 7: Regulatory policies
Patent and legal issues
Governmental fiscal policy regulations
Grant support and climate control by international community
Factor 8: Marketing potentiality and forces
Price suitability for sales' facilitation
Product aesthetic and packaging
Product positioning and launch
Market potential for end products
New market penetration for new and emerging technology
Factor 9: Organisation's strategic issues and features
Organisational strategic implications
Organisation's business policies
Training and development support for technical and marketing staff
Personnel resources with generic and specialised nanocommercialisation knowledge

(continued)

Table 3 (continued)

Factor 10: Health and safety
Public opinion about toxic issue
Safety issue

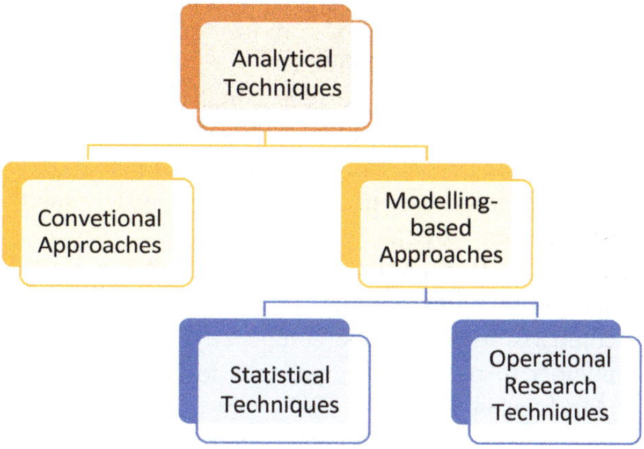

Fig. 7 Analytical approaches providing support for quality control and commercialisation decision-making process. [Adapted by the researcher from Hernandez and Bennison (2000), Kuo et al. (2002) and Priya and Venkatesh (2012)]

relationship between the critical factors for commercialisation decision-making. The deficiency of these approaches is resolved by using the operational research techniques. An *operational research technique* is the analytical hierarchy process (AHP). This process effectively helps the organisation in commercialisation decision-making through the prioritisation of the available decision factors. As a result of this limitation of the conventional approaches, the modelling-based approaches could be adopted; while the Delphi method is used to identify the weight coefficients of the evaluation factors, the AHP is an optimal approach to identify the collection of the weights for priority scaling.

Meanwhile, the Delphi method (DM) and AHP have their strengths and weaknesses (Ho and Wang 2002). Identifying the strengths and weaknesses, Yousuf (2007) elucidated that DM is sufficient for gathering data for a wide and complex problem with a history of insufficient information. He further identified that the methods of summarising and presenting the responses are rather necessitous.

The AHP has the following strengths and weaknesses:

Strength 1: The advantages of AHP over other multifactor methods are its flexibility, intuitive appeal to the decision-makers and its ability to check inconsistencies (Ramanathan 2001). Generally, users find the pairwise comparison form of data input straightforward and convenient.

Strength 2: The AHP method has the distinct advantage that it decomposes a decision problem into its constituent parts and builds hierarchies of factors. Here, the importance of each element becomes clear (Macharis et al. 2004).

Strength 3: AHP helps to capture both subjective and objective evaluation measures. While providing a useful mechanism for checking the consistency of the evaluation measures and alternatives, AHP reduces bias in decision-making (Parcon 2006:98).

Strength 4: The AHP method supports group decision-making through consensus by calculating the geometric mean of the individual pairwise comparisons (Zahir 1999).

Strength 5: AHP is uniquely positioned to help model situations of uncertainty and risk since it is capable of deriving scales where measures ordinarily do not exist (Millet and Wedley 2002).

Weakness 1: AHP can be considered as a complete aggregation method of the additive type. The problem with such aggregation is that compensation between good scores on some factors and bad scores on other factors can occur. Detailed, and often important, information can be lost by such aggregation (Levi-Jakšić and Rakoćevićm 2012).

Weakness 2: With AHP the decision problem is decomposed into a number of subsystems, within which and between which a substantial number of pairwise comparisons need to be completed. This approach has the disadvantage of the number of pairwise comparisons to be made and thus becomes a lengthy task (Macharis et al. 2004).

Weakness 3: Another important disadvantage of the AHP method is the artificial limitation of the use of the 9-point scale. Sometimes, the decision-maker might find it difficult to distinguish among them and which is more important than another (Levi-Jakšić and Rakoćevićm 2012).

Despite the several weaknesses of these methodologies, they have been popularly used in many applications of the public and private sectors. Meanwhile, triangular fuzzy numbers may be introduced into the Delphi method and analytical hierarchy process to address these limitations. Triangular fuzzy numbers are known to eliminate the bottlenecks of ambiguities and uncertainties associated with biasness in the judgements of the experts. The fuzzy Delphi method (FDM) combines the traditional Delphi method and the fuzzy theory and can be used to gather useful and relevant variables from opinions of experts for the purpose of being integrated in decision-making process (Detcharat et al. 2013). Kahraman (2008) posits that fuzzy logic is essential in approaching multifactor decision-making problems with often intrinsic vague, imprecise or fuzzy data in cases of real-life decision-making situations. Empirically providing an approach towards technology selection, Hsu et al. (2010) reported that the FDM was christened by Murray et al. (1985) by conceptualising the combination of the fuzzy set theory and the Delphi method in an attempt to resolve some linguistic ambiguity in opinions and judgements of the experts interviewed. These analytical methods applied in future study will seek to resolve all forms of inexactness encountered in the responses of experts in the process of data collection.

The FDM was proposed by Ishikawa, Amagasa et al. (1993). It is an analytical method based on the Delphi method and fuzzy theory. It is a type of collective decision-making method (Linstone and Turoff 2002), with several rounds of anonymous written questionnaire surveys conducted to ask for experts' opinion. The goal of FDM is to achieve a forecast without influence (Jeremy et al. 2003). It possesses the following properties: anonymity, feedback, statistical and convergence (Ho and Wang 2002). In order to integrate decision-makers' opinion with technology evaluation factors, the fuzzy Delphi method is an effective, economical tool (Ishikawa et al. 1993). Noorderhaben (1995) in Hsu et al. (2010) indicated that applying the fuzzy Delphi method to group decision can solve the fuzziness of common understanding of expert opinions. Future study could therefore apply the fuzzy Delphi method to the selection process of system variables to increase the confidence of the model.

Lan and Sheng (2014) identified the fuzzy analytic hierarchy process (FAHP) and adopted it in analysing the relative weight and ranking of the crucial factors influencing the consumers' purchase The authors aimed at making up for the insufficiency of former research which had scantly been researched by a few savants in a study that could be used as a reference for governmental policies for marketing strategy. Lan and Sheng (2014) gave consecutive accounts of the algorithms of the FAHP as variously reported by Van Laarhoven and Pedrycz (1983), Buckley (1985) and Csutora and Buckley (2001). Lan and Sheng (2014) reported as stating the concept of fuzzy in AHP to solve the values in the pairwise comparison matrix (Van Laarhoven and Pedrycz 1983); geometric mean is adopted to integrate the experts' opinions for precise and consistent factor judgement in the buildup of the fuzzy positive reciprocal matrix (Buckley 1985), and the λ_{max} method is applicable in calculating the fuzzy weights (Csutora and Buckley 2001). The FAHP is used to solve the values in the pairwise comparison matrix obtained which are essential for a large number and wide scope of organisational decision-making for quality control and commercialisation of technologically processed materials for food packaging.

7 Model Development for Nanotechnology Commercialisation

There is need for an improved understanding and applications of nanotechnology commercialisation models for successful transfer of NEMs. This is true for every emerging technology. Oriented towards solving real-life problems, modelling has been seen by Will et al. (2002) as the basis for most of the early researches in the management of operations. The word 'modelling' comes from the Latin word *modellus* (Schichl 2004). Through the process of modelling, scientists represent ideas about the real world to one another and then simultaneously make changes to these depictions over time in response to new evidence and understandings (Cummins 2007).

Schichl (2004) describes a model as a simplified version of something that is real with its functions illustrated to include explaining a phenomenon, making predictions, decision-making and communication. According to the USEPA (2009), models are used for the purpose of data analyses which are based on the scientific understanding of physical and chemical processes by providing a structure for

Fig. 8 A typical linear model. [Adapted by the Researcher from Mahdjoubi (1997)]

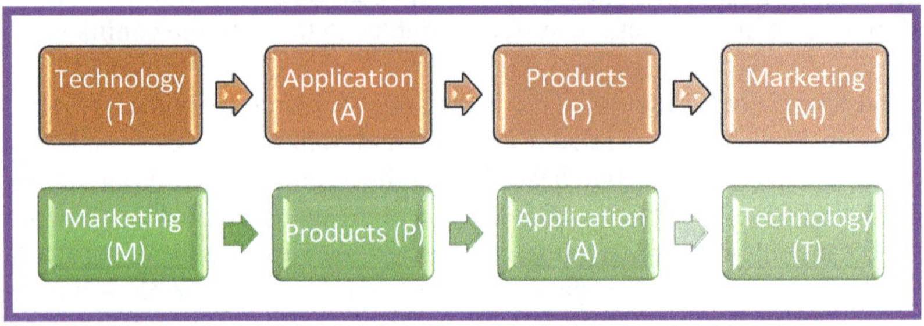

Fig. 9 TAPM/MPAT framework of the first- and second-generation linear models. [Adapted by the Researcher from Amadi-Echendu and Rosetlola (2011)]

understanding a task's behaviour and predicting the effects of actions taken. They are categorised into quantitative and qualitative models.

Being a non-irrational knowledge-generation approach, quantitative model-based research is structured so that the model developed is (partly) a function of the behaviour of real-life operational processes (Meredith et al. 1989). These are also structured to (partly) include decisional problems in real-life operational processes being faced by decision-makers (Will et al. 2002). Observably, quantitative model explains the nature of the causal relationship between the dependent variable and the independent variable. Hence, through this event, it is possible to make observational explanations and also forecast the future state of the modelled processes. This is supported by Will et al. (2002). These categories of models have their relevance and possible contributions in the statistical methods for the development, manipulation and commercialisation of carbon nanotubes. Meanwhile, the work of Amadi-Echendu and Rosetlola (2011) on these discourses provides the point of departure for this section.

Amadi-Echendu and Rosetlola (2011) confirmed that a number of commercialisation models were reported and critiqued in Rosa and Rose (2007). They gave a summary heuristically as linear models and functional models. The linear model, according to Mahdjoubi (1997), is a R&D model also referred to as the assembly-line model, bucket model, ladder model or pipeline model. Figure 8 presents the process from research to product sales as the innovation pipeline to commercialisation success. The linear model identifies development, production and marketing activities as intermediaries between research and product sales.

Stefanovska et al. (2016) categorised the linear model into the first generation of innovation models (technology push) and the second generation of innovation models (market pull). Amadi-Echendu and John (2008) hypothesised model frameworks for the two generations of innovation models, which are illustrated in Fig. 9.

Unfortunately the current days' reality for technology commercialisation does not support the linear models as both lack the feedback loops. Rothwell (1992) and Marinova and Phillimore (2003) presented six generations of innovation models. According to Stefanovska et al. (2016), the last four generations are the functionally integrated innovation models which are supplemented by several feedback loops. The functional models are non-linear models which are the current days' reality for emerging technology commercialisation process. Zuniga and Correa (2013) iterated that there are feedback loops at every stage throughout the process of emerging technology commercialisation which are necessary for quality control.

Corroborating Zuniga and Correa (2013) on the need for feedback loops for quality control in an invention-commercialisation process, Stefanovska et al. (2016) suggested that consumers' feedback is an essential element of any technology innovation. This discovery by Zuniga and Correa (2013) reveals that feedback is essential for the purpose of quality control of the commercialisation process engineered materials from nanotechnology.

Meanwhile, Aithal and Aithal (2016) remarked that commercialisation of nanotechnology from the research stage to profitable material and product is highly risky at the 'Valley of Death' point of commercialisation. Comparing nanotechnology with other emerging technologies, Craig et al. (2013) confirmed that the reasons for this high risk are related to the product focus, product development and market engagement. The 'Valley of Death' point of commercialisation, as indicated in Fig. 10, is a phrase that is used to illustrate the gap between academic-based innovations and their commercial applications within the marketplace.

The articulation of the identified critical factors is necessary for successful commercialisation of engineered nanomaterials in bridging this 'Valley of Death' for improved nanotechnology commercialisation process.

8 Conclusion

The chapter presented salient and critical factors that could be considered in the management decision-making of commercialisation of applications on emerging technology, particularly essential elements for the successful commercialisation of engineered nanomaterials. It focused on bridging the link between the innovation, development and the markets in the commercialisation of food packaging products from bio-nanocomposites. It is expected that, within a global market, a thorough understanding and adoption of these identified factors for successful commercialisation of active packaging technologies will yield system-based solution.

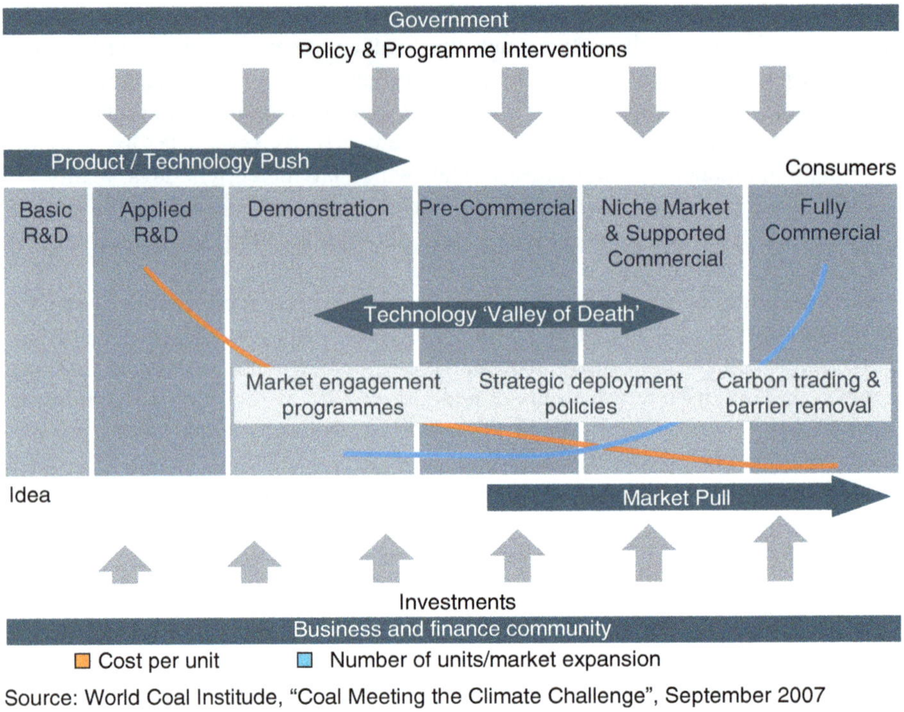

Fig. 10 'Valley of Death' point of commercialisation. (Amadi-Echendu and Rosetlola 2011)

References

Aithal P, Aithal S (2016) A new model for the commercialisation of nanotechnology products and services. Int J Comput Res Dev 1(1):84–93

Alfadul S, Altahir O, Khan M (2017) Application of nanotechnology in the field of food production. Acad J Sci Res 5(7):143–154

Amadi-Echendu J, John A (2008) On the commercialization of trailing knowledge and IP. In: Proceedings of PICMET'08 conference, Cape Town, PaperID 08R0018

Amadi-Echendu J, Rosetlola R (2011) Technology commercialization factors, frameworks and models. In: Proceedings of first international technology management conference, pp 144–148

Amagasa M, Ishikawa A, Shiga T, Tomizawa G, Tatsuta H, Mieno R (1993) The max-min Delphi method and fuzzy delphi method via fuzzy integration. Fuzzy Sets Syst 35:241–253

Andreev G, Minashkin V, Nevskii I, Putilov V (2009) Nanotechnology-derived materials: potential risk in preparation and use. Russ J Gen Chem 79(9):1974–1981

Azerendo H, Mattoso L, McHugh T (2011) Nanocomposites in food packaging – a review. In: Reddy B (ed) Advances in diverse industrial applications of nanocomposites. InTech Publisher, Rijeka

Bandarian R (2007) Evaluation of commercial potential of a new technology at the early stage of development with fuzzy logic. J Technol Manag Innov 2(4):73–85

Bisgaard S (1997) Designing experiments for tolerancing assembled products. Technometrics 39 (2):142–152

Borzouei H, Mirdamadi S, Hosseini S (2011) Affective factors in commercialisation of nanotechnology in Iran's agricultural sector. Ann Biol Res 2(6):56–61
Box E, Bisgaard S (1987) The scientific context of quality improvement. Qual Prog 6:54–61
Bradley L, Castle L, Chaudhry Q (2011) Applications of nanomaterials in food packaging with a consideration of opportunities for developing countries. Trends Food Sci Technol 22:604–610
Buckley J (1985) Fuzzy hierarchical analysis. Fuzzy Sets Syst 17:233–247
Bumbudsanpharoke N, Ko S (2015) Nano-food packaging: an overview of market, migration research, and safety regulations. J Food Sci 80(5):910–923
Chaudhry Q, Castle L (2011) Food applications of nanotechnologies: an overview of opportunities and challenges for developing countries. Trends Food Sci Technol 22:595–603
Craig B, Richard M, Grant E, Darren M (2013) The commercialization of nanotechnology: the five critical success factors to a nanotech-enabled whole product. In: Tsuzuki T (ed) Nanotechnology commercialization. Pan Stanford Publishing, Canberra, pp 171–204
Csutora R, Buckley JJ (2001) Fuzzy hierarchical analysis: the lambda-max approach. Fuzzy Sets Syst 120:181–195
Cummins P (2007) Models for the 1762 Arakan earthquake and tsunami: the potential for giant tsunamigenic earthquakes in the northern Bay of Bengal. Nature 449:75–78
Detcharat S, Pongpun A, Tarathorn K (2013) A hybrid multi-factors decision model for technological innovation capability assessment: research on Thai automotive parts firms. Int J Eng Technol Innov 3(1):20–37
Echeveste M, Amaral C, Rozenfeld H (2007) A support tool for the selection of statistical techniques for industrial product development and improvement processes. In: Loureiro G, Curran R (eds) Complex systems concurrent engineering. Springer, London. https://doi.org/10.1007/978-1-84628-976-7_28
Garg P, Ghatmale P, Tarwadi K, Chavan S (2017) Influence of nanotechnology and the role of nanostructures in biomimetic studies and their potential applications. Biomimetics 2:7
Gholami-Rostam N, Mahdavinejad M, Gholami-Rostam M (2015) Commercializing usage of nano-insulating materials in building industry and future architecture. Procedia Mater Sci 11:644–648
Glen JC (2006) Nanotechnology: future military environmental health considerations. Technol Forecast Soc Chang 73(2):128–137
Hansen SL, Heggelund R, Besora P, Mackevica A, Boldrin A, Baun A (2016) Nanoproducts: what is actually available to European consumers. Environ Sci Nano 3(1):169–180
Hassanali A, Zahra (2015) Prioritization effective factors on commercialization of products using fuzzy AHP APPROACH (Empirical evidence: knowledge-based business of incubators centers of Iran north region). Surv Methodol 5(2):1746–1751
Hernandez T, Bennison D (2000) The art and science of retail location decision. Int J Retail Distrib Manag 28(8):357–367
Ho Y, Wang H (2002) Applying Fuzzy Delphi method to select the variables of a sustainable urban system dynamics model. J Architect 41(2):453–468 Taipei
Hobson D (2009) Commercialization of nanotechnology. WIRES Nanomed Nanobiotechnol 1 (1):2–5
Honarvar Z, Hadian Z, Mashayekh M (2016) Nanocomposites in food packaging applications and their risk assessment for health. Electron Phys 8(6):2531–2538
Hosseini S, Esmaeeli S, Ansari B (2011) Challenges in commercialisation of nano and biotechnologies in agricultural sector of Iran. Afr J Biotechnol 10(34):6516–6521
Hsu Y, Lee C, Kreng V (2010) The application of Fuzzy Delphi method and Fuzzy AHP in lubricant regenerative technology selection. Expert Syst Appl Elsevier 37:419–425
Inigo F, Siavash H, Mika S (2017) Current and future business models for 3D printing applications. In: Ballardini M, Norrgard M, Partanen J (eds) 3D Printing, intellectual property and innovation: insights from law and technology. Wolters Kluwer, Alphen aan den Rijn, pp 33–62
Ishikawa A, Amagasa M, Shiga T, Tomizawa G, Tatsuta H, Mieno R (1993) The max-min delphi method and Fuzzy Delphi method via fuzzy integration. Fuzzy Sets Syst 35:241–253

Jayaramudu J, Reddy G, Varaprasad K, Sadiku A, Ray S, Rajulu V (2013) Preparation and properties of biodegradable films from sterculiaurens short fibre/cellulose green composites. Carbohydr Polym 93:622–627

Jeremy C, Glenn V, Theodore J (2003) Futures research methodology ver. 2.0. American Council for the United Nations University the Millennium Project, Washington, DC

Kahraman C (2008) Multi-factors decision making and fuzzy sets. In: Kahraman C (ed) Fuzzy multi-factors decision making. Springer Theory and applications with recent developments, vol. 16, pp 1–18

Kamolkittiwong, Phruksaphanrat (2015) An analysis of drivers affecting green supply chain management implementation in electronics industry in Thailand. J Econ Bus Manag 3 (9):864–869

Kanmani P, Rhim J (2014) Physiochemical properties of gelatin/silver nanoparticle antimicrobial composite films. Food Chem 148:162–169

Keller-McNulty S, Nakhleh CW, Singpurwalla ND (2005) A paradigm for masking (camouflaging) information. Int Statist Rev 73:331–349

Kim S, Lee J, Lim H, So D, Kim K (2010) Korean experience in nanotechnology industrialization. Tech Monitor 1(1–2):21–29

Koç E, Burhan H (2015) An application of analytic hierarchy process (AHP) in a real world problem of store location selection. Adv Manage Appl Econ 5(1):41–50

Kumar S, Luthra S, Haleem A, Mangla S, Garg D (2015) Identification and evaluation of critical factors to technology transfer using AHP approach. Int Strateg Manage Rev 3:24–42

Kumar V, Jain PK (2003) Commercialization of new technologies in India: an empirical study of perceptions of technology institutions. Technovation 23:113–120

Kuo J, Chi S, Kao S (2002) A decision support system for selecting convenience store location through integration of fuzzy AHP and artificial neural network. Comput Ind 47:199–214

Lan S, Sheng T (2014) The study on key factors of influencing consumers' purchase of green buildings: application of two-stage fuzzy analytic hierarchy process. Int Bus Res 7(6):49–62

Lee C, Lee S, Jhon M, Shin J (2013) Factors influencing nanotechnology commercialisation: an empirical analysis of nanotechnology firms in South Korea. J Nanopart Res 15:1444–1459

Levi-Jakšić M, Rakoćevićm S (2012) Innovative management and business performance. In: Proceedings of the XIII international symposium, Serbia

Linstone HA, Turoff M (2002) The Delphi method: techniques and applications. Addison-Wesley Pub. Co, Reading ISBN 0-201-04294-0

Lo C, Wang C, Chien P, Hung C (2012) An empirical study of commercialisation performance on nanoproducts. Technovation 32(3–4):168–178

Lopez R (2013) Applications and issues in the fields of nanotechnology, information technology, neurotechnology, and biotechnology. Int J Bus Soc Sci 4(8):39–50

Lu J, Jeng S, Wang K (2009) Review of statistical methods for quality improvement and control in nanotechnology. J Qual Technol 41(2):148–164

Ludhiyani A, Pathak R, Joshi S (2011) Business implications of recent developments in nanotechnology for micro scale devices. Intellect Econ 1(9):117–134

Luthra S, Kumar S, Garg D, Haleem A (2015a) Barriers to renewable and sustainable energy technologies adoption: Indian perspective. Renew Sust Energ Rev 41:762–776

Luthra S, Garg D, Haleem A (2015b) An analysis of interactions among critical success factors to implement green supply chain management towards sustainability: an Indian perspective. Resour Policy. https://doi.org/10.1016/j.resourpol.2014.12.006i

Macharis C, Verbeke A, De Brucker K (2004) The strategic evaluation of new technologies through multicriteria analysis: the advisors case. Res Transport Econ 8:443–462

Mackevica A, Revilla P, Brinch A, Hansen S (2016) Current uses of nanomaterials in Biocidal products and treated articles in the EU. Environ Sci Nano 3(5):1195–1205

Mahboudi M, Ananthan B (2010) Effective factors in technology transfer in the pharmaceutical industries of iran: a case study. J Knowl Manag 3(1&2):98–110

Mahdjoubi D (1997) The linear model of technological innovation: background and taxonomy, UTexas working paper
Majeed K, Jawaid M, Hassan A, Bakar A, Khalil A, Salema A, Inuwa A (2013) Potential materials for food packaging from nanoclay/natural fibres filled hybrid composites. Mater Des 46:391–410
Mansoori GA, Soelaiman TA (2005) Nanotechnology – an introduction for the standards community. J ASTM Int 2(6):1–21
Marinova R, Phillimore J (2003) Models of innovation. In: Shavinina V (ed) International handbook on innovation. Elsevier, Amsterdam
Maurizio A, Jan J, Errico M, Sabine F, Paolo V, Volpe M (2005) Biodegradable starch/clay nanocomposite films for food packaging applications. Food Chem 93:467–474
Mazzola L (2003) Commercializing nanotechnology. Nat Biotechnol 21(10):1137–1143
Mensitieri G, Di Maio E, Buonocore G, Nedi I, Oliviero M, Sansone L, Iannace S (2011) Processing and shelf life issues of selected food packaging materials and structures from renewable resources. Trends Food Sci Technol 22(2–3):72–80
Meredith J, Raturi A, Amoako-Gyampah K, Kaplan B (1989) Alternative research paradigms in operations. J Oper Manag 8:297–326
Millet I, Wedley W (2002) Modelling risk and uncertainty with the analytic hierarchy process. J Multi-Criteria Decis Anal 11:97–107
Mohannak K, Samtani A (2014) A factors based approach for evaluating innovation commercialisation. In: DRUID society 2014 conference on entrepreneurship – organization – innovation, 16–18 June 2014, Copenhagen Business School, Denmark. Accessed 15 Apr 2017
Murray T, Pipino L, Gigch J (1985) A pilot study of fuzzy set modification of Delphi. Hum Syst Manag 5:76–80
Nogueira P, Paino M, Zucolotto V (2013) Nanosilver: properties, application and impacts on health and environment. Vigilância Sanitária em Debate 1(4):57–68
Noorderhaben N (1995) Strategic decision making. Addison-Wesley, Workingham
Oladimeji H, Singh S (2013) Operational research modeling approaches for commercialization of nano-engineered materials. In: Kanny M (ed) Advances in composites, biocomposites and nanocomposites. Manipal University Press, Manipal, pp 395–411
Othman S (2014) Bionanocomposite materials for food packaging applications: types of biopolymer and nano-sized filler. Agric Agric Procedia 2:296–303
Parcon P (2006) Develop your decision making skills. Lotus Press, New Delhi
Priya P, Venkatesh A (2012) Integration of analytic hierarchy process with regression analysis to identify attractive locations for market expansion. J Multi-Criteria Decis Anal 19(3–4):143–153
Rafieian F, Shahedi M, Keramat J, Simonsen J (2014) Thermomechanical and morphological properties of nanocomposite films from wheat gluten matrix and cellulose nanofibrils. J Food Sci 79:100–107
Ramanathan R (2001) A note on the use of the *analytic hierarchy process* for environmental impact assessment. J Environ Manag 63:27–35
Reynolds G (2007) FDA recommends nanotechnology research, but not labelling. www.foodproductiondaily.com. Accessed 22 Nov 2017
Rhim J, Park H, Ha C (2013) Bio-nanocomposites for food packaging applications. Prog Polym Sci 38:1629–1652
Rockefeller Foundation (2005) Nanotechnology and the poor: opportunities and risks. Meridian Institute, Dillon
Roco M (2004) Nanotechnology's future. Forum Sect Sci Am 295(2):21
Romig AD, Baker AB, Johannes J, Zipperian T, Eijkel K, Kirchhoff B, Mani HS, Rao CNR, Walsh S (2007) An introduction to nanotechnology policy: opportunities and constraints for emerging and established economies. Technol Forecast Soc Chang 74(9):1634–1642
Rosa J, Rose A (2007) Report on interviews on the commercialization of innovation. Statistics. Canada. ISBN 978-0-662-46359-7

Rossi M, Passeri D, Sinibaldi A, Angjellari M, Tamburri E, Sorbo A, Carata E, Dini L (2017) Nanotechnology for food packaging and food quality assessment. Adv Food Nutr Res 82:149–204

Rothwell R (1992) Successful industrial innovation critical factors for the 1990s. R&D Manag 22 (3):221–239

Rouhi J, Mahmud S, Naderi N, Raymond C, Mahmood M (2013) Physical Properties of fish gelatin-based bio-nanocomposite films incorporated with ZnO nanorods. Nanoscale Res Lett 8:1–6

Schichl H (2004) Models and the history of modeling. In: Kallrath J (ed) Modeling languages in mathematical optimization. Applied optimization, vol 88. Springer, Boston

Sekhon B (2010) Nanotechnology. Sci Appl 3(1):1–15

Sekhon B (2014) Nanotechnology in agri-food production: an overview. Nanotechnol Sci Appl 7:31–53

Shapira P, Gök A, Salehi F (2016) Graphene enterprise: mapping innovation and business development in a strategic emerging technology. J Nanopart Res 18:269–281

Shen Y, Chang S, Lin G, Yu H (2010) A hybrid selection model for emerging technology. Elsevier Technol Forecast Soc Chang 77:151–166

Singh T, Shukla S, Kumar P, Wahla V, Bajpal V (2017) Application of nanotechnology in food science: perception and overview. Front Microbiol 8:1501–1508

Soleimanpour M, Hosseini S, Mirdamadi S, Sarafrazi A (2011) Challenges in commercialisation of nanotechnology in agriculture sector of Iran. Ann Biol Res 2(4):68–75

Sonker AS, Richa R, Pathak J, Rajneesh R, Pandey A, Chatterjee A, Sinha RP (2017) Bionanotechnology: past, present and future. In: Sinha R, Richa JP (eds) New approaches in biological research. Nova Science Pub Inc, New York

Stefanovska L, Polenakovikj R, Dzidrov M (2016) Summary of innovation models on a company level – creating a framework for an innovation model that will increase a company's innovation activity. Sci Proc Sci Tech Union Mech Eng 26(212):47–50

Tang X, Kumar P, Alavi S, Sandeep P (2012) Recent advances in biopolymers and biopolymer-based nanocomposites for food packaging materials. Crit Rev Food Sci Nutr 52:426–442

Uldrich J, Newberry D (2003) The next big thing is really small: how nanotechnology will change the future of your business. Crown Business/Random House Inc, New York

USEPA (2009) Environmental protection agency, Washington, DC, EPA/600/R-08/139F

Van Laarhoven P, Pedrycz W (1983) A fuzzy extension of Saaty's priority theory. Fuzzy Sets Syst 11:229–241

Werner B, Koontz J, Goddard J (2017) Hurdles to commercial translation of next generation active food packaging technologies. Curr Opin Food Sci 16(8):40–48

Wikle C, Royle J (2004) Spatial statistical modeling in biology, in modern biometry. In: Wilson S (ed) Encyclopedia of Life Support Systems (EOLSS). Eolss Publishers, Oxford

Will J, Bertrand M, Fransoo C (2002) Operations management research methodologies using quantitative modeling. Int J Oper Prod Manag 22(2):241–264. https://doi.org/10.1108/01443570210414338

Yousuf M (2007) Using experts' opinions through Delphi technique. Pract Assess Res Eval 12 (4):1–8 http://pareonline.net/getvn.asp?v=12&n=4

Zahir S (1999) Clusters in group: decision making in the vector space formulation of the analytic hierarchy process. Eur J Oper Res 112:620–634

Zhigljavsky A (2011) Stochastic global optimization. In: International encyclopedia of statistical science. Springer, Berlin/Heidelberg, pp 1521–1524

Ziamou PZ (2002) Commercialization of new technologies: consumers' response to a new interface. J Prod Manage 19:365–374

Zuniga P, Correa P (2013) Technology transfer from public research organizations: concepts, markets, and institutional failures. World Bank, Washington, DC

Environment-Friendly Biopolymers for Food Packaging: Starch, Protein, and Poly-lactic Acid (PLA)

Sai Pyae Sone Aung, Hnin Htet Htet Shein, Kyaw Nyein Aye, and Nitar Nwe

Abstract Currently, petroleum-based plastics are widely used for food packaging in all over the world. Due to their environmental pollution problems, most of the researchers are interested to replace the petroleum-based plastics with environment-friendly biodegradable polymers for packaging materials, in which biodegradable packaging materials based on starch, protein, and poly-lactic acid (PLA) have been produced for packaging of food and medicine. However, these materials are weaker than petroleum-based plastics in using as packaging materials. In order to produce strong packaging materials, these bio-based materials are composed of biopolymers and synthetic polymers. They are already commercialized mainly for single-use disposal packaging applications such as bottles, cold drink cups, thermoformed trays and lid containers, blister packages, overwrap as well as flexible films. Therefore, this chapter focuses on production of starch, protein, and PLA; production and characterization of composite materials based on starch, protein, and/or PLA; preparation of packaging materials using composite materials based on starch, protein, and/or PLA; advantage and disadvantage of these packaging materials used in food packaging; commercial usage of packaging materials; and environmental impact of these packaging materials.

Keywords Food packing · Starch packaging · Protein packaging · PLA packaging · Composite

---

S. P. S. Aung (✉)
Biotechnology Laboratory, Ecological Laboratory, Advancing Life and Regenerating Motherland, Yangon, Myanmar

Department of Chemistry, University of Yangon, Yangon, Myanmar

H. H. H. Shein · K. N. Aye
Biotechnology Laboratory, Ecological Laboratory, Advancing Life and Regenerating Motherland, Yangon, Myanmar

Department of food engineering, Yangon Technological University, Yangon, Myanmar

N. Nwe
Biotechnology Laboratory, Ecological Laboratory, Advancing Life and Regenerating Motherland, Yangon, Myanmar

S. Ahmed (ed.), *Bio-based Materials for Food Packaging*,
https://doi.org/10.1007/978-981-13-1909-9_8

1 Introduction

Today, petroleum-based polymers like polyethylene (PE), polypropylene (PP), polystyrene (PS), etc. have been widely used in many areas such as packaging, decoration, production of automobile parts, and others. These plastics have excellent long-life properties and take years to degrade under normal condition because they are highly resistant to natural process including microbial and enzymatic reactions and also chemical reactions. Currently these disposal plastics materials are destroyed using burning or landfilling process. Burning of these plastics releases toxic gases like dioxins, furans, mercury, and polychlorinated biphenyl in the atmosphere (Verma et al. 2016). Landfill is not also the suitable process for these problems. Therefore, plastic problems are now the hot issues in our world. As the population is increasing year by year, the food and its packaging materials become one of the environmental concerns. To maintain the environmental safety, the demand for the sustainability has increased. In order to solve these problems, packing for the food made from biopolymer (renewable and biodegradable source) has now been of interest (Fahma et al. 2017; Malathi et al. 2014; Ferreira et al. 2016). The general cycle of bio-based packaging is shown in Fig. 1.

Biopolymers such as protein, carbohydrate, etc. are the long-chain molecules derived from plants and animals. Biopolymers are ensuring biodegradability and biocompatibility and provide the opportunities to eco-friendly usage (Muller et al. 2017; Othman 2014; Hu 2014). Nowadays, biopolymers such as starch, protein, cellulose, chitosan, PLA, and their derivatives are used in food packaging. These biopolymers are used as edible film or coating on the food with the aim of reducing the loss of moisture, prevention of oxidation, decreasing the migration of lipids, and so on (Bourtoom 2009; Pawar and Purwar 2013).

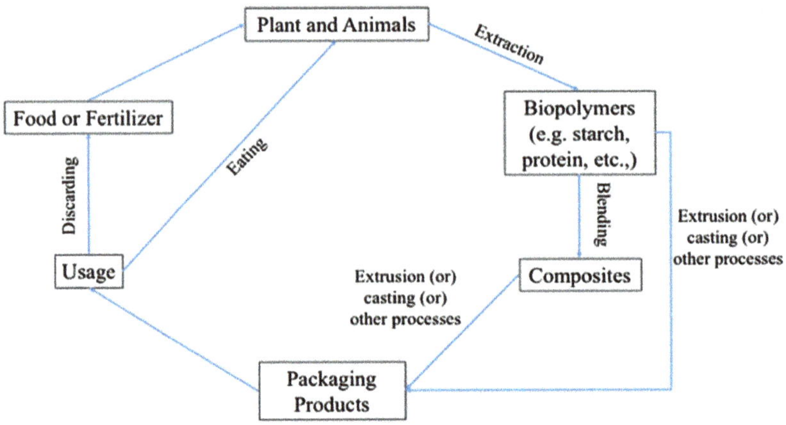

Fig. 1 General cycle of bio-based packaging. (Idea got from Muller et al. 2017)

2 Biopolymers: Starch, Protein, and Poly-lactic Acid (PLA)

2.1 *Starch*

Starch is composed of D-glucose units linked with α-(1,4′)-glycosidic bonds and branched α-(1,6′)-glycosidic bonds. Depending on the sources, starch consists of 10–30% amylose (D-glucose units linked with α-(1,4′)-glycosidic bonds) and 70–90% amylopectin (D-glucose units linked with α-(1,4′)-glycosidic bonds and α-(1,6′)-glycosidic bonds). Starch is one of the most abundant biopolymers, which can be naturally found in tuber like potato; cereal seed like corn, rice and wheat; and root like cassava (tapioca) (Murgić et al. 2015; Nwe 2015).

The production of starch powder is shown in the Fig. 2.

Starch has the film-forming properties and is commonly produced by casting or extrusion (Vartiainen et al. 2014). The film strength depends on amylose content and amylopectin generally contributes to film with low mechanical properties (Ezeoha and Ezenwanne 2013). The large amylose content provides the better mechanical properties when there is no plasticizer. However, when plasticizer is used, the high amylopectin is favorable since it is more reactive and has larger plasticity (Lourdin et al. 1990, and 1995 cited in Vilpoux and Avérous 2004). As starch is widely available and low cost, starch-based biomaterials have received great attention. However, starch-based biomaterial has poor moisture barrier and mechanical properties compared with synthetic polymer, so they become drawbacks in food packaging (Avella et al. 2005). In order to improve properties of starch film, bio-composite films have been investigated.

2.2 *Protein*

Protein is made up of linear repeating units called amino acids linked together by the peptide bonds. There are 22 amino acids (selenocysteine and pyrrolysine with 20 standard amino acids) that can be found in protein (Ambrogelly et al. 2007).

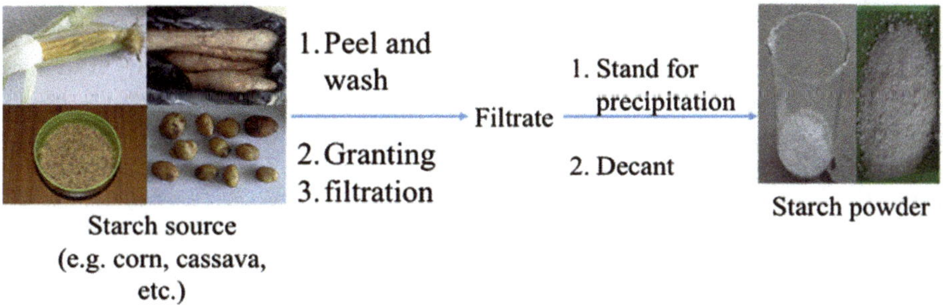

Fig. 2 The production of starch powder. (Nwe 2015)

Protein has been produced by the fermentation or chemical synthesis (Aung et al. 2017). Consequently, proteins obtained from different sources have different physicochemical properties. Proteins are naturally in globular form like wheat gluten and corn zein or fibrous form like casein and gelatin (Wittaya 2012; Lacroix and Vu 2014; Coltelli et al. 2015). Proteins for making film and coating can be obtained from both animals (like casein and whey from milk protein) and plants (like corn zein protein and soy protein) (Dangaran et al. 2009).

Protein films are generally better than polysaccharide film in mechanical and barrier properties and also provide higher nutritional value (Cuq et al. 1998; and Khwaldia et al. 2004 cited in Zink et al. 2016). There are still limitations in packaging with protein (film or coating) such as low water vapor barrier due to being hydrophilic in nature and poor mechanical properties compared with synthetic polymers (Bourtoom 2009; Wittaya 2012).

2.3 Poly-lactic Acid (PLA)

Lactic acid has a formula of $C_3H_6O_3$. Lactic acid can be obtained from the microbial fermentation of carbohydrate or chemical synthesis from the petrochemical. Lactic acid has stereoisomers of L(+) and D(−) lactic acid. The suitable microbe should be used to produce only one of isomers, L-lactic acid or D-lactic acid. Racemic mixture of lactic acid can be obtained from the chemical synthesis (Wee et al. 2006; Avinc and Khoddami 2009; Vuković-Kwiatkowska and Kaczmarek 2014).

PLA is aliphatic polyester of lactic acid (2-hydroxypropanoic). PLA is the synthetic polymer because it is not directly got from the nature but synthesized from the renewable monomer, lactic acid. There are many ways to obtain PLA from lactic acid. The three main methods are (1) direct condensation of D-lactic acid and/or L-lactic acid, (2) azeotropic dehydration polymerization, and (3) ring opening polymerization of lactide. PLA can have three different structures – (1) poly (D-lactide) (PDLA), (2) poly(L-lactide) (PLLA), and (3) poly(D, L-lactide) (PDLLA) – and have different properties (Byun and Kim 2014; Avérous 2008; Pawar and Purwar 2013; Hu 2014).

Production of PLA is shown in Fig. 3.

PLA has film-forming properties. Extrusion is the most common technique used for the casting of PLA film. PLA-based materials are usually brittle and rigid so that plasticizer is added in order to enhance their mechanical properties (Muller et al. 2017).

General structure and composition of starch, protein, and PLA are summarized in Table 1.

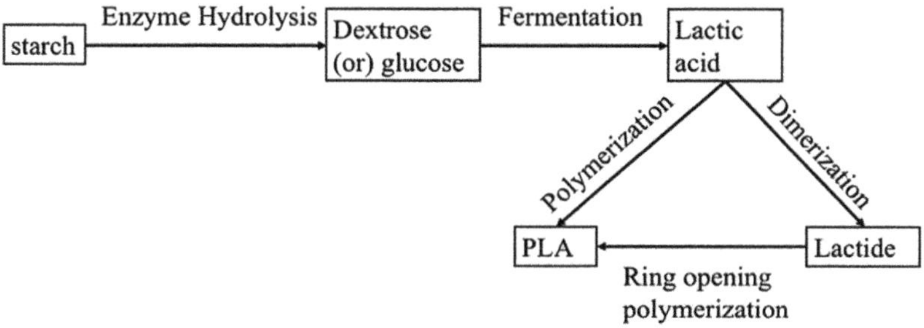

Fig. 3 The production of PLA. (From Avérous 2008; Avinc and Khoddami 2009)

3 Packaging Materials Using Starch, Protein, and Poly-lactic Acid (PLA) and Their Composites

Biopolymers have been produced from various natural sources and have been used as substitution of non-biodegradable plastics (petroleum-based plastics) in order to make more sustainable image which is renewable, rapidly degrades, and reduces the waste in environment (Vartiainen et al. 2014; Othman 2014; Borges et al. 2015). When used in food packaging, biopolymer films have some benefits to improve the quality and to extend the shelf life of the food products in food packaging. However, there are limitations of biopolymers because of their low mechanical properties and poor water vapor barrier properties. Therefore, many researches have been done to improve the mechanical, thermal, and physical properties of these biopolymers by combining with other materials and producing the composite (Rhim and Kim 2014).

Starch, protein, and PLA can themselves be used in production of food packaging material. However, due to their low properties, plasticizer is added to increase the elongation and elastic modulus as has plasticity effect (Said et al. 2016).

The packaging material formed from the starch, protein, PLA, and their composite is shown in Table 2.

4 Advantages and Disadvantages of Packaging Food Products with Packaging Materials Using Biopolymers

Packaging materials from starch, protein, and poly-lactic acid (PLA) and their composite have been used for packaging of fruits, vegetables, meats, and seafoods. They prevent the food products from oxidation, extend shelf life, and improve the quality of food. However, due to their low mechanical properties, there are some disadvantages in comparison with the packaging materials using synthetic polymers. Different properties are shown by the same packaging material when the packaging product is changed (Talens et al. 2010). Their advantages and disadvantages of packaging food products with materials using biopolymers are shown in Table 3.

Table 1 Structure and composition of starch, protein, and PLA

Biopolymer	Source	Compositions	References
Starch	Cassava	19–22% amylose	Cuq et al. (1997), Pérez et al., and Gutierrez et al. cited in García et al. (2015)
		28–81% amylopectin	
	Potato	17–24% amylose	Cuq et al. (1997) cited in García et al. (2015)
		76–83% amylopectin	
	Rice	15–35% amylose	
		65–85% amylopectin	
	Wheat	20–25% amylose	
		75–80% amylopectin	
	Corn	17–25% amylose	Cuq et al. (1997) and Sandhu et al. (2007) cited in García et al. (2015)
		75–83% amylopectin	
Protein	Milk protein	Compose of casein and whey proteins	Wittaya (2012) and Lacroix and Vu (2014)
	Whey protein	Contain five main parts: α-lactalbumin, β-lactoglobulin, bovine serum albumin (BSA), immunoglobulin, proteose peptone	Nwe (2015) and Lacroix and Vu (2014)
	Casein	Consist of three main components α-casein, β-casein, and κ-casein that form micelles which are stabilized by calcium-phosphate bridging	Khwaldia et al. (2004)
	Collagen	Fibrous and structural protein in animal tissue	Alizadeh and Behfar (2013) and Hames and Hooper (2011)
		Main amino acids: glycine, proline, hydroxyproline, and alanine	
		Collagen contains a repeating tripeptide sequence of Gly-X-Y, in which X is often Pro and Y is often Hyp	
	Gelatin	Partial hydrolysis of native collagen	Nwe et al. (2010)
	Wheat gluten	Compose of two water-insoluble proteins: gliadins (low molecular weight) and glutenin (high molecular weight)	Wittaya (2012), Lacroix and Vu (2014), and Šuput et al. (2015)
PLA		Polymers of lactic acid	

5 Commercial Production of Biopolymer-Based Packaging Material Using Starch, Protein, and Poly-lactic Acid (PLA)

Packaging materials based on starch, protein, and PLA have been produced commercially. Some of the products and their trade names are shown in Table 4.

Table 2 Packaging materials using starch, protein, poly-lactic acid (PLA), and their composites

Biomaterials	Plasticizers	Processes	Packaging material	Properties	References
Cassava starch	Glycerol	1. Mix 12% moisture cassava starch, polyvinyl alcohol, talc powder, and urea	Film	Tensile strength is 24.87 N/mm^2	Ezeoha and Ezenwanne (2013)
		2. Add glycerol			
		3. Stir to semi-dry powder			
		4. Extrude with blown film extruder			
Cornstarch	Glycerol	1. Prepare 4% w/v starch solution with distilled water	Film	Low WVP and O_2P	Aghazadeh et al. (2017)
				~0.89 MPa tensile strength, ~53.52% elongation at break	
Rice starch		2. Heat to 80 °C for 30 min by stirring		Low WVP and O_2P	
				~0.278 MPa tensile strength, ~141.16% elongation at break	
Wheat starch		3. Add glycerol to make 2% w/v during stirring		Low WVP and O_2P	
		4. Pour on plastic plate		~0.84 MPa tensile strength, ~86.494% elongation at break	
		Dry in oven at 60 °C for 20 h			
Starch	Glycerol	1. Prepare 4% w/v starch solution with distilled water	Coating solution		Aghazadeh et al. (2017)
		2. Heat to 80 °C for 30 min by stirring			
		Add glycerol to make 2% w/v during stirring			
Starch-chitosan	Glycerol	1. Prepare 1% w/v chitosan solution	Coating solution		Aghazadeh et al. (2017)

(continued)

Table 2 (continued)

Biomaterials	Plasticizers	Processes	Packaging material	Properties	References
		2. Stir with a magnetic stirrer for 24 h			
		3. Add equal volume starch solution			
		Stir with a magnetic stirrer for 10 min			
Starch-chitosan	Glycerol (30% weight of starch)	1. Prepare 4% starch gelatinous solution at 90 °C for 10 min	Film	Decrease WVP, moisture absorption, and solubility, and increase tensile strength with increasing chitosan and rosemary oil contents' antimicrobial activity	Sayyahi et al. (2017)
		2. Cool to 50 °C			
		3. Add glycerol, chitosan solution, and rosemary oil			
		4. Homogenize for 2 min at 120000 rpm			
		Cast and dry at 50 °C for 20 h			
LDPE-tapioca		1. Add LDPE to high-speed mixer at 30 rev/min	Film	Increasing tapioca, decreasing strength and stiffness, but increasing ductility	Ali et al. (2013)
		2. Add glycerol, maleic anhydrate, stearic acid, and wax, and mix for 10 min for each material added			
		Inject into twin-screw extruder at 70–80 rpm			
Whey protein		1. Prepare 10% (w/w) WPI with deionized water	PET film coating		Müller et al. (2017)
		2. Homogenize in electric stirrer for 30 min at 200 rpm at room temperature			
		3. Heat at 90 °C for 30 min at 200 rpm			

		4. Place in ultrasonic bath for 15 min at 23 °C and 37 kHz/s			
		5. Add glycerol			
		6. Stir for 30 min at 200 rpm with Thermomix			
		7. Place in ultrasonic bath for 15 min			
		8. Coat on PET film (for nanosuspensions, rpm and time must be increased)			
Whey gluten-methylcellulose binary blend film	25% wt glycerol	Casting method	Film	Water vapor permeability	Zuo et al. (2009) cited in Gupta and Nayak (2014)
				Tensile strength 1.7–44 MPa	
Keratin-chitosan		Blending	Film (nanofiber)	Water insoluble	Tanabe et al. (2002) cited in Gupta and Nayak (2014)
				Strong and flexible	
				Softness judging	
				Antimicrobial activity	
				Tensile strength 27–34 MPa	
				4–9% ultimate elongation	
Keratin-PEO blend		Blending	Film (nanofiber)	Stress at break 6 MPa	Tonin et al. (2007) cited in Gupta and Nayak (2014)
				Strain at break 46.3 MPa	
Collagen	Glycerol	1. Pour aqueous solution of fish collagen (1% w/w) on Perspex plate	Film		Alizadeh and Behfar (2013)
		Dry to constant weight (48 h)			
Collagen	80% glycerol	1. Prepare 10% collagen solution	Film	~5.265 MPa tensile strength, ~64.307% elongation at break, water vapor transmission rate 1.170 $g.H_2O.m^{-2}.h^{-1}$	Said et al. (2016)
		2. Heat in water bath at 70 °C for 45 min with stirring			

(continued)

Table 2 (continued)

Biomaterials	Plasticizers	Processes	Packaging material	Properties	References
		3. Pour on the cast in hot condition			
	90% glycerol	1.Dry in oven at 55 °C for 18–20 h		~2.586 MPa tensile strength, ~69.619% elongation at break, water vapor transmission rate 0.387 $g.H_2O.m^{-2}.h^{-1}$	
	100% glycerol			~2.277 MPa tensile strength, ~75.037% elongation at break, water vapor transmission rate 1.247 $g.H_2O.m^{-2}.h^{-1}$	
Collagen-alginate (glutaraldehyde as cross-linking agent, glycerol as humectant)		1. Prepare 6.25% of sodium CMC transparent gel with water by stirring	Film	~75 kPa tensile strength, 76% elongation at break (at 0.3 g alginate	Yang et al. (2014)
		2. Prepare 25% starch paste with water by heat at 70 °C for 4 min			
		3. Prepare 4% collagen (with water, glutaraldehyde, and glycerol)			
		4. Mix with 6.25% sodium CMC transparent gel and 25% starch paste			
		5. Stir at 40 °C to homogenize			
		6. Add sodium alginate			
		7. Stir 15 min			
		8. Pour on a template			
		Freeze-dry for 24 h			
Gelatin		1. Prepare 50 wt% gelatin solution with water	Fiber	Soluble in water within 1 day	Chaochai et al. (2016)

		2. Squeeze out the solution through nozzle under slight pressure			
		3. Roll up fiber at constant speed by windup roller			
		Retain fiber on windup roller for 24 h			
Gelatin (formaldehyde as reticulant agent)	4.5% w/w glycerol	1. Dissolve 10% w/w gelatin in distilled water at 50 °C 15 min with shaking bath	Film	Decrease WVP and solubility	Carvalho and Grosso (2006)
		2. Add reticulant agent		Increase in tensile strength	
Gelatin (glyoxal as reticulant agent)		3. Shake at 50 °C 15 min		Decrease WVP and solubility	
		4. Cast the obtained filmogenic solution			
		Dry at room temperature for 24–48 h			
Gelatin-chitosan	Sorbitol	Low-temperature process		Tensile strength 130 MPa	Lacroix and Vu (2014)
		High-temperature process		Tensile strength 34.4 MPa	
				44.7% elongation	
Lignin-gelatin	0.6% w/v glycerol	1. Dissolve 4% w/v gelatin powder in Millipore, deionize water at 60 °C for 30 min	Film	Partially soluble in water after 24 h at 25 °C	Aadil et al. (2016)
		2. Stir at 60 °C for 1 h		Excellent UV light barrier at range of 280–350 nm	
		3. Cast the obtained filmogenic solution		Good antioxidant activity	
		Dry in oven at 60 °C for 2 h		Flexible, stable, and eco-friendly	
Peanut protein		1. Add D.W to make 3% protein content	Film		Jangchud and Chinnan (1999)
		2. Add 1N NaOH to adjust pH 9.0 during dissolution and stirring			

(continued)

Table 2 (continued)

Biomaterials	Plasticizers	Processes	Packaging material	Properties	References
		3. Add glycerine ratio 3:5			
		4. Heat to 90 °C with stirring			
		5. Filter the solution through polyester screen			
		6. Cool for 10 min			
		Pour on nonstick plate and leave for 16 h			
Soy protein-nanocellulose		1. Mix 10% soy protein gel and 9% NFC suspension	Aerogel		Arboleda et al. (2013)
		2. Stir with spatula			
		3. Cast into plastic mold			
		4. Cool to 13 °C for 1 h			
		5. Freeze at −10 °C			
		6. Dry in Edwards Micro Modulyo Freeze Dryer			
		Use cutter to remove protrusion			
Wheat gluten	30% glycerol	1. Prepare 7.5% gluten solution	Film	WVP – 2.43 gmm/m^2dkPa	Tanada-Palmu et al. (2000)
		2. Add glycerol		O_2P 2.14 cm^3μm/m^2dkPa	
		3. Heat at 75 °C with magnetic stirrer about 10 min (pH adjust to 10)		Tensile strength decreased and % elongation at break increased with increased water activity	
		4. Centrifuge at 7000 rpm for 6 min			
		5. Pour on Teflon-covered granite surface and dry for 20 h			

Sodium caseinate	Glycerol	1. Dissolve sodium caseinate in distilled water	Coating solution		Shariatifar and Jafarpour (2013)
		2. Add glycerol			
		3. Heat and constantly stir at 60 °C			
		4. Degas under vacuum			
PLA		Extrusion or casting	Film	Low water vapor permeability but high oxygen permeability	Muller et al. (2017)
Untreated GSP-PLA		1.1. Dissolve GSP and PLA in dichloromethane	Fiber	Tensile strength −0.855 MPa at 5% wt filler	Adeosun et al. (2016)
				Absorption is high	
Treated GSP-PLA		2. Fiber spinning (rate of flow is 1 mL/min)		Best water resistance at 7%wt filler	
				Tensile strength −0.7 MPa at 3% wt filler	
PLA-starch film		Blending	Film	Good barrier capacity for water vapor and gases	Muller et al. (2017)
		Combination (multilayer)			
PLA-PHB blends		Blending	Fiber	Antioxidant and antimicrobial activity to extend a food product's shelf life	Arrieta et al. (2017)
		Electrospinning technique			
PLA-PCL blend		Casting methods	Film	Antimicrobial activity	Muller et al. (2017)

CMC carboxymethylcellulose, *GSP* groundnut shell particle, *LDPE* low-density polyethylene, *NFC* nanofibrillar cellulose, *O_2P* oxygen permeability, *PCL* polycaprolactone, *PEO* poly(ethylene oxide), *PET* polyethylene terephthalate, *PHB* poly(hydroxybutyrate), *SPI* soy protein isolate, *WPI* whey protein isolate, *WVP* water vapor permeability

Table 3 Advantages and disadvantages of packaging materials in products

Packaging materials	Food products	Advantages	Disadvantages	Reference
Potato starch	Guavas	Increase shelf life at least 3 days at 25 °C and 50–70% RH		Gallo et al. (2003)
		Increase preservation of the sensorial characteristic (size, yellow, color, and aroma) for 15 days		
Cassava starch	Strawberries	No significant effect on soluble solid, titratable acidity, pH, and color of strawberries	Coating with potassium sorbate reduces shelf life because of microbial spoilage	Garcia et al. (2012)
		Reduce the respiration rate		
		Delay weight and firmness loss		
Yam starch	Cherry tomato	Promote stability for loss in mass, soluble solid, titratable acidity ratio, phenolic compounds, antioxidant activity, lycopene content		Reis et al. (2015)
Cornstarch-beeswax	Raspberry	Decrease respiration rate	Indicate hypoxic condition which can affect shelf life quality attributes in cold storage	Pérez-Gallardo et al. (2012)
		Permit the preservation of color without affecting anthocyanin content		
Cornstarch-methylcellulose-soybean oil	Crackers, low water activity-type cereal food	Resistance to water vapor transmission, extend shelf life		Talens et al. (2010)
Starch-based composite coating	Strawberries	Reduce weight loss		García et al. (2001)
		Maintain surface color		
		Delay senescence of fruit		
		Show antimicrobial activities		
		Extend shelf life (10–14 days)		
Yam starch-chitosan coating	Carrot slices	Reduce the growth of microbial		Durango et al. (2006)

(continued)

Table 3 (continued)

Packaging materials	Food products	Advantages	Disadvantages	Reference
Starch-*N. sativa* oil	Pomegranate	Reduced softening of arils, weight loss and % of browning index, loss of vitamin C, loss of anthocyanin and delayed microbial decay; maintain the quality of fruits		Oz and Ulukanli (2012)
Pea starch-whey protein isolate-carnauba wax	Walnuts and pine nuts	Prevent oxidative and hydrolytic rancidity at 25 °C storage	Combined film reduced the tensile strength and elongation compared with single-component film	Mehyar et al. (2012)
		Improved their smoothness and taste and improved sensory characteristics	Coatings imparted unacceptable yellowish color on walnuts	
Chitosan-whey protein	Ricotta cheese	Decrease microbial contaminants growth		Pierro et al. (2011)
		Extend shelf life		
		Lower oxygen and carbon dioxide permeability		
		Delayed titratable acidity development		
Whey protein concentrate-rice bran oil	Kiwi fruit	Preserved the total soluble solid material, color, firmness, taste, and the overall acceptability of the fruits		Hassani et al. (2012)
		Higher sensory attributes		
		Delay ripening of fruit		
		Slow down the increase of acidity and weight loss		
SPI	Cut apples	Show high sensory score	Do not keep appearance (become yellow after 2 weeks)	Ghavidel et al. (2013)
Whey protein concentrate				

(continued)

Table 3 (continued)

Packaging materials	Food products	Advantages	Disadvantages	Reference
		Show significant effect on keeping texture		
Zein film	Fresh fruits	Reduce moisture, oxygen, and carbon dioxide transmission	Brittle, insoluble in water at pH 7	Wittaya (2012)
		Reduce loss of firmness and delay color change		
Corn zein film	Nuts, fruits, candies, and oxygen-sensitive food	Oxygen and moisture barrier		Gupta and Nayak (2014)
	Packaging for tomatoes, cooked turkey, popcorn, and shell eggs	Resistance to grease and microbial grow		
Corn zein film	Tomatoes	Delayed color changes and ripening	Higher water vapor transmission rate than shrink wrap film	Park et al. (1994, 1994a)
		Reduced firmness loss and weight loss	Show alcohol fermentation due to anaerobic fermentation	
		Lower oxygen and carbon dioxide transmission rate than shrink warp film at 0% RH		
		Extend the shelf life		
Collagen film	Edible sausage casing	Excellent oxygen barrier at 0% moisture		Lacroix and Vu (2014)
Collagen fiber film	Meat and fish, sausage casing	Good antioxidant property, maintain good taste and high transparency	Poor mechanical properties	Yang et al. (2014)
Galactomannans-collagen coating	Apple and mango	Reduce oxygen consumption rate and carbon dioxide releasing rate	Fruit surface shows low energy	Lima et al. (2010)
Gelatin film	Encapsulate low moisture or oil-phase food ingredients	Reduce oxygen and moisture and transport of oil		Wittaya (2012)

(continued)

Table 3 (continued)

Packaging materials	Food products	Advantages	Disadvantages	Reference
Casein film coating	Fruit and vegetable	Transparent and flexible	Poor moisture barrier	Wittaya (2012)
		Protect moisture absorption and oxidation		
Milk protein film	Slices of apple and potato	Transparent after complete drying		Tien et al. (2001)
		Delay browning		
SPI film	Fresh ground beef	Natural microbials for food preservation		Lacroix and Vu (2014)
SPI-beeswax coating	Cut eggplants	Prevent softening of the tissues		Ghidelli et al. (2010)
		Reduce browning of the tissue		
Wheat gluten film	Fruit (strawberries) and vegetable	Antimicrobial effects depend on plasticizers		Lacroix and Vu (2014)
Gluten film	Dry roasted peanuts and fried chicken pieces	Good barrier to oxygen and carbon dioxide	Highly permeable to water vapor	Gupta and Nayak (2014)
Sodium caseinate coating	Apples	Reduce weight loss and firmness loss		Shariatifar and Jafarpour (2013)
		Delay softening of fruit and ripening		
		Give better sensory characteristics (at 4 °C for 112 days)		
PLA/nisin film	Liquid foods (orange juice and liquid egg white)	Antimicrobial activity		Jin and Zhang (2008)
		Significantly inhibited growth of *L. monocytogenes*		
Poly-lactic acid, lactic acid, nisin solution	Vacuum-packaged fresh beef	Extend the shelf life by antimicrobial effects		Ariyapitipun et al. (1999)

RH relative humidity

Table 4 Commercial food packaging products

Biopolymer	Brand (trade name)	Product	Reference
Starch	Mater-Bi	Compostable and biodegradable bags, laminated paper, container cup, packaging paper	http://www.novamont.com/eng/mater-bi (20 Nov 17) Byun and Kim (2014)
	Bioplast	Compostable and biodegradable bags, films	http://en.biotec.de (20 Nov 17)
	Renature (Storopack)	Compostable loose fill packaging	http://www.storopack.us/en/products-solutions/flexible-protective-packaging/loose-fill-flowable-packaging-peanuts.html (29 Nov 17)
Protein	Opta Glaze	Coating to prevent microbial growth	Pavlath and Orts (2009)
	Z*Coat	Coating to nut meats, pecan, and chocolate-covered peanut	
PLA	ecovio and EcoFlex	Compostable and biodegradable bags, films	http://product-finder.basf.com/group/corporate/product-finder/en/brand/ECOVIO (24 Nov 17)
		Foam packaging, injection molding products	
	Biomax	Compostable container cups, films	http://www.dupont.com (20 Nov 17)
	Ingeo	Bottles, cups, bags,	Byun and Kim (2014)
	BioFoam	Trays, ice-cream containers, boxes	www.synbratechnology.com/biofoam/ (29 Nov 17)

6 Environmental Impacts of Packaging Materials Prepared Using Starch, Protein, and Poly-lactic Acid (PLA)

Starch-, protein-, and PLA-based packaging material can be degradable by microbial, enzymatic, or chemical process. Most of bio-based packaging materials are biodegradable and biocompatible (Ali et al. 2013; Marsh and Bugsu 2007). These make biodegradable food packaging reduce the plastic waste as well as plastic problem in the world (Muller et al. 2017). These can also save marine life (Marsh and Bugsu 2007). Biodegradable rate depends on environmental conditions and packaging material properties (Pikoń and Czop 2013). However, when the biopolymer is composited, depending on the additional materials, the biodegradable property can be changed, for example, grafting starch with PE (polyethylene). Starch itself is biodegradable and biocompatible; however, in the case of starch-PE film, only starch is biodegradable (Vilpoux and Averous 2004).

On the other hand, the sufficient quantities of biodegradable packaging required the sufficient renewable sources (plants and animals). Therefore, the wider use of biodegradable packaging will threaten our food production (Marsh and Bugsu 2007).

7 Conclusion

Replacement of biodegradable and renewable polymers in food packaging brings greener environment by reducing non-biodegradable petroleum-based polymers and waste disposal cost. However, the use of biopolymers in food packaging has some drawbacks in comparison with the non-biodegradable petroleum-based polymers like low mechanical property and low barrier property. Therefore, some researches have been done with nano-sized particles in order to improve the properties. On the other hand, the single-layer coating is not competitive to the synthetic polymers, and the multilayer coating processes have also been used and improved the properties of the packaging materials. The development of new technologies in biopolymers and food packaging leads to more suitable products for consumers' usage.

Acknowledgment The authors would like to thank Mr. Win Myo Thu and the members of Advancing Life and Regenerating Motherland for their encouragement. We also like to thank Win Le Nu and Htun Inzali for sharing their information.

References

Aadil KR, Barapatre A, Jha H (2016) Synthesis and characterization of Acacia lignin-gelatin film for its possible application in food packaging. Bioresour Bioproc 3:27. https://doi.org/10.1186/s40643-016-0103-y

Adeosun S, Taiwo O, Akpan E et al (2016) Mechanical characteristics of groundnut shell particle reinforced polylactide nano fibre. Matéria (Rio J) 21(1):482–491. https://doi.org/10.1590/S1517-707620160002.0045

Aghazadeh M, Krim R, Sultan MT et al (2017) Comparison of starch films and effect of different rice starch-based coating formulations on physical properties of walnut during storage time at accelerated temperature. J Food Process Eng. https://doi.org/10.1111/jfpe.12607

Ali RR, Rahman WAWA, Kasmani RM et al (2013) Tapioca starch biocomposite for disposable packaging ware. Chem Eng Trans 32:1711–1716. https://doi.org/10.3303/cet1332286

Alizadeh A, Behfar S (2013) Properties of collagen based edible films in food packaging: a review. Ann Biol Res 4(2):253–256

Ambrogelly A, Palioura S, Söll D (2007) Natural expansion of the genetic code. Nat Chem Biol 3 (1):29–35. https://doi.org/10.1038/nchembio847

Arboleda JC, Hughes M, Lucia LA et al (2013) Soy protein–nanocellulose composite aerogels. Cellulose 20:2417–2426. https://doi.org/10.1007/s10570-013-9993-4

Ariyapitipun T, Mustapha A, Clarke AD (1999) Microbial shelf life determination of vacuum-packaged fresh beef treated with polylactic acid, lactic acid, and nisin solutions. J Food Prot 62 (8):913–920

Arrieta MP, Samper MD, Aldas M et al (2017) On the use of PLA-PHB blends for sustainable food packaging applications. Materials. https://doi.org/10.3390/ma10091008

Aung SPS, Nu WL, Win KC et al (2017) Enzymes for production of chitin, chitosan and chitooligosaccharide and determination of activities of enzymes using chitinous substrates. In: Ahmed S, Ikram S (eds) Chitosan: hydrogels and composites for their applications. Wiley, New York, pp 71–88

Avella M, Vlieger JJD, Errico ME et al (2005) Biodegradable starch/clay nanocomposite films for food packaging applications. Food Chem 93:467–474. https://doi.org/10.1016/j.foodchem.2004.10.024

Avérous L (2008) Polylactic acid: synthesis, properties and applications. In: Belgacem MN, Gandini A (eds) Monomers, polymers and composites from renewable resources. Elsevier, Oxford, pp 433–450

Avinc O, Khoddami A (2009) Overview of poly(lactic acid) (PLA) fibre. Fibre Chem 41 (6):391–401

Borges JA, Romani VP, Cortez-Vega WR et al (2015) Influence of different starch sources and plasticizers on properties of biodegradable films. Int Food Res J 22(6):2346–2351

Bourtoom T (2009) Edible protein films: properties enhancement. Int Food Res J 16:1–9

Byun Y, Kim YT (2014) Bioplastics for food packaging: chemistry and physics. In: Han JH (ed) Innovations in food packaging, 2nd edn. Academic, London, pp 353–368

Carvalho RAD, Grosso CRF (2006) Properties of chemically modified gelatin films. Braz J Chem Eng 23(1):45–53

Chaochai T, Imai Y, Furuike T et al (2016) Preparation and properties of gelatin fibers fabricated by dry spinning. Fibers. https://doi.org/10.3390/fib4010002

Coltelli M, Wild F, Bugnicourt E et al (2015) State of the art in the development and properties of protein-based films and coatings and their applicability to cellulose based products: an extensive review. Coatings. https://doi.org/10.3390/coatings6010001

Dangaran K, Tomasula PM, Qi P (2009) Structure and function of protein-based edible films and coatings. In: Embuscado ME, Huber KC (eds) Edible films and coatings for food applications. Springer, New York, pp 25–56

Durango AM, Soares NFF, Andrade NJ (2006) Microbiological evaluation of an edible antimicrobial coating on minimally processed carrots. Food Control 7(5):336–341. https://doi.org/10.1016/j.foodcont.2004.10.024

Ezeoha SL, Ezenwanne JN (2013) Production of biodegradable plastic packaging film from cassava starch. IOSR J Eng (ISORJEN) 3(10):14–20

Fahma F, Sugiarto, Sunarti TC et al (2017) Thermoplastic cassava starch-PVA composite films with cellulose nanofibers from oil palm empty fruit bunches as reinforcement agent. Int J Polym Sci. https://doi.org/10.1155/2017/2745721

Ferreira ARV, Alves VD, Coelhoso IM (2016) Polysaccharide-based membranes in food packaging applications membranes. Membranes. https://doi.org/10.3390/membranes6020022

Gallo JAQ, Amaro MRD, Cabrera DMBG et al (2003) Application of edible coating to improve shelf life of mexican guava. Acta Hortic 599:589–594. https://doi.org/10.17660/ActaHortic.2003.599.76

García MA, Martino MN, Zaritzky NE (2001) Composite starch-based coatings applied to strawberries (*Fragaria ananassa*). Nahrung/Food 45(4):267–272. https://doi.org/10.1002/1521-3803(20010801)45:4<267::AID-FOOD267>3.0.CO;2-3

Garcia LC, Pereira LM, Sarantopoulos CIGDL et al (2012) Effect of antimicrobial starch edible coating on shelf-life of fresh strawberries. Packag Technol Sci 25(7):413–425. https://doi.org/10.1002/pts.987

García NL, Famá L, D'Accorso NB et al (2015) Biodegradable starch nanocomposites. In: Thakur V, Thakur M (eds) Eco-friendly polymer nanocomposites, Advanced structured materials, vol 75. Springer, New Delhi, pp 17–77

Ghavidel RA, Davoodi MG, Asl AFA et al (2013) Effect of selected edible coatings to extend shelf-life of fresh-cut apples. Intl J Agric Crop Sci 6(16):1171–1178

Ghidelli C, Sanchís E, Argudo CR et al (2010) Application of soy protein-beeswax edible coating with antioxidants: effect reducing enzymatic browning of fresh-cut eggplants. Acta Hort 877:591–596. https://doi.org/10.17660/ActaHortic.2010.877.77

Gupta P, Nayak K (2014) Characteristics of protein-based biopolymer and its application. Polym Eng Sci. https://doi.org/10.1002/pen.23928

Hames D, Hooper N (2011) Biochemistry. Garland Science, New York

Hassani F, Garousi F, Javanmard M (2012) Edible coating based on whey protein concentrate-rice bran oil to maintain the physical and chemical properties of the kiwifruit (*Actinidia deliciosa).* Trakia J Sci 10(1):26–34

Hu B (2014) Biopolymer-based lightweight materials for packaging applications. In: Yang Y, Xu H, Yu X (eds) Lightweight materials from biopolymers and biofibers, vol 1175. Am Chem Soc, Washington, DC, pp 239–255

Jangchud A, Chinnan MS (1999) Peanut protein film as affected by drying temperature and pH of film forming solution. J Food Sci 64(1):153–157

Jin T, Zhang H (2008) Biodegradable polylactic acid polymer with nisin for use in antimicrobial food packaging. J Food Sci 73(3):127–134. https://doi.org/10.1111/j.1750-3841.2008.00681.x

Khwaldia K, Perez C, Banon S et al (2004) Milk proteins for edible films and coatings. Crit Rev Food Sci Nutr 44(4):239–251. https://doi.org/10.1080/10408690490464906

Lacroix M, Vu KD (2014) Edible coating and film materials. In: Han JH (ed) Innovations in food packaging, 2nd edn. Academic, London, pp 277–304

Lima ÁM, Cerqueira MA, Souza BWS et al (2010) New edible coatings composed of galactomannans and collagen blends to improve the postharvest quality of fruits- influence on fruits gas transfer rate. J Food Eng 97:101–109

Malathi AN, Santhosh KS, Udakumar N (2014) Recent trends of biodegradable polymer: biodegradable films for food packaging and application of nanotechnology in biodegradable food packaging. Curr Trends Technol Sci 3(2):73–79

Marsh K, Bugsu B (2007) Food packaging-roles, materials, and environmental issues. J Food Sci 72:39–55. https://doi.org/10.1111/j.1750-3841.2007.00301.x

Mehyar GF, Al-Ismail K, Han JH et al (2012) Characterization of edible coatings consisting of pea starch, whey protein isolate, and carnauba wax and their effects on oil rancidity and sensory properties of walnuts and pine nuts. J Food Sci 77(2):52–59. https://doi.org/10.1111/j.1750-3841.2011.02559.x

Muller J, González-Martínez C, Chiralt A (2017) Combination of poly(lactic) acid and starch for biodegradable food packaging. Materials. https://doi.org/10.3390/ma10080952

Müller K, Jesdinszki M, Schmid M (2017) Modification of functional properties of whey protein isolate nanocomposite films and coatings with nanoclays. J Nanomater. https://doi.org/10.1155/2017/6039192

Murgić ZH, Rešček A, Siročić AP et al (2015) Polymers for food packaging. In: Nanoparticles in active polymer food packaging. Smithers Pira Technology Ldt., United Kingdom, pp 9–71

Nwe N (2015) Biopolymers based micro- and nano-materials. ASME Press, Momentum Press, New York

Nwe N, Furuike T, Tamura H (2010) Selection of a biopolymer based on attachment, morphology and proliferation of fibroblast NIH/3T3 cells for the development of a biodegradable tissue regeneration template: alginate, bacterial cellulose and gelatin. Process Biochem 45:457–466. https://doi.org/10.1016/j.procbio.2009.11.002

Othman SH (2014) Bio-nanocomposite materials for food packaging applications: types of biopolymer and nano-sized filler. Agric Agric Sci Procedia 2:296–303. https://doi.org/10.1016/j.aaspro.2014.11.042

Oz AT, Ulukanli Z (2012) Application of edible starch-based coating including glycerol plus *Oleum nigella* on arils from long-stored whole pomegranate fruits. J Food Process Preserv 36(1):81–95. https://doi.org/10.1111/j.1745-4549.2011.00599.x

Park HJ, Chinnan MS, Shewfelt RL (1994) Effect of corn-zein film coating on the storage life of tomatoes. J Food Process Preserv 18:317–331. https://doi.org/10.1111/j.1745-4549.1994.tb00255.x

Park HJ, Chinnan MS, Shewfelt RL (1994a) Edible coating effects on storage life and quality of tomatoes. J Food Sci 593:568–570. https://doi.org/10.1111/j.1365-2621.1994.tb05563.x

Pavlath AE, Orts W (2009) Edible films and coatings why, what and how? In: Embuscado ME, Huber KC (eds) Edible films and coatings for food applications. Springer, New York, pp 1–23

Pawar PA, Purwar AH (2013) Bioderadable polymers in food packaging. Am J Eng Res 2 (5):151–164

Pérez-Gallardo A, Mattinson SD, Lazcano-Peralta A et al (2012) Effect of native and acetylated-crosslinked waxy corn starch-beeswax coatings on quality attributes of raspberries during storage. Starch 64(8):665–673

Pierro PD, Sorrentino A, Mariniello L et al (2011) Chitosan/whey protein film as active coating to extend Ricotta cheese shelf-life. LWT-Food Sci Technol 44(10):2324–2327. https://doi.org/10.1016/j.lwt.2010.11.031

Pikoń K, Czop M (2013) Environmental impact of biodegradable packaging waste utilization. Pol J Environ Stud 23(3):969–973

Reis RCD, Devilla IA, Correa PC et al (2015) Postharvest conservation of cherry tomato with edible coating. Afr J Agric Res 10(11):1164–1170. https://doi.org/10.5897/AJAR2014.9238

Rhim J, Kim YT (2014) Bioolymer-based composite packaging materials with nanoparticles. In: Han JH (ed) Innovations in food packaging, 2nd edn. Academic, London, pp 413–442

Said MI, Erwanto Y, Abustam E (2016) Properties of edible film produced using combination of collagen extracts of bligon goatskin with glycerol. Am J Anim Vet Sci 11(4):151–159. https://doi.org/10.3844/ajavsp.2016.151.159

Sayyahi Z, Beigmohammadi F, Shoaiee S (2017) Optimization of starch biopolymer enriched with chitosan containing rosemary essential oil and its application in packaging of peanuts. Nutr Food Sci Res 4(3):19–28. https://doi.org/10.18869/acadpub.nfsr.4.3.19

Shariatifar M, Jafarpour E (2013) Edible coating effects on storage life and quality of apple. J Basic Appl Sci Res 3(6):24–27

Šuput DZ, Lazić VL, Popović SZ et al (2015) Edible films and coatings-sources, properties and application. Food Feed Res 42(1):11–22. https://doi.org/10.5937/ffr1501011s

Talens P, Fabra MJ, Chiralt A (2010) Edible polysaccharide films and coatings. Nova Science Publishers, New York

Tanada-Palmu P, Helén H, Hyvönen L (2000) Preparation, properties, and applications of wheat gluten edible films. Agric Food Sci Finl 9(1):23–35

Tien CL, Vachon C, Mateescu MA et al (2001) Milk protein coatings prevent oxidative browning of apples and potatoes. J Food Sci 66(4):512–516. https://doi.org/10.1111/j.1365-2621.2001.tb04594.x

Vartiainen J, Vähä-Nissi M, Harlin A (2014) Biopolymer films and coatings in packaging applications—a review of recent developments. Mater Sci Appl 5:708–718. https://doi.org/10.4236/msa.2014.510072

Verma R, Vinoda KS, Papireddy M et al (2016) Toxic pollutants from plastic waste. Procedia Environ Sci 35:701–708

Vilpoux O, Avérous L (2004) Starch-based plastic. In: Cereda MP, Vilpoux O (eds) Technology, use and potentialities of Latin American starchy tuber. NGO Raízes and Cargill Foundation, Sao Paolo, pp 521–553

Vuković-Kwiatkowska I, Kaczmarek H (2014) Innovative composites of poly (lactic acid) for the production of packaging foils. CHEMIK 68(2):135–140

Wee Y, Kim J, Ryu H (2006) Biotechnological production of lactic acid and its recent applications. Food Technol Biotechnol 44(2):163–172

Wittaya T (2012) Protein-based edible films: characteristics and improvement of properties. In: Eissa AA (ed) Structure and function of food engineering. InTech, Rijeka, pp 43–70

Yang H, Guo X, Chen X et al (2014) Preparation and characterization of collagen food packaging film. J Chem Pharm Res 6(6):740–745

Zink J, Wyrobnik T, Prinz T et al (2016) Physical, chemical and biochemical modifications of protein-based films and coatings: an extensive review. Int J Mol Sci. https://doi.org/10.3390/ijms17091376

Web References

BASF we creat chemistry (2017) http://product-finder.basf.com/group/corporate/product-finder/en/brand/ECOVIO. Accessed 24 Nov 17
Biotec plastic for a better live (2017) http://en.biotec.de. Accessed 20 Nov 17
Dupont (2017) http://www.dupont.com. Accessed 20 Nov 17
Novamon (2017) http://www.novamont.com/eng/mater-bi. Accessed 20 Nov 17
Storopack Perfect Protective Packaging (2015) http://www.storopack.us/en/products-solutions/flexible-protective-packaging/loose-fill-flowable-packaging-peanuts.html. Accessed 29 Nov 17
Synbra Technology bv (2017) www.synbratechnology.com/biofoam. Accessed 29 Nov 17

Biodegradable Smart Biopolymers for Food Packaging: Sustainable Approach Toward Green Environment

Roohi, Pallavi Srivastava, Kulsoom Bano, Mohammed Rehan Zaheer, and Mohammed Kuddus

Abstract The advantage of biodegradable plastics over conventional synthetic plastics in respect to complete and safe disposal makes a way for its use in broad range of usage including agriculture, biomedical, and food packaging. In present scenario, biodegradable materials used for wrapping food products are mostly synthesized renewable raw materials like polylactide, starch-/cellulose-based polymers, polybutylene succinate, and polybutylene succinate-co-adipate. Spoiling of food by microbial contamination along with lipid and protein oxidation is a major concern in terms of food safety and quality, and it may occur in packaged foods even after using traditional food preservation techniques such as freezing, drying, heating, salting, and fermentation. All these problems can be overcome by using bio-based active and smart packaging that provides great opportunities for enhancing quality, tenderness, and freshness along with antimicrobial assets. Recently, antioxidant packaging is developed with bioactive molecules in encapsulated form that also improves aroma characteristics. This chapter gives an overview of different types of biodegradable polymers that are applied for covering of foodstuff and what consequences it confers on environment after biodegradation and composting.

Keywords Biodegradable polymers · Food packaging · Smart packaging · Biodegradation · Environmental impact

Roohi · P. Srivastava · K. Bano
Protein Research Laboratory, Department of Bioengineering, Integral University, Lucknow, India

M. R. Zaheer
Department of Chemistry, Gagan College of Management and Technology, Aligarh, India

M. Kuddus (✉)
Department of Biochemistry, University of Hail, Hail, Saudi Arabia

S. Ahmed (ed.), *Bio-based Materials for Food Packaging*,
https://doi.org/10.1007/978-981-13-1909-9_9

1 Introduction

The basic application of packaging on food products is meant for safeguard from external environment and destruction and also to provide information to consumers regarding ingredient and nutrition (Coles et al. 2003). A few secondary purposes are also important such as flexibility, accessibility, and interference. Other advantages of packaging are that they delay the corrosion of the product, maintain the valuable impact of development, improve existence, and enhance the excellence of food. Food packaging also helps in transportation across long distances securely and can be stored for long periods of time. Airtight packaging must increase the food quality; it is essential for food to maintain its marketable properties. Ideally the material is recyclable and can be used again and again. Due to tampering of food and pharma products, special packaging is used to provide extra protection so that packages cannot be tampered with in stores. Tamper-evident features, which are usually comprised of additional packing materials, may involve special bottle liners or closures. A heat seal that changes color upon opening is an example of an additional packaging feature. It is important for manufacturer to maintain balance between safety issues and other factors such as cost and eco-friendly perspective. Many consumers are paying closer attention to environmental issues that are involved with packaging technology.

2 Types of Material Used for Packaging

As per the literature reports, 35% of the packaging materials used were paper and board (cartons) worldwide, while 30% accounts for plastics including polyethylene terephthalate (PET), polyvinyl chloride (PVC), and polystyrene (PS) (Tonjes and Greene 2012). Plastics have swiftly added share 10% more in the market in the early 1960s. The rest of the packaging material includes metals like aluminum, steel, and glass. Increased consumption in European countries results in increased amount of packaging used, which is much harmful to the environment. The area of landfills is increasing and waste filling them is not spontaneously disintegrated (Kale et al. 2006). Food packaging refuse accounts for major portion of municipal waste. Thus, for proper choice, it is necessary to remain updated with different packaging materials actually used for food packaging. Food and beverage packaging with regard to raw substance required in the production can be classified as follows: metal, glass, plastic, paper, and cardboard in addition to packaging from assorted material. All mentioned materials have their own advantages and disadvantages.

2.1 Glass

Glass can take different shapes and undergo recycling without changing its mechanical properties (but glass processing needs high temperature), has efficient barrier properties (lack of permeability of gasses and water vapor), but are disadvantageous in terms of ratio of mass to volume (glass packaging are heavy). Glass is also brittle, and is nondegradable (but it does not damage the environment); it is mainly used for manufacturing of bottles and jars.

2.2 Metals (Steel and Aluminum)

Metals possess very good barrier properties, are quite expensive, can be recycled, and are mainly used for canned food (food in cans are pasteurized) and in producing temperature-resistant metallic trays that are probably used for frozen and hot meals; thin aluminum foils are mostly used for wrapping of sweets, cheese, coffee, tea, etc.; bottle caps and jar lids are also metallic.

2.3 Paper and Cardboard

Paper and cardboard based packaging are usually cheap cellulose-based material (from wood); paper waste often involves recycling, incineration, and biodegradation while decaying in the environment. It is light; is easily printable; shows permeability for breezing, aqueous vapor, and oxygen; has reduced tearing resistance; is applied on different fields in the form of boxes and belongings; and is mainly employed for packing of dry stuffs like sugar, salt, and baked foods. Papers are also employed in labeling of packages made of different materials such as glass and plastics.

2.4 Synthetic Plastic

Synthetic polymers like PET, PS, and PVC are categorized by low production cost, good mechanical and barrier properties, and easy processability and modification for getting required properties; it is light, esthetic, unbreakable, elastic, transparent or colored, and nondegradable in nature and can be recycled.

2.5 Mixed Materials (Laminates)

Packaging composed from few thin layers, e.g., metallic, plastic, and paper films, has very good barrier properties; laminate packaging are sealed owing welding possibility; it is difficult for utilization because of problems with layer separation. Layering packaging are active and intelligent packaging, which contains substances extending shelf-life, that do not interact with food article; these substances are mainly oxygen and ethylene absorbers, compounds emitting or binding oxygen dioxide, are water regulators, and also have antioxidants and antibacterial substances.

2.6 Nanocomposites

New-generation packaging with specific properties, containing small amount of mineral fillers (such as glass or carbon fibers, calcium carbonate, or silicates) with very small size particles, which improve mechanical and barrier properties; can be used for manufacturing of bottles or films with very low permeability of oxygen and water vapor; though it is expensive, it has the potential to be recycled.

2.7 Non-biodegradable Materials

Non-biodegradable packaging materials are inorganic and man-made packaging material that will not decompose or break down into simpler forms of matter. PP and PVC packaging material generally belong to non-biodegradable packaging materials.

We cannot use non-biodegradable packaging material much because it remains in the environment as pollutants. The commercial packaging materials do not degrade naturally to the monomers, and so the conserved energy could not be regenerated in the environment. Dependence on synthetic polymers for food packaging is an indefensible and non-ecological practice toward earth pollution. The energy stored in these persistent polluting substances could not be reclaimed due to the nondegradable behavior of synthesized polymer. Recycling is thus an alternative toward the reuse of such nondegradable materials in order to make desired products. This approach is very economical to save the natural resources with the reduction in the gathered persistent pollutants. A sustainable development for recycling and reuse of packaging materials like glass, metals, and pulp has been achieved although the reduction in availability of toxic pollutant as a waste in landfill is not so successful (Northwood and Okaley-Hill 1999). Synthetic packaging materials are copolymerizing with different chemical substances to enhance the mechanical properties. Most of the chemical substances include different additives like plasticizers and colorants. Thin layer coatings of food packaging films are an essential need to create a functional film, which act as a barrier toward the impermeability of oxygen, water, and moisture content of food. The process of recycling of plastic wastes involves several steps such as collection, identification, and transportation of plastic material from the dump site that frequently renders the recycling as non-economical, making incineration or landfill more convenient.

3 Biodegradable Materials

A distinction between biopolymer and biodegradable polymer is very important for its application in packaging industry. Biodegradable polymers are the materials that undergo degradation that result in inorganic substances with the generation of CO_2, CH_4, and H_2O in aerobic as well as anaerobic conditions by the enzymatic action of microorganisms, while biopolymer synthesis is based on renewable substrate. Based on the route of production, the degradable bio-derived polymers are characterized into the following groups (Valdés-García et al. 2014):

(i) *Polymers derived from organic waste*: substances from agricultural waste as polysaccharides (chitin, corn, galactomannan) and protein (zein, casein, gluten, whey)
(ii) *Polymers synthesized by microorganisms naturally*: such as polyhydroxybutyrate (PHB)
(iii) *Polymer produced by using agro-resource substances*: such as polylactic acid (PLA)
(iv) *Polymers where monomers are produced by conventional method* from fossil resources

Bioplastics synthesized naturally by utilizing the natural resources are the recent developments in new-generation packaging materials and very effective in reducing the greenhouse effect on the earth. The biodegradable plastic produced through the sustainable method follows the composting cycle of degradation by the action of microorganism. Biodegradable plastic is a choice of materials while recycling is neither convenient nor economical. Bioplastics are alternatives toward the reduction of hazardous effect on environment caused by conventional plastics comparatively. Biodegradable plastics are disposed of through composting along with food waste comprising 25–30% of total solid waste. Other methods of plastic disposal include sewage treatment plants (hydrolyzed biological plastics) or landfill (agricultural applications). A sustainable development toward the study on bioplastics is a subject of great interest the last 10 years. Biodegradable plastics produced from using natural resources accomplished the necessary needs in application sectors and is an innovative solution toward technical as well as environmental examination. The biodegradable plastics derived from renewable resources have a wide range of application in food packaging. One should always remember that bio-based and biodegradable are not equal terms. Biodegradability is the property that the bio-based material may possess, whereas bio-based is not essentially biologically derived.

The biodegradable polymer accounts a very low percentage of total material waste (e.g., 3% PLA). Since, conventional plastics are not meant to degrade during composting, to retard confusion, all plastic materials are barred out from composting of all polymer in sewage waste. Bioplastics required in food cuisine may assure flow of waste for composting, comprising only in small amount. Composting may come up with different problems. Standards are as follows in commercial composting condition at higher temperature: 58 °C is the standard degradation rate for biological degradation, while for domestic waste it is in the range of 20–30 °C, while in UK 20–25 °C is applied in custom mode of composting (Song et al. 2009). In spite of being degradable and eco-friendly, bioplastics such as PLA are still not contributing a major part in packaging industry. At present, the manufacturing of bioplastics at industrial scale is not so appropriate due to the insufficient facility of production. The polymer film demands some mechanical modifications required to be satisfied taken into account for packaging material. In comparison to the conventional packaging materials, the bioplastics should accomplish the required expectations to function as multilayer packaging film. The biodegradation does not account with the currently available packaging materials. The development of biodegradable packaging materials is an area of vigorous study (Robertson 2013).

4 Biodegradable vs. Non-biodegradable Biopolymers

1. The polymers that acquire decomposition by natural substitutes have been designated as biodegradable. Natural means naturally acquired microorganisms, water content, presence of oxygen and water vapors, etc., while non-biodegradable does not degrade or decompose under natural ecology.
2. Organic materials generated from food waste such as fruit and vegetable inedible skin, deceased animals and plants, etc. are included in the category of biodegradable substances while the polymers synthesized through chemical methods such as polystyrene, polyethylene metal substances, and aluminum products come in non-biodegradable category.
3. A simple decomposed organic substance is assimilated in the soil by natural agents and consequently processed under atmospheric carbon cycle, while non-biodegradable polymers that opposed the degradation by natural ecology and do not break into small constituent consequently cause accumulation of solid waste.
4. Biodegradation usually occurs within a few weeks or months, whereas non-biodegradable substances require thousands of years or even do not decompose and stay in its forms the way it is.

5 The Advantages of Bio-Based Bioplastics

In the process of assimilation of biodegradable products, complex compounds are broken down into monomers. The use of biodegradable plastics is very beneficial to human beings, and they are eco-friendly to the environment as some of them are recyclable while few are synthesized from organic raw materials which are degradable in a particular period of time without the generation of any toxic substances. Bio-based bioplastics synthesized by microbes by utilizing natural organic carbon sources are biodegradable and thus reduce depletion of the natural resources required for synthesis of conventional plastics. Conventional synthetic materials are nondegradable, while print and paper products are degraded through biological process. Products such as goods' packaging and disposable plasticware made up of plant- and/or corn-derived degradable polymers have several ecological advantages over nondegradable products. The development in the production of packaging bio-derived material from live feedstocks is receiving an increasing concern (Mostafa et al. 2015).

A lesser amount of energy is required in the production of biodegradable plastic as compared to the conventional petroleum-derived plastics. One report provided by FoodServiceWarehouse (US) informs that ~65% less energy is required in biodegradable plastic production as compared to the synthetic plastics. Singh et al. (2014) in his review has summarized that cyanobacteria can synthesize the PHB as an intracellular bioplastics in the presence of sunlight that reduces greenhouse gas emission with no generation of CO_2. PLA, bioplastics developed by renewable resources such as cornstarch, has been addressed as a successful replacement of conventional plastics for packaging, produced on low energy requirement basis (Guo and Crittenden 2011). Moreover, cost of bioplastics has significantly dropped in recent years as reported by Khalil et al. (2016), where PLA amounted to be 3 dollars/pound in 1990 and dropped to 90 cents/pound in 2010. This feature signifies that the use of bioplastics is now up to the reach of common man. The price of bio-based plastics is made comparable to the conventional one due to increased oil price. Comparably less amount of energy is required in production of bioplastics than synthetic polymer as 2.20-pound production of PLA requires 27.2 MJ/kg energy based on fossil fuel while PP and HDPE requires 85.9 and 73.7 MJ/kg, respectively. Cellulose nano-fibers have emerged as the most renewable resource material for food packaging applications (Azeredo et al. 2017). Cellulose-based food packaging films will result in the reduction of production cost due to its low-cost availability on a wide range. The decomposition of biodegradable packaging materials is concerned with the impact on environment. The landfill of biodegradable plastics including food packaging material and restaurant and kitchen waste is affected by the generation of greenhouse gas, i.e., methane, much higher to CO_2 under anaerobic digestion, which can be captured as an energy source (Song et al. 2009).

Eco-friendly packaging of food material is also known as green packaging as the materials have some unique properties such as they are biodegradable and can be recycled or reused. Some of these eco-friendly packaging can be reused in their native form while other can be decomposed in an eco-friendly environment. Studies based on environmental impact of cellulose-based packaging film have shown that a comparatively low toxicity as well as environmental risks is associated with it (Ni et al. 2012). Being originated from renewable feedstocks, the bio-based, bio-derived, or blended bioplastics have shown eco-friendly nature of degradation. Decomposition of biodegradable plastics in the soil improves the nutritional efficacy by releasing extracellular compounds and thus enhances the soil fertility. Mixing of biodegradable plastics along with the biodegradable products such as agricultural waste and paper products in composting generates humus material in the soil, which enhances the increase in crop-yield with low requirements of chemical fertilizers (Song et al. 2009). At present, bioplastics can only be decomposed in commercial composting site under the maintenance of high temperature, in contrast to the paper and textile products, which is composted in garden-bin. However, a rise in such composting facility has evolved.

6 Disadvantages of Non-biodegradable Polymers

Synthetic packaging material derived from inorganic resources is nondegradable as well as not decomposed into the smaller molecules or the simpler forms. Packaging materials made of polypropylene, polyethylene, and polyvinyl chloride are included into the non-biodegradable polymers' category, which on the other hand restricts their use as in packaging due to the long time accumulation in the environment. These gathered toxic non-biodegradable polymers are causing pollution to the environment as nature could not decompose it and no energy is thus generated. The production of non-biodegradable packaging materials traps lots of natural resources as well as energy, which cannot be reclaimed due to the long time persistence of these materials, leading to depletion in resources and gathering of polluting substances. The gathered plastic debris has several adverse effects. The accumulated nondegradable plastics enter into marine environment and affect adversely the marine organisms. The floating plastic debris on the seawater serves as mistaken food material and is ingested by seabird species. Moore (2008) has reported that plastic debris alone is affecting the marine organisms. Until now, 267 marine species have been affected by floating plastic debris due to the inevitable ingestion of non-biodegradable accumulates. Synthetic non-biodegradable food packaging materials accumulate in the ocean as well as in the land. In a report of Derraik (2002), 23,000 t of conventional packaging plastic polymers were dumped into the ocean every year. Accumulated food packaging materials made up of polystyrene and polyamides (nylons) have led to severe health hazards such as cancer, allergies, and system dysfunction. Li et al. (2016) has reported that 64.3% of macroplastic debris based on packaging materials was accumulated in marine environment and affected the franciscana dolphins (*Pontoporia blainvillei*).

In the manufacturing stage of conventional plastics, traces of chemical compounds are added in the form of plasticizer or softeners to enhance the softness. These chemical compounds such as bisphenol A (BPA), phthalates, and polybrominated di-phenyl ether (PBDE) remain intact or cause several inevitable effects on human beings and to the environment. In the previous studies, it is observed that BPA affects the animal reproductive system while the styrene monomer is found as the most emerging carcinogens. Sahu et al. (2016) has reported from a study based on exposure to BPA that an increased level of chemicals was observed in the urine, which leads to diabetes and heart diseases. The accumulated plastic wastes reach to ocean through many ways like in the form of leaches leaked from landfill sites of plastic debris. The persistent chemical present in the plastic debris releases toxic compound. In a report by Barnes et al. (2009), about 40–80% of mega plastic debris was from packaging material as most of them are single-use material and usually dumped into landfill sites. This accumulated debris will remain the same for long duration if no other solution is found as these are very durable due to non-biological degradation and will not fade away.

7 Biodegradable Polymers in Food Packaging Industry

The packaging material which remains in close contact with food material includes utensils, salad packs, wrappers, laminating films, dairy products, and meat items. These products thus come in contact with aqueous and nonaqueous condition of stored food at low to high temperature (Conn et al. 1995). Growing concern by the researchers in the last few years forces the industries for the production of recyclable and compostable biopolymers from renewable feedstocks. Modification in the properties (physical and mechanical) of biopolymer by planning and designing could in consequence produce biodegradable polymer as compared to other conventional packaging material like PS and PET.

8 Aliphatic: Aromatic Blending

Aliphatic polyesters comprise a higher fraction in biodegradable bioplastics as compared to the aromatic copolymers due to the favorable hydrolysis of di-carboxylic backbone present in polyesters. These food packaging materials are developed by the blending of conventional polymers with the biologically degradable polymer such as polyethylene terephthalate (PET) with other polyesters. In spite of its good mechanical property, PET has high melting point and is inactive over microbial attack which makes it unfavorable for polymer degradation. The blending of aliphatic polyesters results in the formation of weak points that helps in the hydrolysis. Although it is biologically degradable, produced by utilizing nonrenewable natural resources (oil and fossil fuels), its accumulation affects the total waste dumped worldwide. The degradation of PET thus depends on the way it is disposed off as it breaks down in about 8 weeks after disposal in a controlled way, while the degradation can account for 50 years of time period in an uncontrolled way of disposal. This polymer is generally used in kitchenware and plastic bottles but is very expensive (Siracusa et al. 2008).

9 Aliphatic Polyesters

Aliphatic polyesters have the analogous properties to the conventional polymer like polyethylene (PE) and polypropylene (PP). These biodegradable materials are synthesized by condensation reaction of high molecular weight monomers derived from natural resources. These odorless polymers are utilized in the packaging of beverages. The aliphatic polyesters are biodegradable and produces CO_2 and H_2O as end products. Nodax is the commercially available polymer developed by Procter & Gamble Company that is degradable in aerobic as well as in anaerobic environment.

10 Polycaprolactone (PCL)

Based on the category of biodegradable polymer, the PCL are polymerized from nonrenewable substances. PCL has thermo-softening property and shows resistance toward water and other chlorinated solvents with melting point of around 60 °C. Low-priced biodegradable plastic for food application (trash bags) could be achieved by mixing PCL with starch.

11 Polylactide Aliphatic Copolymer (CPLA)

This polymer is developed by blending D/L-lactide and aliphatic polyesters having hardness like polystyrene (PS) and is elastic like polypropylene (PP) depending upon the mixture ratio. The polymer is found to be stable at higher temperature range (200 °C). As compared to the synthetic polymer, low amount of carbon dioxide could be produced in combustion and no organic pollutants can be produced in burning. Biologically the degradation of CPLA occurs within 5–6 months of disposal, while with food waste the decomposition starts within 2 weeks of disposal.

12 Polylactic Acid (PLA)

PLA is the feasible biopolymer that can be synthesized from renewable feedstocks such as corn and sugarcane under fermentation (Sin 2012) conditions. PLA polymer is degradable, recyclable and translucent, has high average molecular weight, and is sensitive to water solubility. Polymerization of PLA is accomplished with the two existence forms of lactic acid (D/L). The characteristics of PLA depend and vary on the ratio of both the forms of lactide monomers. Various studies on the properties of PLA polymer have been carried out where D-lactide content was in varied concentrations (6–12%). Variation in properties was observed, i.e., from semicrystalline to amorphous state, respectively. However, the higher ratio (12%) of D-lactide is a favorable property for its application in food packaging (Saeidlou et al. 2012) (Table 1).

In a study of Kale et al. (2006), composting of biodegradable PLA materials (bottles, trays, and containers) was carried out under different atmospheric conditions. A different technique (gel permeation chromatography, thermal gravimetric analysis) was applied to study the breaking and visualization of these materials in composting conditions. In the composting, the soil was mixed with cow dung and wood flakes at higher temperature (60 °C). The composting was carried for different times (up to 4 weeks) at a relative humidity of 68%. The effect of temperature on

Table 1 Characteristics of amorphous PLA

Characteristics	Unit	PLA
Glass transition temperature (Tg)	(°C)	62.1 ± 0.7
Melting temperature (Tm)	(°C)	195–245
Enthalpy	($\Delta H^{c}{}_{m}$) (J g^{-1})	93–148
Percent crystallinity	%	
Density (ρ)	(g/cm^3)	1.36
Solubility	–	Dichloromethane, acetonitrile, chloroform
Degradation	–	Hydrolysis (random nonenzymatic, enzymatic by microbes)
Molecular weight (Mw)	(kDa)	66
Stereoisomer	–	L-lactic acid
O_2 permeability	cm^3 mil/m^2	550
CO_2 permeability	cm^3 mil/m^2	3000
Water vapor transmission	$g \cdot \mu m \cdot kPa^{-1} \cdot m^{-2} \cdot d^{-}$	161–237
Tensile modulus (E)	(GPa)	3500
Tensile strength (σ)	(MPa	48–53
Elongation at break	(%)	7
Percent of elongation	(%)	12
Transmission	(230–250 nm)	95%
Thermal conductivity	(190 °C	0.195

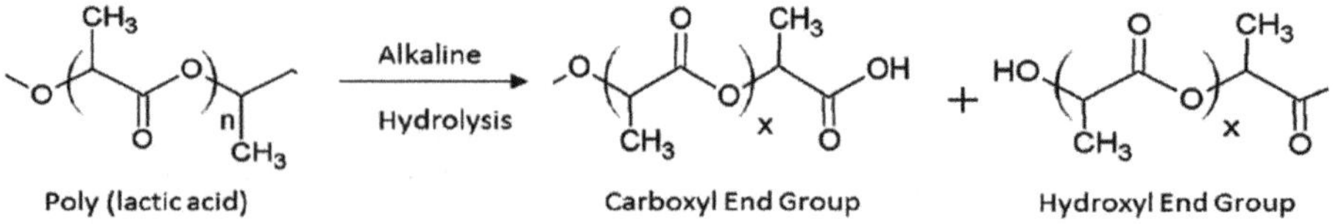

Fig. 1 PLA hydrolysis and molecular cleavage

PLA degradation was also studied as it was well known that PLA is susceptible to water hydrolysis and breaking of ester bonds. In a result, it was found that crystallinity affects the degradation of different PLA materials having different concentrations of D/L-lactide. Percentage of lactide monomers changes the rate of degradation as higher rate of degradation was observed with low content, making the polymer crystalline difficult to degrade. Hydrolysis was observed with decrease in thickness and thus resulting material becomes more fragile (Fig. 1). Measurement in the variation of material property and thickness was calculated. During the first 2 weeks of degradation, a small increase in molecular weight of PLA material was reported due to the cross-linking of subsequent monomer, produced under the effect of UV radiation. In view of the fact that hydrolysis is occurring arbitrarily, PLA chains with longer linear backbone are more susceptible than the shorter ones. The long tails of PLA is more acquire to cleave in comparison to short tails. Decomposition of PLA materials carried out in municipal as well as industrial amenities proceeds with the process of hydrolysis at higher temperature range. Further studies on decomposition techniques of food packaging materials under composting condition is required to be evaluated.

13 Polyhydroxyalkanoates (PHA)

Microorganisms synthesize PHA polymers naturally in rich availability of carbon sources. This class of biopolymer is biodegradable, biocompatible, and thermostable having melting temperature of about 180 °C. Similar to PLA, the crystallinity depends upon the type and number of monomer and hence indirectly affects the melting temperature. In the field of packaging, PHA has various ranges of food application. Amongst PHA, polyhydroxybutyrate (PHB) is the most studied and characterized polymer; polymers alone or in combination with the conventional polymers constitute an excellent source of food packaging films (Tharanathan 2003). PHB accumulates in the bacterial cell as a cell inclusion, which sometime may comprise about 80% of cell dry weight (Khosravi-Darani and Bucci 2015). PHB can be derived by polymerization of hydroxyl-butyrate monomer, showing the comparable properties of polypropylene but with more crystallinity. Besides being insoluble to water, PHB is optically active and has good barrier properties toward gas. Khosravi-Darani and Bucci (2015) has studied a comparison between PP and PHB as food packaging material. Author's reported that 50% lower deformation value of PHB was observed as compared to PP. PHB performs in better way as compared to PP at freezing and higher temperature. The other copolymer named polyhydroxybutyrate-valerate (PHBV) is used as packaging material due to its good mechanical properties (less brittle). In spite of the fact that the production and availability of PHBV polymer are cost-effective but is showing a good rate of degradation of about 1 month in an active microbial environment. Yu et al. (1998) reported the synthesis of different PHA polymers by utilizing different food wastes as carbon sources. Different polymers show diverse mechanical characteristics such as tensile strength, viscosity, etc. Thus, digesting food waste as carbon source can help in the reduction of production price.

14 Polybutylene Succinate (PBS)

The recent development in the techniques of food packaging made from PBS includes retention of food quality, aroma emitters, antimicrobial activity, and gas scavenger/emitters (CO_2/O_2). Consistent change in the flavor of packaged foodstuffs is an important emerging issue toward its acceptance by the customer and hence needs to be resolved. The undesirable change in flavor of foodstuff is generally achieved by the removal of active aroma compounds present in the food items. This technique of aroma scalping could essentially be overcome by the use of biodegradable packaging materials. Cihal et al. (2015) reported that the aroma scalping properties, i.e., sorption and permeation of PBS-based packaging film, were evaluated by using different sets of aroma compounds such as ethyl acetate, ethyl butyrate, and ethyl hexanoate. Litreature results confirmed that biopolymer PBS has comparable permeable properties that are even lower than the nondegradable film and is found to be a replacement for conventional food packaging material due to its production from renewable resources.

15 Polysaccharide-Based Biodegradable Polymer

15.1 Starch

Polysaccharide-based packaging materials are the most renowned form of food packaging as they are the richest macromolecules present in the environment. Among different polysaccharides (chitosan, chitin, galactomannan), starch is the most studied material in food packaging applications derived from plant. Fundamentally, starch is the cheapest renewable carbon source, is completely biodegradable, and can show thermoplastic properties. Commercially the starch is obtained from the corn and potato. The blending of starch with conventional polymers improves its flexibility towards the packaging material and makes it comparable to the oil-based polymer. Different percentages of starch (up to 90%) were added into the petrochemical-derived plastic which enhanced its susceptibility toward water (hot/cold). Decomposition of starch-based material accelerates with the action of microbial enzymes making the thermoplastic surface porous. The addition of starch into nondegradable material exceeds the disintegration up to 60%. Conversion of starch-based thermoplastic material into foam could also be achieved by using water steam in spite of PS as packaging substance. Deformation of thermoplastic into desirable packaging products such as tray and dishes is successful due to its consumption by microorganism within about 10 days of disposal.

16 Cellulose

The cellulose occurs in the earth as the most abundant form and is obtained by removing lignin from woody pulp. Cellophane film is produced by dissolving the degradable polysaccharides into sodium hydroxide solution with carbon disulfide in solvent casting process. Derivatization of cellulose is accomplished by removing hydroxyl groups in esterification. Now derivatization of cellulose is an issue of current study. Cellulose derivatives like cellulose (di-)acetate and cellulose (tri-) acetate involve some plasticizers to construct thermoplastics. Different derivatives have different solubilities toward water such as hydroxypropyl cellulose is water-soluble while ethyl-cellulose is not. Blending of plasticizer with ethyl-cellulose enhances its use in lamination and molding. Derivatization of cellulose increases its use as in food packaging; however, it is cost-effective (Cyras et al. 2007). Few other polysaccharide-based biodegradable polymer materials are listed in Table 2.

17 Bio-Based Packagings

Processing and designing of packaging materials are multifunctional treads which require watchfulness and a variety of considerable engineering processes to make it a proper material having the desired properties. Some characteristics to be taken into account for manufacturing of food packaging films that involves impermeable barrier towards gas and water vapor, waterproof ability, thermo-tolerance, UV light tolerance, transparency, printability, availability, antifogging capacity and costs. More than that, an important thing to think about is the course of degradation which is essential. Because of this, the disposal of used packaged should be taken into a careful consideration.

It is essential to understand that the possible market need and packaging could not be fulfilled by using a single bio-based material. Consequently, rising concerns are seen toward development of packaging material having multiple layers of bio-based polymer and conventional method to produce food packaging film having layers of polymer that act as a barrier and protects food. In conventional method, coating of food packaging materials is achieved with ethylene-vinyl alcohol or polyamides

Table 2 Properties and applications of polysaccharide-based biodegradable polymer

Raw materials	Origin	Solubility	Advantages	Disadvantages	References
Zein	Corn	Ethanol, acetone	Good film properties	Brittle	Ghanbarzadeh et al. (2006) and Sozer and Kokini (2009)
Soy protein isolate (SPI)	Soybean	Water, ethanol	Good gas barrier property, optically active	Water resistance	Zink et al. (2016)
Whey protein	Cheese	Water, ethanol	Mechanically good Low O_2 and CO_2 permeability	Moderate barrier to gas	Schmid et al. (2012)
Wheat (gluten derived films)	Starch	Water	Moderate barrier to oxygen	Moisture resistance, brittle	Zink et al. (2016)
Chitosan	Chitin	Polar solvents	Desired tensile strength and barrier to gas and water, antimicrobial property	Good water retention	Rhim et al. (2007)

blended with low-density polyethylene containing the barrier properties of polyamide as well as the water-resistant and mechanical properties of LDPE. Similar method can also be applied for the development of bio-based packaging material having necessary properties, and so bio-based coating could be developed by means of having gas barrier properties with chitosan, a protein or corn-derived material blended with PLA or PHA, which supports the mechanical strength (Weber et al. 2002).

18 Potential Food Packaging Applications

The packaging films comprise about 25% of usage among the 125 million tons of waste material generated per annum and is an apparent approach toward the possible market need for biologically derived packaging substances, which on the other hand could be enhanced by appropriate procedure at a comparable cost. Particularly the bioplastics will have to contend with conventional packaging materials, which are inexpensive, easy to produce, and have enhanced packaging properties such as lightweight, durable, disposable, and chemically inert to the food stuff. A rise in the production of bio-derived bioplastics would be achieved in the subsequent years. Food packaging comprises the highest growing sector among the plastic packaging field. In a report by Ferreira et al. (2016), plastic food packaging covers about 40% of the European conventional packaging required in the field of transportation, preservation, and storing of food items. In reality, about 50% of European food items are usually packed in the plastic material, and this is very economical due to the good mechanical properties like infirmity, plasticity, strength, etc. These properties make them an excellent source of packaging as well as are very economical for the food packaging industry.

19 Fruits and Vegetables

It is very important to protect fruits and vegetable after harvesting. The presence of ethyl acetate leads to its ripening and affects respiration by changing the concentration of water and gas content in packing material. Choosing a packing material is very tough as different foodstuffs have different rate of transpiration and respiration. Impermeability toward gas and water leads to damage of the fruits as well as vegetable due to fermentation process inside packaging.

Different biopolymers have distinctive gas barrier property. A comparatively low transmittance between carbon dioxide and oxygen rate is obtained in bio-based packaging film with 1:4 by conventional while 1:30 from bio-derived material transmitting carbon dioxide in comparison to oxygen is an appealing demand of packaging material for fruits and vegetables (Sandhya 2010). However, in some cases, transmittance is not required all the time.

20 Cheese

Cheese is protein rich, high water content product that respires continuously, producing carbon dioxide, that makes the use of bio-based packaging material very interesting. Permeability of packaging material for carbon dioxide is very essential requirements to protect the seal from inflation (Auras et al. 2006). Both the biodegradable polymers (PHA and PLA) can be applied for such type of packaging; however, blended materials could also develop.

21 Chilled or Frozen Products

The low gas barrier property of PHA and PLA materials made them extensively a choice of application in food industry. As suggested by Weber et al. (2002), dairy items have very feasible packaging application with PLA. Aliphatic polyesters have wide range of application in packaging of beverages (bottle and cups) as well as non-beverage items. In addition to packing with cardboard, these materials can be applied in the form of thin coating as compared to the conventional one. Haugaard et al. (2001) has found positive assumption in packing of dairy products.

Table 3 Miscellaneous applications of bio-based packaging films in food sector

Applications	Biopolymers	References
Beverages	PLA-coated cups	Jager (2010)
Curdled milk	PLA jars	
Baked items	PLA packs	
Chocolates	Cornstarch trays	Grumezescu (2017)
Organic crop	Cornstarch packs	
Potato chips	Cellulose sheet	
Organic pasta	Cellulose packs	
Sweets	Cellulose films	
Salads	PLA dishes and packs	Haugaard et al. (2003)
Fizzy items	PLA bottles	Sudesh and Iwata (2008)
Kiwi	Cellulose-based trays	Kumar and Thakur (2017)
Fruits and vegetables	PLA packs	Koide and Shi (2007)
Organic bread	PLA bags	Weston (2010)
Chilled fries	PLA sheet	Nieburg (2010)

22 Single-Use Tableware

Application of degradable tableware manufactured from bio-based material opens up an era of disposable material for restaurants. If all the fast-food packing materials and tableware are developed from biodegradable process, then it will lead to an enormous way of waste disposal under the process of composting (Muhammadi et al. 2015). Some more current applications of bioplastics have been listed in Table 3.

23 Main Limitations of Biodegradable Films

Employment of bio-derived plastics as food packages is restricted due to different limitations. Besides high production cost as compared to petrochemical-derived plastics, the use and availability of land are also major limiting factors for its production, which prohibited the functionality of its production at industrial level. Although the properties of bioplastics fascinates its use, several disadvantages such as fragility, thermal volatility, low Tm, complex heat stability, and high accessibility toward water and oxygen restrict them to be a good food packaging material (Peelman et al. 2013).

24 Conclusion

Development of biodegradable packaging materials is one of the recent and innovative processes in the area of food packaging. To enhance biodegradable polymers, it is necessary that our surroundings should have more concern regarding environmental health. Such packaging needs more and more study so that the property, storage period, dietary standards, and biosafety may be improved. Besides these properties, the barrier properties also required improvement. Starch- and cellulose-derived degradable bioplastics are mostly building its firm growth toward application in food packaging. Since starch is renewable, low-priced, and easily accessible biopolymer, blending with a few plasticizers additionally with water is required to make it accessible as a distortable thermoplastic called TPSs. Other nondegradable materials are also employed in packaging that exists as a waste in its original form for long duration and releases harmful effects to the environment. Landfilling and incineration are the methods usually applied to wipe out the packaging waste; however, it is not appropriate for plastics degradation.

Microbe-derived biopolymers found to be attarctive in several ways but they are not so economical. Presently biodegradable packages are in use for such foodstuffs that do not necessitate high impermeability for oxygen and water vapor and have short storage period such as fresh green groceries and fruits, or for long-storage products, like dumplings and fries, that do not require excessive oxygen- and/or water-impermeable properties. Yet, the armory of films depicts extensive diversity in properties that could design them to be used as packages meant for further food products with more strict conditions. Tests for preservation and other investigation on the industrialized packaging equipment should be done to confirm that these films could use commercially. Thus, it could be summarized that bio-derived materials have a wide variety of applications in the packaging industry. However, it is quite significant to understand that analysis and assessment of the operative assets of a bio-based polymer is very important prior to use as an alternative for conventional film materials.

Acknowledgment Authors are grateful to Integral University for providing necessary infrastructure and manuscript communication number IU/R&D/2017-MCN000214.

Conflict of Interest The authors declare that they have no conflict of interest.

References

Auras R, Singh SP, Singh J (2006) Performance evaluation of PLA against existing PET and PS containers. J Test Eval 34:530–536

Azeredo HM, Rosa MF, Mattoso LHC (2017) Nanocellulose in bio-based food packaging applications. Ind Crop Prod 97:664–671

Barnes DK, Galgani F, Thompson RC et al (2009) Accumulation and fragmentation of plastic debris in global environments. Philos Trans R Soc Lond Ser B Biol Sci 364:1985–1998

Cihal P, Vopicka O, Pilnacek K et al (2015) Aroma scalping characteristics of polybutylene succinate based films. Polym Test 30(46):108–115
Coles R, McDowell D, Kirwan MJ (2003) Introduction. In: Coles R, McDowell D, Kirwan MJ (eds) Food packaging technology. Blackwell Publishing, London, pp 1–31
Conn RE, Kolstad JJ, Borzelleca JF et al (1995) Safety assessment of polylactide (PLA) for use as a food-contact polymer. Food Chem Toxicol 33(4):273–283
Cyras VP, Commisso MS, Mauri AN et al (2007) Biodegradable double-layer films based on biological resources: polyhydroxybutyrate and cellulose. J Appl Polym Sci 106(2):749–756
Derraik JG (2002) The pollution of the marine environment by plastic debris: a review. Mar Pollut Bull 44(9):842–852
Ferreira AR, Alves VD, Coelhoso IM (2016) Polysaccharide-based membranes in food packaging applications. Membranes 6(2):22
Ghanbarzadeh B, Oromiehie AR, Musavi M et al (2006) Effect of plasticizing sugars on rheological and thermal properties of zein resins and mechanical properties of zein films. Food Res Int 39:882–890
Grumezescu A (2017) Nanotechnology in the agri-food industry food packaging. Elsevier, Cambridge, MA
Guo Q, Crittenden JC (2011) An energy analysis of polylactic acid (PLA) produced from corn grain and corn stover integrated system. In: Sustainable systems and technology (ISSST), 2011 I.E. international symposium on IEEE, pp 1–5
Haugaard VK, Udsen AM, Mortensen G et al (2001) Potential food applications of biobased materials, an EU-concerted action project. Starch 53(5):189–200
Haugaard VK, Danielsen B, Bertelsen G (2003) Impact of polylactate and poly(hydroxybutyrate) on food quality. Eur Food Res Technol 216:233–240
Jager A (2010) Ingeo polylactide een natuurlijke keus. Presentation given at VMT conference: green packaging
Kale G, Auras R, Singh SP (2006) Degradation of commercial biodegradable packages under real composting and ambient exposure conditions. J Polym Environ 14:317–334
Khalil HA, Davoudpour Y, Saurabh CK et al (2016) A review on nanocellulosic fibres as new material for sustainable packaging: process and applications. Renew Sust Energ Rev 64:823–836
Khosravi-Darani K, Bucci DZ (2015) Application of poly (hydroxyalkanoate) in food packaging: improvements by nanotechnology. Chem Biochem Eng Q 29(2):275–285
Koide S, Shi J (2007) Microbial and quality evaluation of green peppers stored in biodegradable film packaging. Food Control 18(9):1121–1125
Kumar S, Thakur KS (2017) Bioplastics-classification, production and their potential food applications. JHA 8(2):118–129
Li WC, Tse HF, Fok L (2016) Plastic waste in the marine environment: a review of sources, occurrence and effects. Sci Total Environ 566:333–349
Moore CJ (2008) Synthetic polymers in the marine environment: a rapidly increasing long-term threat. Environ Res 108:131–139
Mostafa NA, Farag AA, Abo-dief HM et al (2015) Production of biodegradable plastic from agricultural wastes. Arab J Chem
Muhammadi S, Afzal M et al (2015) Bacterial polyhydroxyalkanoates-eco-friendly next generation plastic: production, biocompatibility, biodegradation, physical properties and applications. Green Chem Lett Rev 8(4):56–77
Ni H, Zeng SQ, Wu J et al (2012) Cellulose nanowhiskers: preparation: characterization and cytotoxicity evaluation. Bio Med Mater Eng 22:121–127
Nieburg O (2010) Unique sustainable deep freeze packaging set for European launch. http://www.foodproductiondaily.com/Packaging/Unique-sustainable-deep-freezepackaging-set-for European-launch. Accessed 10 Oct 2012
Northwood T, Oakley-Hill D (1999) Wastebook. Luton friends of the earth, environment agency and the building research establishment
Peelman N, Ragaert P, De Meulenaer B et al (2013) Application of bioplastics for food packaging. Trends Food Sci Technol 32(2):128–141

Rhim JW, Lee JH, Perry KW (2007) Mechanical and barrier properties of biodegradable soy protein isolate-based films coated with polylactic acid. Food Sci Technol Int 40(2):232–238

Robertson GL (2013) Food packaging: principles and practice, 3rd edn. CRC Press, Boca Raton

Saeidlou S, Huneault MA, Li H et al (2012) Poly (lactic acid) crystallization. Prog Polym Sci 37 (12):1657–1677

Sahu CK, Dave U, Sukrutha S et al (2016) A case study and perspectives of human wisdom on eco-friendly food packaging materials in India. Biotechnol (Rajkot) 12(7)

Sandhya (2010) Modified atmosphere packaging of fresh produce: current status and future needs. LWT – Food Sci Technol 43(3):381–392

Schmid M, Dallmann K, Bugnicourt E et al (2012) Properties of whey-protein-coated films and laminates as novel recyclable food packaging materials with excellent barrier properties. Int J Polym Sci 2012:562381

Sin LT (2012) Polylactic acid: PLA biopolymer technology and applications. William Andrew, New York

Singh SB, Grewal A, Kumar P (2014) Biotechnological production of polyhydroxyalkanoates: a review on trends and latest developments. Chin J Bio 2014:1–18

Siracusa V, Rocculi P, Romani S et al (2008) Biodegradable polymers for food packaging: a review. Trends Food Sci Technol 19(12):634–643

Song JH, Murphy RJ, Narayan R, Davies GBH (2009) Biodegradable and compostable alternatives to conventional plastics. Philos Trans R Soc B364:2127–2139

Sozer N, Kokini JL (2009) Nanotechnology and its applications in the food sector. Trends Biotechnol 27(2):82–89

Sudesh K, Iwata T (2008) Sustainability of biobased and biodegradable plastics. Clean (Weinh) 36 (5–6):433–442

Tharanathan RN (2003) Review e biodegradable films and composite coatings: past, present and future. Trends Food Sci Technol 14:71–78

Tonjes DJ, Greene KL (2012) A review of national municipal solid waste generation assessments in the USA. Waste Manag Res 30(8):758–771

Valdés García A, Ciller M, Cristina A et al (2014) Natural additives and agricultural wastes in biopolymer formulations for food packaging

Weber CJ, Haugaard V, Festersen R et al (2002) Production and applications of biobased packaging materials for the food industry. Food Addit Contam 19:172–177

Weston S (2010) Packaging- assessing its environmental credentials. http://www.foodbev.com/article/packaging–assessing–its-environmental–credentials. Accessed 10 Oct 2012

Yu PH, Chua H, Huang AL et al (1998) Conversion of food industrial wastes into bioplastics. Appl Biochem Biotechnol 70–72:603–614

Zink J, Wyrobnik T, Prinz T et al (2016) Physical, chemical and biochemical modifications of protein-based films and coatings: an extensive review. Int J Mol Sci 17(9):1376

Biopolymers, Nanocomposites, and Environmental Protection: A Far-Reaching Review

Sukanchan Palit and Chaudhery Mustansar Hussain

Abstract Environmental engineering science today is witnessing drastic challenges. The domains of engineering and science are moving from one visionary paradigm toward another. In the similar manner, biopolymers and nanocomposites are the ecomaterials of today's research and development forays. Mankind and the vast vision of human scientific endeavor need to be envisioned and readdressed as science and engineering surges forward. Today, climate change, ecological biodiversity, and frequent environmental catastrophes are challenging the vision of science and the vast scientific fabric. The future of human civilization lies in the hands of environmental engineers and scientists. Here comes the scientific truth of application of biopolymers and nanocomposites in environmental pollution control. Mankind's immense scientific grit, the vast technological profundity, and the need of environmental sustainability will lead a long and effective way in the true realization of environmental engineering today. In this chapter, the authors deeply address the importance of biopolymers and nanocomposites in ameliorating the burning issue of water contamination of heavy metals, environmental protection, and industrial wastewater treatment. In this chapter, the authors with cogent insight address the scientific research pursuit in the field of both biopolymers and nanocomposites with a clear objective toward effective proliferation of science and engineering. Human ingenuity, vast scientific ardor, and the vast world of scientific and technological validation are the hallmarks of research endeavor today. In this chapter, the authors deeply focus on the scientific success, the potential, and the technological motivation in the field of smart materials, material science, and environmental engineering. Polymer science and technology and its wide and vast research forays are the other pillars of this well-researched chapter. The challenge and the vision of biopolymers, composites, and nanotechnology are the other facets of this chapter. This chapter

S. Palit (✉)
Department of Chemical Engineering, University of Petroleum and Energy Studies, Dehradun, Uttarakhand, India

C. M. Hussain
Department of Chemistry and Environmental Sciences, New Jersey Institute of Technology, Newark, NJ, USA

S. Ahmed (ed.), *Bio-based Materials for Food Packaging*,
https://doi.org/10.1007/978-981-13-1909-9_10

will open new windows in the domains of nano-science, nanotechnology, and the relevant fields of environmental engineering in decades to come.

Keywords Biopolymers · Nanocomposites · Environment · Protection · Vision · Remediation · Pollution

1 Introduction

Human scientific endeavor and human scientific paradigm today are in the midst of deep scientific introspection and vast innovation. Technology, engineering, and science are in the path of newer scientific regeneration and vast scientific rejuvenation. Environmental and energy sustainability are the cornerstones of research and development initiatives globally. In the similar vein, environmental engineering and chemical engineering are in the wide avenue of deep scientific emancipation. Technology and engineering are highly challenged today as human scientific endeavor surges forward toward a newer eon and a newer paradigm. Material science and polymer science need to be envisioned and readdressed as science and technology moves forward. The author in this chapter pointedly focuses on the need of polymer science and nanotechnology to human society. Modern science is a pillar with a vast vision of its own. This chapter is an eye-opener toward the world of scientific ardor and the scientific intricacies in scientific research endeavor in the field of biopolymers and nanocomposites. Water purification and industrial wastewater treatment are the other visionary areas of scientific research today. Arsenic and heavy metal drinking water contamination is challenging the vast scientific fabric and the deep scientific landscape in present-day human civilization. Modern science and human society thus stand in the critical juncture of scientific difficulties and scientific introspection. In this chapter, the author deeply comprehends the need of biopolymers, nanocomposites, and nanotechnology in tackling the immense scientific enigma of environmental pollution control. Human society and human civilization today stand in the vast abyss of introspection and scientific fortitude. This chapter unravels scientific innovation and scientific instinct in the field of nanotechnology, material science, and environmental engineering in years to come (Palit 2017b).

2 The Aim and the Objective of This Study

Human mankind and human civilization are today in the path of rejuvenation and deep scientific regeneration. Technology and engineering science have few answers to the scientific intricacies and the barriers to innovation and vision. The primary objective of this study is to target the latent areas of applications of biopolymers and nanocomposites in tackling the scientific enigma of environmental protection.

Modern technology needs to be readdressed as global water shortage assumes immense importance in our present-day human mankind. Scientific barriers, scientific ingenuity, and vast scientific needs of the human society will lead an effective way in the true realization of environmental engineering. Environmental engineering and environmental sustainability are the true needs of human society and human civilization today. In this chapter the author with immense insight addresses the vast scientific farsightedness, the intense acuity, and the vast vision of science in the research pursuit in nanotechnology, polymer science, and the vast world of environmental engineering. The pivots of this chapter are biopolymers, nanocomposites, and environmental protection. Environmental engineering science and human scientific endeavor today stand in the midst of scientific introspection and scientific comprehension. The author in this chapter reiterates the scientific divination and the vision behind scientific validation in environmental pollution control today. The authors also delve deep into traditional and nontraditional environmental engineering techniques with the main purpose of furtherance of science and engineering. This chapter surpasses scientific imagination and deep scientific thoughts as science and technology of environmental protection moves forward. Sustainable development and holistic infrastructural development are the needs of human society and modern science today. These areas need to be more explored and envisioned as human scientific research pursuit moves forward. The primary aim and objective of this study are to unravel the hidden scientific and technological truth and the vast scientific profundity in nanotechnology and polymer science applications in tackling global environmental crisis (Palit 2017b).

3 The Vast Scope of This Study

Modern science and scientific research initiatives today stand tall in the midst of vision and scientific fortitude. Environmental protection and scientific revelation are in the path of deep rejuvenation. The scope of this study is vast and versatile. Material science, polymer science and technology, and nanotechnology are in the avenues of newer vision and widespread innovation. Today technology and engineering science have few answers to global water crisis and heavy metal groundwater and drinking water contamination. Human scientific progress and human scientific stature need to be readdressed and reenvisioned with the passage of scientific vision and time. The scope of this study goes beyond material science and environmental engineering toward heavy metal groundwater remediation. Today the human mankind and sustainable development are in the midst of deep anguish and vast vision. Scientific research pursuit is at a state of immense catastrophe and vast failures. Environmental protection and water purification are the frontiers of science today and are covered in this chapter. Thus, the emancipation of modern science is truly realized and human civilization surges into a new era. This chapter also addresses the burning issue of arsenic groundwater contamination of developed and developing economies in our planet (Palit 2017b). The author also poignantly ponders into traditional and nontraditional environmental engineering techniques (Palit 2017b). Technology and science of nanotechnology are surpassing vast and versatile scientific frontiers today. This hidden scientific truth of nanotechnology is deeply elaborated with similar vision in this chapter (Palit 2017b).

4 What Do You Mean by Biopolymers?

Biopolymers are polymers produced by living beings; in other words, they are polymeric macromolecules. Since they are polymers, biopolymers contain monomeric units that are bonded to form large molecules. There are three main classes of biopolymers, classified according to monomeric units used and the structure of the biopolymer formed: polynucleotides, which are long polymers composed of 13 or more nucleotide monomers; polypeptides, which are short polymers of amino acids; and polysaccharides which are often linear bonded polymeric carbohydrate structures. Biopolymer is a polymer that is obtained from human beings. It is a biodegradable chemical compound that is regarded as the most found organic compound in the ecosphere and the human ecosystem. The name “biopolymer” envisages that it is a biodegradable polymer (Palit 2017b). Today scientific ingenuity, scientific vision, and scientific divination are in the path of immense regeneration. Polymer science, nanomaterials, and material science are today facing immense scientific difficulties and scientific intricacies (Palit 2017b). Biopolymers and composites are smart materials which have immense applications in environmental protection and the holistic world of environmental engineering. In this treatise, the authors reiterate the scientific success, the needs of engineering science in human progress, and the futuristic vision of polymer science and material science (Palit 2017b).

5 What Do You Mean by Nanocomposites?

Nanocomposite or polymer composites are a multiphase solid material where one of the phases has one, two, or three dimensions of less than 100 nanometers (nm) (Palit 2017b). In the broadest sense, this definition and the vast description can include porous media, colloids, gels, and copolymers. Scientific vision, the vast scientific cognizance, and the scientific revelation in biopolymer applications are the pallbearers toward a new scientific knowledge in the field of nanotechnology. The technology, engineering, and science of nanotechnology and the science of composites are overcoming vast and versatile visionary frontiers (Palit 2017b). The authors in this paper deeply comprehend and thus restructure the need of composite science in the true emancipation and proliferation of science and engineering (Palit 2017b)

6 The Scientific Doctrine and the Scientific Vision of Environmental Protection

Human scientific research initiatives and human civilization are today in the avenue of new scientific rejuvenation and vast vision. Environmental protection today stands in the midst of vast scientific introspection and deep scientific articulation. Global climate change, the loss of biodiversity, and the frequent environmental disasters are the torchbearers toward a newer scientific order in environmental pollution control (Palit 2017b). Industrial wastewater treatment, drinking water and groundwater treatment paradigm, and water purification technologies are changing the face of environmental engineering science. In this paper, the author deeply focuses on the need of applied science such as polymer science and nanotechnology in the proliferation of science and engineering. Technology, engineering, and science are today in a state of distress with the growing concerns for global climate change and frequent environmental disasters. Human scientific candor, the engineering travails, and the vast scientific ardor will all lead an effective way in the true realization of environmental sustainability and sustainable development today (Palit 2017b). The scientific doctrine and the scientific vision of environmental protection are today in the midst of immense revival and regeneration (Palit 2017b). Global climate change is challenging the vast scientific firmament of human scientific ingenuity and scientific acuity (Palit 2017b). The future of environmental engineering and environmental protection is bright and pathbreaking as science and technology moves forward. The face of human scientific endeavor today is highly challenging and needs to be envisioned and restructured with the passage of scientific vision and time (Palit 2017b).

7 Environmental and Energy Sustainability and the Visionary World of Environmental Management

Environmental and energy sustainability are today the visionary arenas of scientific research initiatives. The status of global environment is absolutely dismal (Palit 2017b). Environmental engineering science and environmental sustainability are today linked to each other. Along with the advancement of technology, sustainability is an absolute necessity. In this chapter, the author deeply focuses on the human acuity and ingenuity in the application of sustainability in human society. Sustainable development as regards energy and environment is the immediate need of the hour. The state of global environment and water technology is absolutely dismal. In such a critical juncture of vast scientific vision and time, technology and innovations are the necessities of human scientific progress (Palit 2017b). Water science and technology also needs to be restructured as mankind steps into a new era of scientific rejuvenation. Frequent environmental catastrophes and lapses in environmental sustainability are challenging the scientific firmament today (Palit 2017b). Global climate change and disastrous loss of ecological diversity are transforming the face of human civilization. In this chapter, the author comprehends the scientific success, the scientific regeneration, and the futuristic vision in the application of engineering tools such as polymer science and nanotechnology in tackling environmental pollution control. Today energy and environmental sustainability are linked to social and economic sustainability (Palit 2017b). Human society will be surely elevated and move toward a newer paradigm if provision of basic needs such as water, food, and energy is met. Thus, the need of also energy sustainability (Palit 2017b). Energy security is also a cornerstone of scientific endeavor today. Energy engineering and the application of renewable energy technology will all lead a visionary way in the true realization of energy sustainability today. In this chapter, the authors deeply elucidate the human success and the vast insight in the application of material science, polymer science, and nanotechnology to tackling environmental pollution control (Palit 2017b).

Scientific vision and the world of scientific validation are the repertoire of global environmental management today. Environmental management and sustainable development are two sides of the visionary coin. Environmental management today encompasses the scientific success and the vast scientific vision of environmental pollution control (Palit 2017b).

Barrow (2005) deeply elucidates on the vast domain of environment management and development. The science of sustainability whether it is energy or environment is today the utmost needs of mankind and human scientific progress. The author elucidates in this treatise theory and approaches in environmental management in developing countries, resource management issues by sector, and the vast domain of environmental tools and policies (Barrow 2005). Environmental management is evolving rapidly and is increasingly applied in developing as well as developed nations and to the wide world of transboundary and global issues. Human scientific progress in applied science and basic areas is witnessing immense challenges, and the scientific vision in environmental management is highly retrogressive (Barrow 2005). Environmental management is being applied by many sectors of human activity, and its practitioners are diverse (Barrow 2005). This area of scientific research initiative needs to be restructured and reorganized as science and engineering surges forward. This well-researched treatise explores and investigates environmental management with the focus on developing countries and the development processes throughout the world. The author (Barrow 2005) with immense scientific conscience and divination pronounces the need of environmental management as well as environmental sustainability toward progress of developing nations (Palit 2017b).

8 Significant Scientific Endeavor in the Field of Biopolymers

Polymer science, material science, and the world of nanomaterials are today in the path of newer scientific vision and deep fortitude. Bioengineering is a frontier science today. The world of validation in polymer science and biopolymers needs to be reframed and restructured with the passage of scientific history and time. Today polymer science is a huge scientific edifice with a purposeful vision of its own. In the similar vein, material science and nanotechnology are changing the face of scientific research pursuit (Palit 2017b).

Rao et al. (2014) deeply discussed with cogent insight biopolymers and its vision. The environmental impact of non-biodegradable plastic wastes is raising a widespread global and societal concern (Rao et al. 2014). This resulted in a scientific surge in the research and development forays in biopolymers (Palit 2017b). Human scientific advancements, the vision of polymer science, and the vast world of scientific validation will lead a visionary way in the true emancipation and true scientific truth of material science today (Palit 2017b). Nanotechnology is another area of visionary pursuit globally. This paper depicts poignantly the success of engineering science, the scientific validation in the application of biopolymers, and the vast scientific profundity in the application of nanocomposites and nanotechnology (Rao et al. 2014). Biopolymers (polysaccharides) have vast applications in medicine, food, and petroleum industries. Approximately, 140 million tones of synthetic and organic polymers are produced worldwide every year (Rao et al. 2014). Since polymers are highly stable, their degradation cycles in the environment and the biosphere are extremely limited (Palit 2017b). Environmental pollution by synthetic polymers has been widely recognized as a major global problem (Rao et al. 2014). Plastics and polymers are an integral part of our daily human life and scientific progress. Man's immense scientific grit and determination, mankind's vast scientific prowess, and the futuristic vision of polymer science and material science will all lead a visionary way in the true emancipation of biopolymer application today (Rao et al. 2014; Palit 2017b). Due to stability and resistance to degradation, these are accumulated in the environment (Rao et al. 2014). Mankind and human scientific genre are today in the midst of vast scientific farsightedness. Biopolymers are defined as polymers formed under natural conditions (Palit 2017b). Therefore, they are also defined as natural polymers (Rao et al. 2014; Palit 2017b). For material applications, cellulose and starch are highly interesting and immensely promising (Rao et al. 2014). In this review paper, the author deeply ponders on the functions and synthesis of biopolymers, production of biopolymers, intracellular versus extracellular production of biopolymers, genetic engineering and biopolymer technology, and genetically engineered biopolymer production system. The authors also discussed in minute details various types of biopolymers (Rao et al. 2014). The needs of scientific research pursuit in polymer science are vast and versatile. This paper is a veritable eye-opener to the scientific intricacies of biopolymer applications to human scientific progress (Rao et al. 2014; Palit 2017b).

Chauhan and Chauhan (2013) discussed with lucid and cogent insight the vast scientific world of natural fibers and biopolymers. In recent decades, there have been a tremendous increase in scientific interest, research, and development in the natural fibers, biopolymers, polymer materials, and engineered polymer materials for their extensive use in bio-composite, packaging, agriculture, medicine, etc. (Chauhan and Chauhan 2013). The challenges and the vision of polymer science and material science in today's world of science and engineering are vast and versatile (Palit 2017b). Ecological engineering and environmental engineering are two branches of engineering which need to be envisioned and reorganized with the passage of scientific vision and visionary timeframe (Chauhan and Chauhan 2013). Today is the world of bioengineering and nanotechnology (Chauhan and Chauhan 2013). The vast vision behind the scientific research forays is to promote the biodegradable polymer materials that would reduce the need for the synthetic polymer production while reducing the environmental pollution. The vast vision of material science, the futuristic vision of biopolymers, and the vast scientific ingenuity are all the torch-bearers toward a newer era in the field of science and engineering (Chauhan and Chauhan 2013). This paper provides a deep and cogent insight to the progress in the utilization of natural fiber and biodegradable polymer (Palit 2017b). The authors with deep scientific conscience also discussed the usefulness of natural fibers and polymer for environmental protection applications (Chauhan and Chauhan 2013). There has been a tremendous increase in scientific attention, research, and development initiatives for the natural fiber, biopolymers, and polymer materials for their extensive and evergrowing use in bio-composite, packaging, agriculture, medicine, etc. (Chauhan and Chauhan 2013). Technology and engineering science of natural fibers and biopolymers are today surpassing vast and versatile visionary frontiers. Many research and development initiatives are today targeted toward modifying traditional natural fibers and polymers to more user-friendly and novel polymers, polymer composites, nanocomposites, and advanced materials (Chauhan and Chauhan 2013). Fiber-reinforced composites are the smart and ecomaterials of today. Today the other side of the visionary coin is the importance of environment friendliness of the natural fibers, biopolymers, nanocomposites, and fiber-reinforced composites (Chauhan and Chauhan 2013; Palit 2017b). This is the upshot of this entire scientific endeavor. The authors deeply focus and investigate the need of scientific and technological emancipation in material science and polymer science today (Chauhan and Chauhan 2013; Palit 2017b).

Gabor and Tita (2012) discussed with deep and cogent foresight in a review biopolymers used in food packaging. The vision of this review is to reveal with deep scientific conscience the most relevant biopolymers used in food packaging. Most research and development forays that are investigated are classified in three categories: biopolymers from renewable sources, chemically obtained biopolymers, and microbial obtained biopolymers. Biopolymers, polymer science, and material science are today in the midst of immense scientific regeneration and deep scientific ingenuity (Gabor and Tita 2012). The actual inclination in packaging research is to develop and promote the use of "bioplastics" which are useful in reducing waste disposal and environment friendly and are good alternative of petroleum, a nonrenewable resource (Gabor and Tita 2012). Today the scientific world is widely challenged with the monstrous issue of depletion of fossil fuel resources and global climate change. In this paper, the authors deeply discussed biodegradable polymers in food packaging and the vast and definite vision behind it (Gabor and Tita 2012). The areas of biopolymers touched upon are polysaccharides, alginate, cellulose, chitin/chitosan, curdlan, gellan, pectin, starch, and xanthan. The other areas envisioned in this paper are proteins, collagen, gelatin, soy protein, whey protein, and aliphatic polyesters such as polylactic acid and polyhydroxybutyrate (Gabor and Tita 2012). Scientific research pursuit, the needs of biodegradable plastics, and the futuristic vision of polymer and material science will all lead an effective way in the true realization of science and engineering (Gabor and Tita 2012; Palit 2017b).

Valdes et al. (2014) investigated with deep insight natural additives in bioformulations for food packaging. The main research directions in food packaging are targeted toward improvements in food quality and food safety engineering (Valdes et al. 2014). For this purpose, food packaging with a deeper vision toward longer product shelf life as well as stringent environmental regulations is the immediate need (Valdes et al. 2014). Scientific vision, vast scientific repertoire, and the needs of scientific and technological validation are the necessities of research and development forays in polymer science and material science today. The innovations in new materials, and particularly innovative biopolymer formulations, can definitely help to deeply address these requirements and also the vast packaging applications (Valdes et al. 2014). Human scientific vision in biopolymer science is at its scientific helm today (Palit 2017b). The vastly innovative upshot of this paper is the use of natural extracts in biodegradable and eco-friendly polymers (Valdes et al. 2014). The other avenue of research endeavor of this paper is revalorization of agricultural residues as reinforcement in biopolymers. Legislative issues and environmental regulations in biopolymer packaging are the other salient features of this paper (Palit 2017b).

9 Recent Significant Research Pursuit in Application of Nanocomposites

Scientific vision, deep scientific discernment, and vast scientific motivation are the needs of modern science today. Nanotechnology, material science, and composite science are the areas of immense scientific regeneration today (Palit 2017b). Nanocomposite application and its vast scientific emancipation are the necessities of research endeavor today. In this section, the authors point out toward the vast scientific vision, the futuristic vision of technology, and the immense prowess behind nanocomposite application in modern society (Palit 2017b).

Modi et al. (2014) deeply elucidated with vast scientific conscience in a review on green polymer nanocomposites and their vast applications. Technological vision, scientific ingenuity, and deep scientific and engineering prowess are the cornerstones of research endeavor in nanotechnology and composite science today (Modi et al. 2014). Nanocomposites and fiber science are the newer areas of scientific and engineering innovation today. This paper addresses the domain of green polymer nanocomposites. Through this research endeavor, the authors study bio-reinforced composites in automotive, construction, packaging, and medical applications (Palit 2017b). From the deep and investigative study of nanocomposites, the authors studied the unique properties of natural fillers and organic polymers (Modi et al. 2014). Man's immense scientific prowess, the vision of composite science, and the needs of environmental protection will all lead a visionary way in the true emancipation of material science, nanotechnology, and composite science today. This paper investigates techniques of production, trends, and prospects in the field of green polymer nanocomposites (Modi et al. 2014). Technological acuity and the vast world of scientific validation are the utmost need of the moment as science and engineering surges forward. Today, there are immense research and development initiatives for the development and application of environmentally friendly and sustainable bio-reinforced composites for wide use in automotive, construction, packaging, and medical fields (Modi et al. 2014; Palit 2017b). The authors discussed in minute details processing methods of nanocomposites, review of green polymers, thermoplastic starch-based composites, polylactic acid-based composites, cellulose-based composites, and other biopolymer-based composites. The progress of mankind and the utmost needs for green engineering are the veritable torchbearers toward a greater scientific emancipation of polymer science and material engineering today. The authors also poignantly depict the application of green polymer nanocomposites (Modi et al. 2014). The veritable needs of green technology and green engineering are the pallbearers toward a newer age in the field of material science today (Modi et al. 2014; Palit 2017b).

Camargo et al. (2009) discussed with lucid insight synthesis, structure, properties, and new application opportunities of polymer nanocomposites. Human scientific vision, the vast scientific fortitude, and the human scientific foresight are the pallbearers toward a greater scientific emancipation in the field of nanocomposites (Camargo et al. 2009; Palit 2017b). Nanocomposites, a high-performance material, exhibit unusual property combinations and unique design properties. In this comprehensive review, the three types of matrix nanocomposites are elucidated underpinning the vast need for these materials, their important processing methods, and some recent results on structure and potential applications. Scientific vision in the field of nanotechnology combined with polymer science is deeply investigated in this well-researched paper (Camargo et al. 2009; Palit 2017b). Nanocomposites can be classified into three broad categories: (1) ceramic matrix nanocomposites, (2) metal matrix nanocomposites, and (3) polymer matrix nanocomposites (Camargo et al. 2009). Conducting polymer nanocomposites are novel and smart materials of today's modern-day human civilization and scientific research pursuit. Potentials and opportunities in nanocomposites with deep scientific vision are the other salient features of this paper. The authors also touched upon processing of nanocomposites, structure and properties, and the vast future perspectives of composite science (Camargo et al. 2009).

Adeosun et al. (2012) discussed and investigated in minute details green polymer nanocomposites. Green technology and green engineering are the areas of scientific endeavor which are in the path scientific regeneration (Adeosun et al. 2012; Palit 2017b). Today is the need of environmental protection and green technology. Bio-reinforced composites have extensive applications in automotive, construction, packaging, and medical applications due to the evergrowing concerns for environmental sustainability. Sustainability science and sustainability engineering are today the needs of human scientific research pursuit. Green polymer nanocomposites show unique properties of combining the advantages of natural fillers and organic polymers. Science of polymer and material engineering are today linked with the vast world of environmental protection. This paper with vast scientific divination and vision investigates current research trends, the techniques of production, opportunities, challenges, and prospects in the field of green nanocomposites (Adeosun et al. 2012). Environmental engineering science and environmental protection are today in the state of immense scientific comprehension and scientific vision. The authors in this paper discussed with deep scientific conscience green polymer nanocomposites, fillers, and nanocomposite processing methods. This paper definitely opens a newer scientific paradigm in the field of polymer and material science (Adeosun et al. 2012; Palit 2017b).

Fawaz and Mittal (2015) with deep scientific insight and conscience reviewed various techniques in the synthesis of polymer nanocomposites (Fawaz and Mittal 2015). Technological validation, the vast scientific conscience, and profundity will definitely lead a visionary way in the true realization of composite science today. Polymer nanocomposites are hybrid organic-inorganic materials with at least one dimension of the filler phase less than 100 nm (Fawaz and Mittal 2015). Polymer nanocomposites are synthesized via various techniques that can be categorized into four major routes: melt intercalation, template synthesis, exfoliation adsorption, and in situ polymerization intercalation (Fawaz and Mittal 2015). The authors discussed in minute details synthesis methods of nanocomposites (Fawaz and Mittal 2015).

Huang et al. (2003) elucidated with vast scientific foresight polymer nanofibers by electrospinning and their vast applications in nanocomposites. Scientific validation and deep technological vision are the needs of human society and modern science today. Various polymers have been successfully electrospun into ultrafine fibers in recent times mostly in solvent solution and some in melt form. Technology needs to be envisioned and reorganized with the passage of scientific vision and the visionary timeframe (Palit 2017b). Information of these polymers together with their immensely visionary processing conditions for electrospinning of ultrafine fibers has been summarized in details in this well-researched paper (Huang et al. 2003).

10 Significant Research Endeavor in Environmental Protection and the Vision for the Future

Environmental protection and the vast domain of environmental engineering and chemical process engineering today stand in the midst of deep scientific vision and foresight. The success of science needs to be reenvisioned, reframed, and readdressed with the passage of scientific vision and visionary timeframe. The application of polymer science in environmental protection is a veritable challenge to scientific progress today. Human scientific research pursuit today stands in the midst of deep scientific divination and vast revelation (Palit 2017b). The author deeply invokes in this paper the need of environmental protection and environmental engineering tools with the sole objective of emancipation and proliferation of science and technology (Palit 2017b). Global water shortage, global climate change, and frequent environmental disasters are challenging the vast scientific firmament of modern civilization today. In this paper, the author reiterates with deep scientific conscience research endeavor in the field of material science, polymer science, and nanotechnology. Today nanotechnology is linked with every visionary avenues of scientific pursuit such as chemical engineering and polymer science (Palit 2017b). The world of environmental management needs to be envisioned with the passage of scientific history, scientific vision, and time. Environmental management is certainly evolving rapidly and is increasingly applied in developing countries and to transboundary and global issues (Palit 2017b).

Palit (2017a) discussed with lucid and cogent insight advanced environmental engineering separation processes, environmental analysis, and application of nano-science and nanotechnology in a far-reaching review. The world of challenges and the vast vision behind environmental protection are moving steadfastly toward a newer vision (Palit 2017a). The vision of science, the scientific ingenuity, and the futuristic vision of conventional and nonconventional environmental methods will all lead a successful way toward environmental engineering realization today (Palit 2017b). The success of environmental engineering separation processes is delineated in details along with deliberation of advanced environmental analysis. The author with deep scientific conscience addresses membrane separation processes and nontraditional environmental engineering treatment techniques such as advanced oxidation processes and integrated advanced oxidation processes (Palit 2017b). The author in this chapter pointedly focuses on global ecological balance, provision of potable drinking water, and the scientific progress of human civilization. The authors also touched upon water process engineering, environmental separation processes, and the vast vision of tomorrow's scientific pursuit. A review of relevant technologies for wastewater treatment and oxidation technologies is deeply and lucidly pondered upon in this chapter. Novel separation processes are wonders of science today. This treatise opens up new windows of restructuring of science and vast scientific profundity in the field of environmental engineering techniques (Palit 2017a, b).

Hussain and Palit (2018) elucidated with deep and cogent foresight nanomaterials, ecomaterials, and the vast vision of material science and engineered nanomaterials. Human scientific regeneration and the vast scientific profundity are the torchbearers of scientific endeavor in material science today (Palit 2017b). The world of nanotechnology is moving toward one paradigm toward another. In the similar vein, environmental engineering science is treading a weary path toward a visionary scientific boundary. The main aim and objective of this treatise is the discussion of scientific pursuit in the field of nanomaterials, ecomaterials, and engineered nanomaterials and the vast domain of material science (Hussain and Palit 2018). Human scientific regeneration, the vast vision of material science, and the visionary aisles of environmental engineering will all lead a visionary way in the true realization of environmental protection (Palit 2017b). The authors discussed nanomaterials for environmental applications. The authors depicted with deep scientific conscience and profundity the significant scientific endeavor in the field of nanomaterials (Hussain and Palit 2018; Palit 2017b).

Palit (2018) discussed with vast scientific farsightedness nanomaterials for industrial wastewater treatment and water pollution control. Industrial wastewater treatment, drinking water treatment, and water purification are the needs of human society today (Palit 2017b). Application of nano-science and nanotechnology in water purification is the challenge and the vision of modern science today. In this chapter, the author depicts with profound conscience novel separation processes and the traditional and nontraditional environmental engineering techniques (Palit 2017b, 2018).

11 Scientific Doctrine of Traditional Environmental Engineering Tools

Traditional environmental engineering tools today are in the midst of scientific vision and deep scientific introspection. Activated sludge treatment, flocculation, and coagulation are traditional environmental engineering tools. Today there are few answers to the monstrous issue of global climate change and loss of ecological biodiversity. Man's immense scientific grit and determination, mankind's vast scientific knowledge, and the vision of human research pursuit will all lead a long way in the true emancipation of science of environmental sustainability. Traditional environmental engineering tools are today replaced by nontraditional tools such as membrane separation processes and advanced oxidation processes (Palit 2017b). Novel separation processes include membrane separation processes. Scientific research endeavor, vast scientific prowess, and immense scientific discernment are the backbones of research pursuit in traditional and nontraditional environmental engineering techniques today. Man's immense futuristic vision, the vast world of scientific and technological validation, and the utmost needs of the basic needs such as energy and water will today truly lead a long way in the true realization of the science of sustainability. Water purification and drinking water treatment are the absolute necessities of human scientific progress today (Palit 2017b). The science of membrane separation processes and desalination is widely expanding as global water shortage moves from one visionary frontier toward another (Palit 2017b).

12 Scientific Vision and Scientific Cognizance of Membrane Science

Membrane science and global water research and development initiatives are today linked by an umbilical cord. The vision and the challenge of novel separation processes are immense and pathbreaking and are replete with vast scientific grit and determination (Palit 2017b). Today membrane science and desalination are veritably linked with global water crisis and heavy metal groundwater remediation. The crisis is extremely severe in South Asian countries as well as developed world. Scientific vision and vast scientific cognizance are today in the crucial juxtaposition of comprehension, scientific might, and vast discernment. Technology, engineering, and science of membrane processes need to be reframed as human civilization and global research and development initiatives move forward (Palit 2017b). Stringent water quality regulations can be achieved efficiently with the help of membrane separation processes. The most important advantages of using membrane science are the ease of operation and little requirement of chemicals. Depending on pore size, the membrane processes can be classified into microfiltration (MF), ultrafiltration (UF), and nanofiltration (NF) (Palit 2017b). The membranes are also subdivided into organic, inorganic, and composite membranes. The primary disadvantage of using membranes is fouling. Global water research and development initiatives are today linked with membrane technology. The needs for scientific endeavor are immense as human scientific pursuit moves forward. In this chapter, the author reiterates the deep scientific understanding and the scientific vision behind membrane applications in tackling global water issues (Palit 2017b).

13 Advanced Oxidation Processes and the Scientific Barriers Behind It

Advanced oxidation processes (AOP) are alternatives to the incineration of wastes which have numerous disadvantages (Munter 2001). The AOPs have proceeded along one of the two routes:

- Chemical oxidation with oxygen in temperature ranges between ambient conditions and those found in incinerators (Palit 2017b).
- The use of oxidants such as ozone and hydrogen peroxide and/or photons that are able to generate highly reactive intermediates – OH radicals (Munter 2001).

Advanced oxidation processes and integrated advanced oxidation techniques are the technologies of tomorrow (Palit 2017b). Nowadays there is a continuously evergrowing worldwide concern for development of water reuse and water treatment techniques mainly focused on agriculture and industry. Advanced oxidation processes (AOPs) are considered as visionary processes to treat industrial wastewater. These processes are used for the removal of those organic pollutants not treatable by conventional environmental engineering techniques due to their high chemical stability and/or high biodegradability (Palit 2017b). Chemical oxidation for complete mineralization is highly expensive today. The main and primary routes for destroying toxic compounds in natural water are biodegradation, bioremediation, and photodegradation (Palit 2017b). Today the scientific intricacies and the scientific barriers of advanced oxidation processes can be removed, and efficiency can be increased with the combination of biological treatment. Oller et al. (2011) discussed with immense lucidity wastewater treatment by combined AOPs (advanced oxidation processes)/biotreatment technology. Scientific perseverance, scientific grit, and vast scientific vision are the needs of scientific research pursuit in advanced oxidation processes (Oller et al. 2011; Palit 2017b). This challenge and the vast vision of science are elucidated in details in this paper (Oller et al. 2011; Palit 2017b). Industrial wastewater treatment and water purification today are in the midst of deep scientific introspection and the vast futuristic vision (Palit 2017b). Technology and engineering science has few answers to the evergrowing concerns for arsenic and heavy metal groundwater contamination and subsequent remediation (Palit 2017b). Nontraditional environmental engineering separation processes are vastly envisioned and need to be reorganized as regards zero-discharge norms and strict environmental regulations (Oller et al. 2011; Palit 2017b).

14 Arsenic and Heavy Metal Groundwater Remediation

Arsenic and heavy metal groundwater contamination and subsequent remediation are burning issues in modern science and today's human civilization (Palit 2017b). Technological vision, the vast futuristic vision of science, and the world of challenges in environmental engineering science will all lead an effective way in the true realization of environmental sustainability (Palit 2017b). Today, groundwater remediation is the need of the hour. In South Asia and developing and developed countries throughout the world, arsenic groundwater contamination has resulted in a devastating crisis. Human scientific ingenuity and vast scientific vision are the pallbearers toward a newer era in water purification (Hashim et al. 2011). Technology and engineering science has veritably no answers to this burning and enigmatic issue of groundwater contamination. In this paper the author points out toward the vast scientific success, the scientific profundity, and the challenges behind heavy metal remediation of groundwater (Palit 2017b). South Asia particularly Bangladesh and India is in the threshold of a devastating scenario (Hashim et al. 2011). Water shortage and the lack of pure drinking water are a scientific enigma in developing and developed countries around the world (Palit 2017b). In this paper, the authors also point out toward the arsenic groundwater remediation technologies which have interfaced with material science, material chemistry, and nanotechnology. Environmental engineering science today is a huge pillar with a definite and purposeful vision of its own. The imminent need of the hour is innovations and technologies in groundwater remediation. Hashim et al. (2011) with deep and cogent foresight delineated remediation techniques for heavy metal- and arsenic-contaminated groundwater. The world of science and engineering are today witnessing the environmental disaster of groundwater heavy metal contamination in India and Bangladesh (Palit 2017b). Scientific vision has few answers toward the monstrous issue of groundwater poisoning. Scientific enigma of arsenic groundwater remediation is deeply comprehended by scientists throughout the world (Hashim et al. 2011). The contamination of groundwater and drinking water by heavy metal, originating either from natural soil sources or from anthropogenic sources, is an enigma and a matter of immense concern to human progress (Palit 2017b). Keeping the issue of environmental sustainability in mind, the technologies encompassing bioremediation and biosorption are highly recommended and envisaged in many water treatment issues (Hashim et al. 2011; Palit 2017b).

Mandal and Suzuki (2002) reviewed the presence of arsenic in groundwater throughout the world. This comprehensive review delineates environmental origin, occurrence, and impact on health by arsenic-contaminated groundwater (Mandal and Suzuki 2002). Arsenic and its compounds are extremely mobile in the environment. Weathering of rocks converts arsenic sulfides to arsenic trioxide, which enters the arsenic cycle (Mandal and Suzuki 2002). The vast scientific vision of arsenic groundwater remediation technologies, the needs of human society, and the provision of basic human needs such as water will all lead an effective way in the true realization of environmental protection (Mandal and Suzuki 2002). The authors in this paper deeply investigate the scientific innovation and the vast scientific profundity in water purification science.

Ahmed et al. (2006) discussed with deep scientific conscience and with urgent concern an epidemiology report on the assurance of supplying clean drinking water in Bangladesh. Human civilization today stands in the midst of deep scientific introspection and vast scientific revival. In the early 1980s, Dr. K.C. Saha from the School of Tropical Medicine in Kolkata attributed skin lesions in the state of West Bengal in India, to exposure to arsenic in groundwater pumped from shallow tube wells on various districts (Ahmed et al. 2006). Scientific progress had virtually come to a standstill and vision stalled as the world's largest environmental disaster destroyed the scientific firmament. Despite these scientific findings, millions of tube wells have been installed in the Bengal basin. The vast popularity of tube wells in Bangladesh and the state of West Bengal in India reduced incidence of diarrheal disease. This ushered in another great disaster which is a permanent agony in India and Bangladesh. Human sufferings are immense and scientific vision stalled as science and engineering surged forward in many developing and developed nations around the world (Ahmed et al. 2006; Palit 2017b). Today, perhaps 100 million people in India, Bangladesh, Vietnam, Nepal, and Cambodia are drinking water with arsenic concentrations up to 100 times the World Health Organization guideline of 10 μg per liter. Technological innovations and vast scientific validation are at a state of immense catastrophe as there are few answers to the evergrowing concern of groundwater contamination (Ahmed et al. 2006). The world's largest environmental disaster is at a state of immense scientific ingenuity as well as deep introspection (Ahmed et al. 2006; Palit 2017b).

15 Modern Science: Difficulties, Challenges, and Opportunities

Human civilization and human scientific research pursuit are today in the path of newer scientific rejuvenation. The world of science, technology, and engineering today is facing immense challenges and vast scientific vision. The state of modern science is immensely dismal in human planet. Nuclear science and engineering and space technology are today in the process of new regeneration (Shannon et al. 2008; Palit 2015, 2016a, b). Yet climate change, loss of biodiversity, and frequent environmental disasters are totally destroying the visionary scientific firmament (Palit 2017b). Science and engineering are today two huge colossus with a deep vision of its own (Palit 2017b). The difficulties and the challenges of modern science are immense, and the opportunities are groundbreaking. Science today has few answers to environmental disasters and the grave concerns for groundwater and drinking water contamination of heavy metals such as arsenic. In this paper, the authors repeatedly target the vast scientific vision, the scientific fortitude, and the needs of basic human resources such as water, food, shelter, and energy (Palit 2017b). The world of technology has few answers to the grave concerns for energy and environmental sustainability. Modern science and vast research endeavor are the answers to sustainable development today. Material science and polymer engineering are today the frontiers of science along with nanotechnology. Nanotechnology today is vastly encompassed with scientific vision and vast scientific profundity (Palit 2017b). Today material science and nanotechnology are two opposite sides of the coin. In this chapter, the author points toward the human scientific success and the vast scientific ingenuity in the field of biopolymers, nanocomposites, and environmental engineering science (Palit 2017b). This chapter opens new windows of scientific ardor and deep scientific understanding in the field of material science and environmental protection (Palit 2017b).

16 Future Recommendations and Future Flow of Scientific Thoughts

Future recommendations and future flow of scientific thoughts in the field of material science, polymer science, and environmental engineering are vast and versatile. Mankind's vast scientific intellect, man's vast ingenuity, and the futuristic vision of environmental engineering will all lead an effective way in the true emancipation of environmental sustainability and the vast area of sustainable development today (Palit 2017b). Future flow of scientific thoughts and research and development initiatives in polymer science, material science, and nanotechnology should be targeted toward energy and environmental sustainability (Palit 2017b). Scientific regeneration is today in the path of vision and scientific fortitude. Water science and technology and the vast domain of water purification also need to be readdressed and envisioned today with the progress of science and engineering. Industrial wastewater treatment, groundwater treatment, and desalination techniques are the visionary avenues of scientific endeavor today. Technological and scientific advancements are today in the crucial juncture of immense disaster with the growing concerns for climate change. Nuclear science and space technology are the cornerstones of research foray today. Yet, the challenges and targets of science and technology are immense, far-reaching, and evergrowing. The science of biopolymers and nanocomposites needs to be veritably overhauled as polymer science and nanotechnology witness immense and drastic changes. The application of biopolymers and nanocomposites in environmental pollution control is in a similar vein undergoing vast and versatile changes. In this treatise the author rigorously points toward the vast scientific success, the technological validation, and vast scientific motivation in the application of polymer science and nanotechnology in environmental engineering (Shannon et al. 2008; Palit 2015, 2016a, b, 2017b).

17 Conclusion and Scientific Perspectives

Human scientific vision and human scientific forays are today in the path of newer rejuvenation and newer scientific grit and determination (Palit 2017b). Human mankind stands in the midst of deep scientific introspection and vast acuity. The world today is highly challenged with the growing concerns for climate change and industrial pollution. Technology and engineering science has few answers to the scientific advancements in environmental protection (Palit 2017b). The whole gamut of scientific research in polymer science, the science of the composites, and nanoscience needs to be structured and envisioned as mankind tackles environmental issues. The state of global environment is immensely dismal today. Scientific perspectives in environmental protection and environmental engineering science should be targeted toward more innovations, scientific advancements, and the provision of basic human necessities such as water, food, shelter, and energy (Palit

2017b). Energy and environmental sustainability today are in the critical juncture of deep global crisis. The need for pure drinking water and energy security in the human planet will all lead a long way in the true realization of energy and environmental sustainability. Application of nanotechnology in environmental pollution control apart from deep comprehension in material science is the other pillar of this paper. The authors in this treatise focus on the boundless scientific potential and vision behind nanotechnology and material science. The need for renewable energy and energy sustainability is the other cornerstone of holistic scientific pursuit today. Global research and development advancements in water purification and environmental protection are the utmost need of the hour and gaining immense importance with the passage of history and visionary timeframe (Palit 2017b). Polymer science and nanotechnology are the other areas of interdisciplinary research which will have direct impact on environmental protection. In this paper, the authors deeply elucidate on the vast domain of biopolymers and its application to environmental protection. Polymer science and nanotechnology are in the path of newer scientific vision and vast candor. This paper unravels new ideas and new scientific innovations in both nanotechnology and material science.

References

Adeosun SO, Lawal GI, Balogun SA, Akpan EI (2012) Review of green polymer nanocomposites. J Miner Mater Charact Eng 11(4):385–416

Ahmed MF, Ahuja S, Alauddin M, Hug SJ, Lloyd JR, Pfaff A, Pichler T, Saltikov C, Stute M, van Geen A (2006) Ensuring safe drinking water in Bangladesh. Science 314:1687–1688

Barrow CJ (2005) Environmental management and development. Routledge Press (Taylor and Francis), New York

Camargo PHC, Satyanarayana KG, Wypych F (2009) Nanocomposites: synthesis, structure, properties and new application opportunities. Mater Res 12(1):1–39

Chauhan A, Chauhan P (2013) Natural fibers and biopolymer. J Chem Eng Process Technol., (Open Access Journal) S6:1–4

Fawaz J, Mittal V (2015) Synthesis of polymer nanocomposites: review of various techniques. In: Mittal V (ed) Synthesis techniques for polymer nanocomposites, 1st edn. Wiley-VCH Verlag GmBH & Co., KGaA, Germany

Gabor D, Tita O (2012) Biopolymers used in food packaging: a review. Acta Univ Cibiniensis Ser E: Food Technol XVI(2):3–19

Hashim MA, Mukhopadhyay S, Sahu JN, Sengupta B (2011) Remediation technologies for heavy metal contaminated groundwater. J Environ Manag 92:2355–2388

Huang Z-M, Zhang Y-Z, Kotaki M, Ramakrishna S (2003) A review on polymer nanofibers by electrospinning and their applications in nanocomposites. Compos Sci Technol 63:2223–2253

Hussain CM, Palit S (2018) Nanomaterials, ecomaterials and wide vision of material science. In: Martinez LMT, Kharissova OV, Kharisov BI (eds) Springer handbook of ecomaterials. Springer International Publishing AG, Basel, pp 1–29

Mandal B, Suzuki KT (2002) Arsenic round the world: a review. Talanta 58:201–235

Modi VK, Shrives Y, Sharma C, Sen PK, Bohidar SK (2014) Review on green polymer nanocomposite and their applications. Int J Innov Res Sci Eng Technol 3(11):17651–17656

Munter R (2001) Advanced oxidation processes- current status and prospects. Proc Estonian Acad Sci Chem 50(2):59–80

Oller I, Malato S, Sanchez-Perez JA (2011) Combination of advanced oxidation processes and biological treatments for wastewater decontamination- a review. Sci Total Environ 409:4141–4166

Palit S (2015) Advanced oxidation processes, nanofiltration, and application of bubble column reactor. In: Kharisov BI, Kharissova OV, Rasika Dias HV (eds) Nanomaterials for environmental protection. Wiley, New York, United States of America, pp 207–215

Palit S (2016a) Nanofiltration and ultrafiltration- the next generation environmental engineering tool and a vision for the future. Int J Chem Technol Res 9(5):848–856

Palit S (2016b) Filtration: frontiers of the engineering and science of nanofiltration-a far-reaching review. In: Ortiz-Mendez U, Kharissova OV, Kharisov BI (eds) CRC concise encyclopedia of nanotechnology (Taylor and Francis). Wiley, New York, pp 205–214

Palit S (2017a) Advanced environmental engineering separation processes, environmental analysis and application of nanotechnology: a far-reaching review. In: Hussain CM, Kharisov B (eds) Advanced environmental analysis: application of nanomaterials, vol.1. RSC Detection Science Series No.9, Royal Society of Chemistry, London, United Kingdom, pp 317–16

Palit (2017b) Application of nanotechnology, nanofiltration and drinking and wastewater treatment- a vision for the future. In: Grumezescu AM (ed) Water purification. Academic Press (Elsevier), Cambridge, MA

Palit S (2018) Nanomaterials for industrial wastewater treatment and water purification. In: Martinez LMT, Kharissova OV, Kharisov BI (eds) Springer handbook of ecomaterials. Springer International Publishing AG, Basel, pp 1–41

Rao MG, Bharathi P, Akila RM (2014) A comprehensive review on biopolymers. Sci Rev Chem Commun 4(2):61–68

Shannon MA, Bohn PW, Elimelech M, Georgiadis JG, Marinas BJ, Mayes AM (2008) Science and technology for water purification in the coming decades., Nature. Nature Publishing Group, London, United Kingdom, pp 301–310

Valdes A, Mellinas AC, Ramos M, Garrigos MC, Jimenez A (2014) Natural additives and agricultural wastes in biopolymer formulations for food packaging. Front Chem 2., Article 6:1–10

Chitosan-Based Edible Membranes for Food Packaging

Mohamed E. I. Badawy and Entsar I. Rabea

Abstract Chitosan edible membranes provide extra benefits including compatibility with living organism, barrier properties to gases, nontoxic to mammals, nonpolluting, and low cost compared to synthetic polymers. In addition, bioactive membranes and coatings act as food additive carriers, such as antibiotics and antioxidants, and have been specifically considered in preservation of food due to their ability to extend shelf life. This chapter focuses principally on chitosan membranes derived from the naturally occurring chitin. Moreover, the chapter provides modern technological applications and the prospects of chitosan-modified membranes in food preservation.

Keywords Chitosan · Biopolymer · Membranes · Technological applications · Food preservation

1 Introduction

The urgent need to develop environmentally friendly biologically active materials in food packing technology through enhanced barrier permeability characteristics (water, CO_2, O_2, and vapor odor), transparency, biodegradability, prevention and control of pollutants, mechanical and moisture resistance, and other intelligent characteristics leads to focusing on the new trend in biopolymer technologies (Rhim et al. 2013a; Appendini and Hotchkiss 2002; Brockgreitens and Abbas 2016). In particular, chitosan biopolymer is an incessant protocol of demanding investigation in packaging of food and preservation because of its biodegradability, nontoxicity, transparency, hydrophilicity and antimicrobial activity, as well as high

M. E. I. Badawy (✉)
Department of Pesticide Chemistry and Technology, Faculty of Agriculture, Alexandria University, Alexandria, Egypt

E. I. Rabea
Department of Plant Protection, Faculty of Agriculture, Damanhour University, Damanhour, Egypt

S. Ahmed (ed.), *Bio-based Materials for Food Packaging*,
https://doi.org/10.1007/978-981-13-1909-9_11

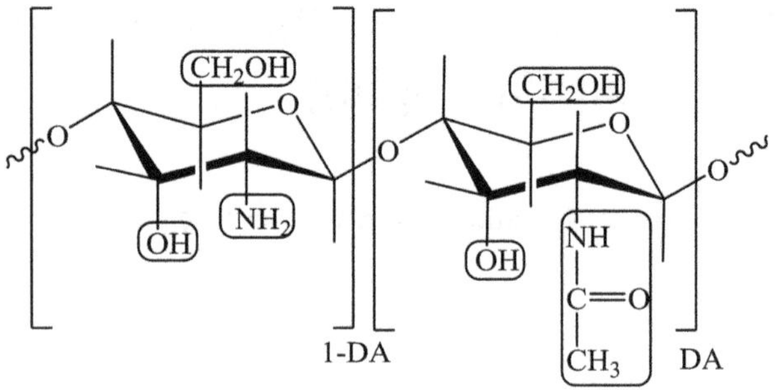

Fig. 1 Chemical structure of chitosan molecule (GlcN and remaining of GlcNAc units). DA is the degree of acetylation

cation and anion exchange capacity (Ferreira et al. 2016; Appendini and Hotchkiss 2002; Miller and Krochta 1997). Chitosan is a common linear polymer of β-(1-4)-D-glucosamine (GlcN) and *N*-acetyl-β-(1-4)-D-glucosamine (GlcNAc), gained by chitin deacetylation, the second abundant polysaccharide in nature (Fig. 1) (Muzzarelli 1983; Rabea et al. 2003). Chitosan plays increasingly significant role as edible coating membranes in food packing and preservation with substituting the nonbiodegradable and nonrenewable polymers (Baldwin 1994; Lin and Zhao 2007; No et al. 2007; Elsabee and Abdou 2013). High hydrophilicity of the chitosan molecule limits mechanical strength that can be associated with the deacetylation degree (DDA) (Clasen et al. 2006). The crystallization and density of chitosan are correlated to the functional characterization of films and membranes (Tsai et al. 2008). In the applications in food preservation, semielastic coatings of chitosan can change the interior atmosphere, delaying the maturing of vegetables and fruits (Elsabee and Abdou 2013; Kowalczyk et al. 2015; Badawy and Rabea 2009). The edible chitosan films can also use as carriers releasing of other antimicrobial agents like essential oils that are affected by vapor contact with pathogens (Avila-Sosa et al. 2012; Zivanovic et al. 2005; Acevedo-Fani et al. 2015). Other characteristics of chitosan edible films are the capacity to decrease the loss of the moisture and respiration rate of fruit, aroma, and solutes, thus increasing the shelf life (Zhang et al. 2011; Barbosa-Cánovas et al. 2000; Badawy et al. 2017).

Hydrophobicity control aids to regulate the rate of chitosan degradation that is of dominant importance in investigation of biomass-based and environmental polymers (Badawy et al. 2016; Clasen et al. 2006). Thus, it is usual to blend chitosan polymer with further biopolymers including gelatin, cellulose, and starch to develop the physicochemical characteristics of the prepared films. Furthermore, plasticizers are generally added to improve the dry flexibility and properties of the films (Badawy et al. 2016). This could develop the sustainability of the fabrication of novel intellectual packaging of food, in covenant with customer mandate for environmental transparency (Conte et al. 2015; Magnier and Schoormans 2015).

This chapter offers an indication of the application of chitosan biopolymer membranes in food packaging and preservation. In this context, this chapter chiefly deals with the recent approaches of preparation and characterizations of chitosan composite membranes with their safe application in food sectors.

2 Preparation and Characterizations of Chitosan Membranes

Several methods have been used to prepare membranes of chitosan and develop the separation efficacy including polymerization, mixing, surface modification, and grafting of selective types onto an inert films. Among them, the solution casting process is one of the best common techniques. Meanwhile the key purpose of the packaging of food and coating is regularly to reduce the moisture between the food and the surrounding atmosphere or between two components of a heterogeneous food product; the water vapor permeability (WVP) has to be as low as possible (Gontard et al. 1993; Salehi et al. 2016).

2.1 *Pure Chitosan Membranes*

A chitosan biopolymer has attracted considerable attention by potentially replacing traditional plastic packaging resources owing to improved awareness in sustainable progress. Chitosan has the capability to produce thin films due to its great hydrophilicity and good film-forming characters (Mi et al. 2001; Xu et al. 2005; Caner et al. 1998; Shu et al. 2001; Huei and Hwa 1996; Baldwin et al. 2011; van den Broek et al. 2015; Goosen 1996). The chitosan film membrane is stable structure, tough, durable, flexible, hard to tear, good permeable, and good water barrier. Some required possessions of chitosan are these form films without adding of additives, display good oxygen and permeability to carbon dioxide, as well as admirable mechanical characteristics (Suyatma et al. 2004). The chitosan film characteristics are contingent on its morphology that is influenced by molecular weight, DDA, solvent systems, solvent evaporation, and mechanism of free amine generation (Rathke and Hudson 1994; Elsabee and Abdou 2013). It has been revealed that the chitosan polymorphs are dictated by the preparative circumstances and play a significant role in the resulting properties of tensile. Clean or partly plasticized chitosan film may be prepared by casting from a solvent. In this procedure, chitosan is liquefied in acidified water, and a plasticizer is added. The solution is decanted onto a flat surface and the solvent is evaporated. Austin described chitosan films that were flocculated in a 5% (w/v) chitosan solution in aqueous 4% (v/v) acetic acid and solidified in aqueous bases. Stretching properties of 23.60 kg/mm^2 (23.60 MPa) were described after 1 day. The tensile properties reduced to 16.90 g/mrn^2 (16.90 MPa) after 16 days of aging the solution, and the viscosity was decreased from 4300 to 850 cPs. (Austin 1982). Auerbach described the casting of chitosan films from solution of 10% on plates and drying at 125 °C. The resulting films had a thickness of 0.002–0.015 inch and were transparent and flexible (Auerbach 1975). There were no reports on washing and drying methods, as well as on the concentrations of solutions and molecular weights. Nevertheless, Auerbach followed this report in 1977 with a topic that dealt with molecular weight and viscosity associations (Auerbach 1978).

In chitosan film forming, plasticizing agents such as glycerol, sorbitol, and fatty acids are important to overawe the weakness of the films. This weakness of the films refers to the weak intermolecular forces of the polymer structure. Plasticizers work to reduce the intermolecular forces that soften the stiffness of the film structure and increase the mobility of the biopolymer chains, thereby improving the mechanical properties (Gontard et al. 1993; Srinivasa et al. 2007; Arvanitoyannis et al. 1997; Boesel 2015; Meng et al. 2014). The key objective of the study was the WVP for the films achieved with mechanical characteristics. WVP is strongly overstated by DDA, acid types (formic, lactic, acetic, and propionic acid), and solvent pH. A greater WVP was measured for the highly disrupted chitosan than for the chitosan with a lesser DDA. The WVP values of the chitosan films were augmented with adding the plasticizers such as glycerol (Arvanitoyannis et al. 1997, 1998). However, Srinivasa and co-authors reported that the addition of glycerol to chitosan film reduced the WVP (Srinivasa et al. 2007). Claasin and others described that the quantity of glycerol was equal to 16–100% of the amount of chitosan in the solution to prepare the flexible chitosan membrane. The solution was stirred for several min and applied to a flat surface of polycarbonate to produce a film of 2 mm after dried at 60 °C (Clasen et al. 2006). Chitosan membranes were also produced by evaporation of solvent and then immersing the membranes in NaOH solution (1 M) for 24 h to neutralize the excess of acid and were finally washed with distilled water (Mladenova et al. 2011).

2.2 *Chitosan Blend Membranes*

Furthermost of the reports on the formation and possessions of chitosan films focused on their application as edible membranes. For packaging applications, chitosan alone has moderately low barrier and mechanical properties, which presently limit its industrialized usage. Particularly problematic is the improvement of moisture barrier belongings owing to its hydrophilic nature. However, blending of chitosan with other biopolymers increases these characteristics (Puttipipatkhachorn et al. 2001; Blair et al. 1987; Zhang et al. 2002; Tual et al. 2000; Singh and Ray 1999; Suyatma et al. 2004). The composites are designed once two or more polymers are mixed physically in the soluble state of a suitable solvent. Polymer mixtures can be in different forms, including the separable phase, compatible, non-compatible, interpenetrating and semi-interpenetrating polymer networks, or molecular composites. Combinations are frequently used to develop the elastic properties of the prepared membranes. Mixtures regularly offer a stronger structural constituent in the separation media that supports the active polymer. Physical possessions of a polymer can also be changed by presenting a second polymer, which enhances the original polymer properties in certain sectors. Composite membranes of chitosan with polysaccharides (starch, gelatin, pectin, and methylcellulose) or proteins have

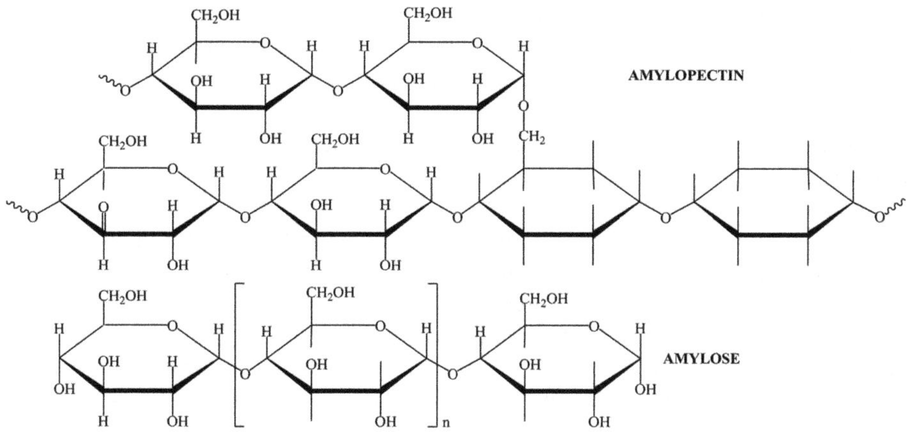

Fig. 2 Chemical structure of starch (amylose and amylopectin)

been widely considered to improve the properties of producing films (Valenzuela et al. 2013; Pereda et al. 2011; Arvanitoyannis et al. 1998; Debeaufort et al. 1998; Elsabee and Abdou 2013; Alishahi and Aïder 2012; Cha and Chinnan 2004).

2.2.1 Chitosan-Starch Membranes

Starch is one of the greatest abundant conventional polysaccharide raw ingredients (Robyt 2008). It consists of two core molecular constituents: amylose (an almost linear polymer of α-1,4-anhydroglucose) and amylopectin (a highly branched short-chain α-1,4 chains linked by α-1,6-glucosidic branching points occurring every 25–30 glucose units) (Fig. 2). The key alterations between chitosan and starch are the α (1,4) glucoside linkage for starch and β (1,4) for chitosan, and the hydroxyl group of the second carbon is substituted by the amine group which seems acetylated in the instance of chitin (Jiménez et al. 2012). Chitosan-starch membranes may be formed from emulsion or a film-forming dispersion, which comprises a high percentage of water. Furthermore, starch films can be obtained by using a drying method where the water content is lower as compared with the wet procedure.

The films formed from chitosan and starch in the existence of sorbitol as plasticizer showed low WVP values. This may be due to the cross-linking caused between chitosan amino groups and the hydroxyl groups of starch molecules by thermal mixing hydrophilic-hydrophilic form in a more hydrophobic material which has relatively low water barrier characteristics (Xu et al. 2005). According to Xu and co-authors, the highest mechanical strength of the films occurred at the starch to chitosan ratio of 1:1 (Xu et al. 2005). Regardless of the starch type, the tensile strength (TS) as well as the elongation (*E*) of the composite films increased in the first place and then declined with the addition of starch. Composite film made with normal thickness exhibited higher TS and *E* than waxy starch (Xu et al. 2005).

Starch/chitosan mixture membranes were prepared by the compression molding method in physical gel state (Zhai et al. 2004). The tensile flexibility and strength of starch membranes were significantly enhanced once the combination of 20% chitosan. Investigations by X-ray and scanning electron microscope (SEM) indicated that interaction and microphase separation occur between the molecules of chitosan and starch.

The properties of thermomechanical of starch-chitosan membranes were examined by dynamic mechanical thermal analysis (DMTA) and large deformation tensile testing. The data indicated that incorporation of sorbitol (10% and 30%) and/or moisture adsorption by chitosan-starch membranes resulted in a substantial decrease in the glassy transition (T_g) of the polysaccharide matrix due to plasticization (Lazaridou and Biliaderis 2002). The rice starch-chitosan films indicated an increase in the TS, WVP, lighter color, and yellowness and a decreasing elongation at the break (*E*) and solubility once combined with chitosan. The addition of chitosan amplified the crystal peak structure of the starch film; nevertheless, too high a concentration of chitosan has produced a phase separation between starch and chitosan (Bourtoom and Chinnan 2008). Blends of chitosan and corn starch were achieved by melt extrusion, and effect of polymer interaction on the thermal, morphology, and mechanical characteristics was examined (Mendes et al. 2016). SEM analyses exhibited a consistent break surface deprived of the existence of chitosan aggregates or starch particles. Though the combination of chitosan resulted in reduce in tensile strength and stiffness, films with enhanced stretch capability and thermal stability were formed.

The thermal mixing between chitosan and starch transformed the hydrophilic-hydrophilic mixture into a more hydrophobic mixture, which exhibited relatively low water barrier characteristics, as previously explained by Jagannath and his co-authors (Jagannath et al. 2003), who reported that the intensity of nonthermal mixing between chitosan and starch was lower than that achieved by intensive thermal mixing. In addition, FT-IR spectrum of chitosan/starch film showed that the peak of amino group of chitosan moved from 1578 to 1584 cm^{-1} by adding starch that confirmed the links between the hydroxyl groups of starch and the amino groups of chitosan occurred (Xu et al. 2005; Meenakshi et al. 2002). Arvanitoyannis and co-authors reported an increase in the WVP with the addition of glycerol (Arvanitoyannis et al. 1997, 1998), which is consistent with our result, where there was an increase in the WVP. However, Srinivasa and co-authors reported that the incorporation of glycerol to chitosan film reduced WVP (Srinivasa et al. 2007). This might be owing to the intrinsic chemical property of chitosan and similarly to the way in which the blend films are prepared. The results also indicated that the WVP correlated positively with the thickness of the films. This finding is consistent with numerous research studies on the WVP of chitosan films that have said that the correlation was an indication that the bulk diffusion may not be the rate-controlling step in the WVP process through edible chitosan films. Instead, interfacial reactions, such as dissolution rate or evaporation rate, may be controlling (Butler et al. 1996; Srinivasa et al. 2007; Kim et al. 2006; Kołodziejska and Piotrowska 2007; Shen and Kamdem 2015; Jridi et al. 2014; van den Broek et al. 2015). Active

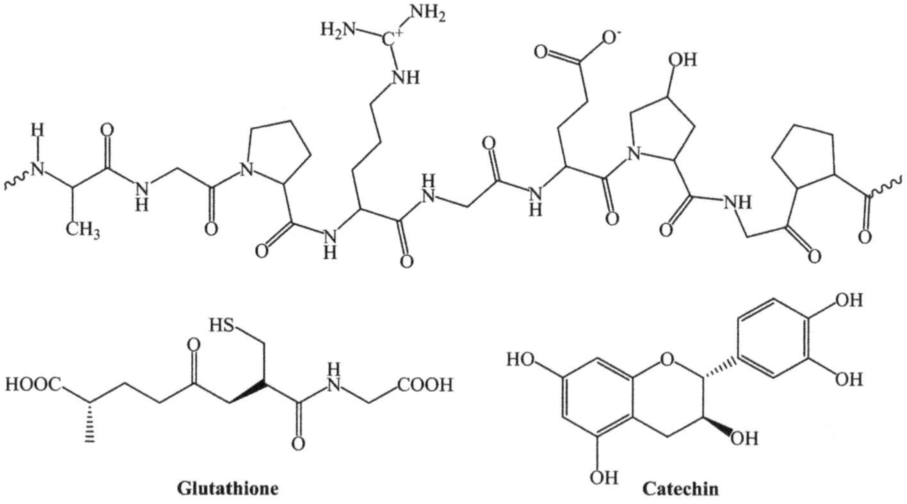

Fig. 3 Chemical structure of gelatin showing abundant amino acids such as glycine, proline, and hydroxyproline in sphere and cylindrical model (-Gly-Pro-Hyp-), glutathione, and catechin

composite chitosan-starch membranes were prepared using molding the composite polymers of 2% chitosan with starch (2% and 5%) through diverse concentrations of glycerol and sorbitol (1%, 2%, 5%, and 10%) as plasticizers (Badawy et al. 2016). The results mentioned that the mixture of starch in combination with plasticizer in chitosan films improves swelling, antioxidant activity, and WVP. Recently, the effect of active combinations such as essential oils and homogenization technology has been studied on the physical belongings of chitosan film-forming solutions and edible films consisting of large-scale chitosan-starch membranes. The data reported that the chitosan starch with such product mixtures displayed an enhanced mechanical response associated to pure chitosan or starch films while providing antimicrobial activity at low cost (Bonilla et al. 2012; Lozano-Navarro et al. 2017; Talón et al. 2017a, b; Siripatrawan and Vitchayakitti 2016).

2.2.2 Chitosan-Gelatin Membranes

Gelatin is transparent, bleak, weak (when dry), flavorless polymer, gained from collagen attained from various animal by-products. It has proline, hydroxyproline, and glycine in its polypeptide chain (Fig. 3). Glycine is responsible for close packing of the chains; however, the presence of proline restricts the conformation (Schrieber and Gareis 2007). Gelatin is very attractive biopolymers owing to its richness, biodegradability, inexpensive, and functional and exceptional film-forming characteristics (Pereda et al. 2011). However, they have low mechanical strength and dissolves relatively quickly in aqueous solution, thus limiting their practical applications. Gelatin is homogeneous due to its good miscibility with chitosan, leading to the development of film properties compared to those obtained from monomer (Kowalczyk et al. 2015; Jridi et al. 2014; Hosseini et al. 2013; Alexandre et al. 2016). In addition, chitosan-gelatin films can carry biologically active compounds that transmit antimicrobial and/or antioxidant activities, for example, to the film. In this case, nanocomposite film can be formed (Ibarguren et al. 2015; Vanin et al. 2014).

Several studies reported that WVP of gelatin films decreased when chitosan was added by ~12%. These effects can be illustrated by the fact that chitosan can develop gelatin binding and decrease the free volume of the polymer matrix, thereby declining the rate of water molecules spreading through the films, resulting in lesser WVP than chitosan gelatin compared with chitosan or gelatin alone (Cao et al. 2007; Arvanitoyannis et al. 1997, 1998; Kowalczyk et al. 2015). Parallel data were also described by Pereda and co-authors in films constructed from bovine hide chitosan and gelatin; however, there was no statistical variance in WVP among chitosan-gelatin membrane and gelatin control one (Pereda et al. 2011).

A chitosan-gelatin membrane sequence was prepared by changing the ratio of the components, and the quantity of the diverse states of water was determined by differential scanning calorimetry. The interface among chitosan and gelatin was confirmed by IR and X-ray investigation and was found to be related with mechanical strength. The results specify that the water intake of a chitosan-gelatin mixture is diminished by sustainable interactions within the networks. Chitosan can increase the tensile strength of complex films, and even with high water content, they can maintain adequate tensile strength and improved elongation (Yin et al. 1999).

Rivero and others prepared composite, bilayer, and laminated biodegradable membranes from chitosan (7.5%) and gelatin (1%) in the presence of glycerol (0.75%) as plasticizer (Rivero et al. 2009). The membranes showed a compact structure and good mechanical properties, and the WVP was independent of the thickness. The gelatin composites with chitosan and chitosan gel to form wound healing constituents were also prepared and showed smooth and standardized surface as revealed by SEM and X-ray measurements.

The gelatin composites were prepared from chitosan and chitosan gel to form wound therapeutic components and showed a smooth and uniform surface as shown in SEM and X-ray measurements (Nagahama et al. 2009). Chitosan-gelatin membranes were prepared by casting the aqueous solutions of both polymers at pH 4.0 and evaporating at 22 or 60 °C. Thermal, mechanical, and permeable water/gas of these films, plasticized with water or polyols, has been investigated (Arvanitoyannis et al. 1998). The results showed that the increase in the plasticizer content resulted in a significant reduction in elasticity and tensile strength with an increase in the percentage of elongation.

The result of glycerol on the mechanical and water barrier possessions of gelatin-chitosan membranes in addition to solubility of water was examined by Kolodziejska and Piotrowska (2007). The ultimate goal of their study was to plan biodegradable constituents with respectable mechanical properties, appropriate for filling several kinds of food through various acids and moisture content. Further study should be focused toward manipulating these mixtures with possible applications in food preservation.

Active composite chitosan-gelatin membranes were organized by molding the composite polymers of chitosan (2%) and gelatin (2% and 5%) with changed concentrations of glycerol and sorbitol (1%, 2%, 5%, and 10%, w/w) as plasticizers (Badawy et al. 2016). The results suggested that the incorporation of gelatin improved swelling, WVP, and the antioxidant action of the films. Films created from gelatin, gelatin-sodium caseinate, and gelatin-chitosan mixtures, with addition of the active ingredients (ethereal garlic oil, cinnamaldehyde, and α-tocopherol), were prepared by casting technique and characterized subsequently (Córdoba and Sobral 2017). Membranes prepared from gelatin-chitosan mixture have the lowest solubility and swelling and the highest hydrophobicity. Analysis of the microscopic structure of membranes obtained from SEM and FTIR displayed a respectable correlation among gelatin-chitosan mixtures and showed that this matrix allowed uniform distribution of active agents through the complex.

2.2.3 Chitosan-Alginate Membranes

Alginate is a linear anionic copolymer composed of (1→4)-linked-ß-D-mannuronate (Block M) and (1→4)-linked-α-L-guluronate (Block G) units. The structure is obtained from the brown seaweeds known as Phaeophyceae containing *Ascophyllum nodosum*, *Laminaria digitata*, *Laminaria hyperborea*, *Laminaria japonica*, and *Macrocystis pyrifera* through conduct with alkaline water solutions (Clark and Green 1936; Lee and Mooney 2012). The chemical structure is composed of successive G residues (GGGGGG), M residues (MMMMMM), and alternating M and G residues (GMGMGM) (Fig. 4). The molecular weight of commercially available alginates ranges from 32,000 to 400,000 g/mol. The factors of the Mark-Houwink relationship ($[\eta] = KM_v^a$) for sodium alginate in 0.1 M NaCl solution at 25 °C are $K = 2 \times 10^{-3}$ and $a = 0.97$, where $[\eta]$ is intrinsic viscosity (mL/g) and M_v is the viscosity-average molecular weight (g/mol) (Rinaudo 1992). The viscosity increases with decreasing pH and reaches a maximum at pH 3–3.5 once the carboxylate groups are protonated in the alginate molecule and form hydrogen bonds. Increasing of the molecular weight of alginate directed to increase the physical properties of the resultant gels. Conversely, high molecular weight alginate solution grows into very viscous, which is frequently unattractive in the treatment (LeRoux et al. 1999; Lee and Mooney 2012).

Alginate has established many applications in numerous fields due to its promising properties, such as low toxicity, biocompatibility, being inexpensive, and mild gelling by the addition of divalent cations such as chitosan (Lucinda-Silva et al. 2010; Mukhopadhyay et al. 2015; Rassu et al. 2016). Alginate is negatively charged while chitosan contains a positive charge equilibrium. Thus, the cationic chitosan molecules can form gels with the nontoxic polyvalent ions of the anionic alginate through ionic cross-linking mechanism. Related to alginate or chitosan gels alone, the polyelectrolyte complex has numerous improved physicochemical properties,

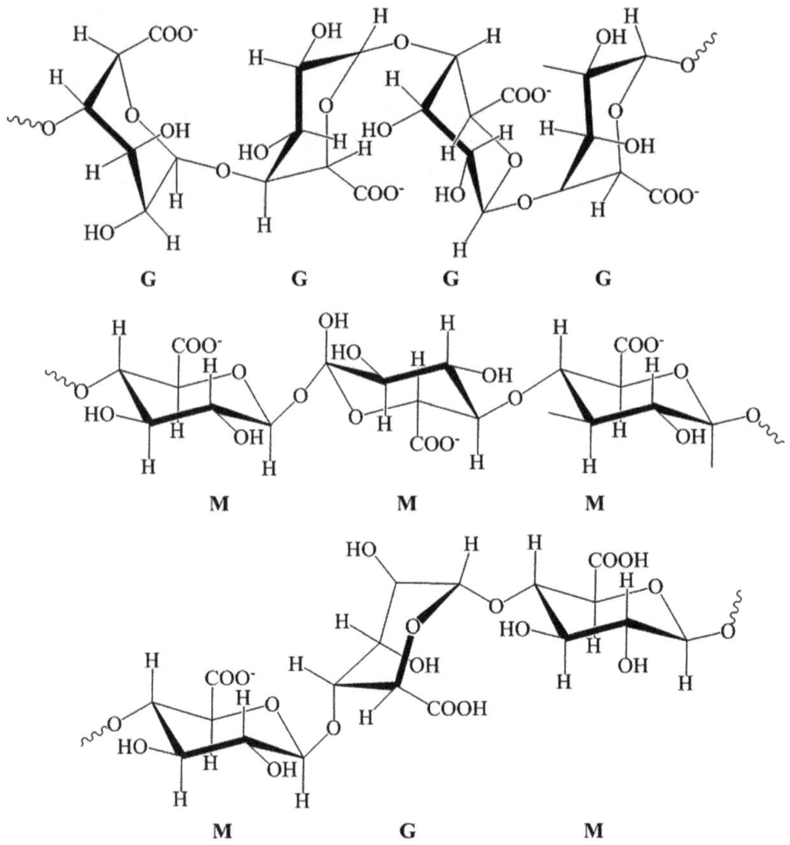

Fig. 4 Chemical structures of G-block, M-block, and irregular block in alginate

such as greater stability against pH changes and improved efficacy than controlled release systems (Khan and Ahmad 2013; Luo and Wang 2014; Bueno et al. 2016). Biologically active molecules can be released from the chitosan-alginate gels or membranes in a controlled manner, depending on the types of cross-linkers and cross-linking methods. Elastic, thin, translucent, new chitosan-alginate polyelectrolyte complex membranes cast from aqueous suspensions of chitosan-alginate coacervates with $CaCl_2$ were prepared and applied to the wound dressing (Wang et al. 2002). Chitosan-alginate blend possesses a worthy film-forming property and produces uniform, transparent, and water-soluble membranes. Membranes can serve as a sacrificial agent, where moisture is lost from the coating before the food dries dramatically. Coating with these membranes can also develop adhesion of the paste to the surface of food products (Yan et al. 2001; Yuan et al. 2007). Sodium alginate and chitosan put onto aminolyzed/charged polyethylene terephthalate to form a successively higher number of nanolayered film with a WVP of 0.85×10^{-11} g m $^{-1}$ s^{-1} Pa^{-1} and a rigidity increase of 0.245 GPa (Carneiro-da-Cunha et al. 2010).

Arzate-Vázquez and others have reported that microscopy techniques, including light scanning, environmental scanning, and atomic microscopies, are effective tools for quantitatively evaluating the exterior morphology of the membranes with chitosan alginate (Arzate-Vázquez et al. 2012). Fractal dimension and entropy were extra valuable for measuring the roughness and complexity of these films, and the higher values were for alginate-chitosan followed by alginate and chitosan films (Arzate-Vázquez et al. 2012). A strong complexation has been confirmed between the chitosan and alginate polymers in their films using titration calorimetry and viscosity studies, which affects the film swelling and the subsequent release of the active substances (Chandrasekar et al. 2016). The results showed that the films containing higher alginate fractions exhibited a significantly lower degree of swelling in the water.

2.3 *Cross-Linked Chitosan Membranes*

Cross-linking of chitosan has been performed by di-/polyfunctional reagents including ethylene glycol diglycidyl ether, epichlorohydrin, and glutaraldehyde (Fig. 5) (Baba et al. 1994; Shim and Ryu 1998; Varma et al. 2004; Tikhonov et al. 1996). Although cross-linking decreases the capacity of adsorption, it improves the chitosan resistance against alkali, acid, and chemicals (Inoue et al. 1993). The cross-linked chitosan is also very stable and retains its strength even in acidic and basic solutions.

Yisong et al. cast chitosan solutions inclosing glutaraldehyde, and the solvent was vaporized under IR light, where cross-linking happened. The membranes had a thickness of 45–50 μm, and the swelling was found to be declined with an increase in the concentration of the cross-linking agent (Yisong et al. 1990). Partially cross-linked chitosan membranes were prepared by adding 20 g of a cooled solution of the relevant quantity of the cross-linker glutaraldehyde or glyoxal (0.6, 0.3, 0.15, and 0.05 g) to chitosan solution (0.75%). The mixed solutions were immediately homogenized and were applied to a flat polycarbonate surface and dried at 60 °C (Clasen et al. 2006). A network structure change was achieved by simple construction of the chitosan with polyanion-sulfoethyl cellulose by introducing artificial pores in the microscale to the center of chitosan with silica particles or polyethylene glycol. The resulting network assemblies and morphologies of these exceptional membranes were described in detail and associated with the molecular parameters of the chitosan as DDA, molecular mass, and charge density (Clasen et al. 2006). The natural chitosan films were cross-linked heterogeneously by dipping in glutaraldehyde solution (0.75%) at 25 °C for 2 h (Vieira and Beppu 2006; Beppu et al. 2007). The quantity of protonable amino groups was measured for each membrane using potentiometric titration. From FTIR spectroscopy, the lost hydroxyl groups, the acetal ring, and the ether bond were formed for interaction between hydroxyl groups of PVA and formaldehyde (Yang et al. 2004). Controlled pore size of macroporous membranes of chitosan and good mechanical properties were prepared and cross-linked with ethylene glycol diglycidyl ether to increase their chemical constancy

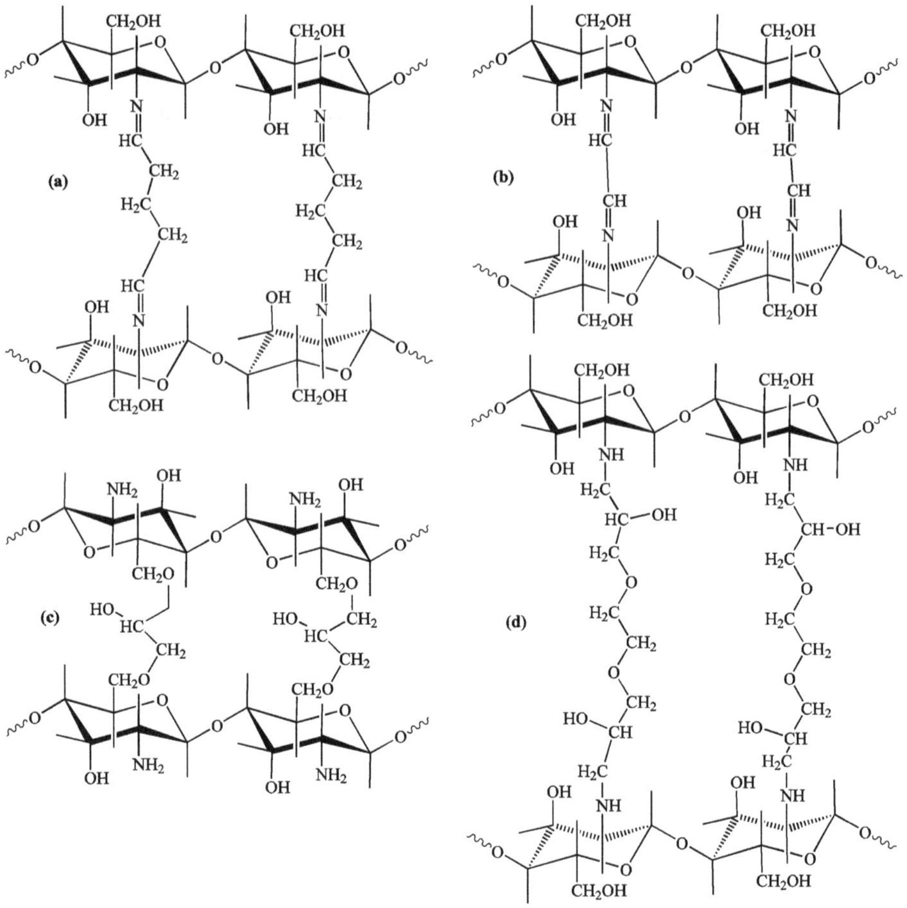

Fig. 5 Chemical structures of cross-linked chitosan-glutaraldehyde (**a**), chitosan-glyoxal (**b**), chitosan-epichlorohydrin (**c**), and chitosan-ethylene glycol diglycidyl ether (**d**)

(Zeng and Ruckenstein 1998). Chitosan cross-linking was approved using γ-glycidoxy propyl trimethoxysilane with in situ formation of chitosan-silica hybrid membranes. The hybrid resources of chitosan and silica formed had improved thermal stability and a low swelling degree in the water (Liu et al. 2004). For improving the physicochemical performance of chitosan-gelatin membranes, cross-linked films were prepared with increasing concentrations of glutaraldehyde. The effect of the concentration of the cross-linking agent on the physicochemical properties including water solubility, hydrophilicity, swelling, moisture content, and transparency was studied (Ahmed et al. 2016). The membranes were found to be thin, very translucent, uniform, and homogeneous and displayed a condensed structure indicating a good compatibility among constituents, which could cooperate by strong hydrogen bonding as was established by FTIR spectroscopy.

Recently, chitosan-graphene membranes have been shown to present excellent physical, thermal, and mechanical properties. These graphene properties, if incorporated in the chitosan-derived films, will potentially overcome the poor mechanical and thermal properties of the chitosan polymer (Han et al. 2011; Shao et al. 2013; Kamal et al. 2017). Furthermore, graphene-based nanomaterials have been established to existent very worthy antimicrobial properties. Chitosan-graphene oxide films will improve their properties as compared to pristine chitosan, because of the good dispersion of the nanofiller (Santos et al. 2012; Akhavan and Ghaderi 2010; Justin and Chen 2014; Ahmed et al. 2017).

Previous studies have combined graphene oxide nanosheets and chitosan at low concentrations to produce strong, biodegradable, and biocompatible chitosan-graphene nanocomposites (Han et al. 2011; Yang et al. 2010). These studies have principally motivated on the mechanical properties of these nanocomposites. Other studies have focused only on the antimicrobial properties of chitosan-graphene oxide nanocomposites and have shown that the antimicrobial property may vary depending on the type of chitosan used (Yadav et al. 2013; Lim et al. 2012). Graphene oxide-chitosan nanocomposite films were recently synthesized by cross-linking graphene oxide with chitosan at high temperatures (120 °C), and their potential application in food packaging was investigated (Grande et al. 2017). The mechanical properties were improved, and the tensile strength at break increased from 22.7 to 6471.6 MPa when 0.1% graphene oxide was added. Thermal studies have shown an increase in endothermic peak values with the increase in graphene oxide content, which is also related to a proliferation in the degree of cross-linking of the films and interaction with water.

2.4 Chitosan Nanocomposite Membranes

Several studies have been used to improve chitosan products for naturally food packaging. The use of these biopolymers and similar was restricted owing to their generally poor mechanical and barrier characteristics, which are able to improve by adding enriched compounds (fillers). Greatest strengthened resources display weak matrix fillings, which tend to develop with reduced load sizes. The use of nano-dimensional fillings at least produces nanoparticles (Rhim et al. 2013a; Sorrentino et al. 2007; Arora and Padua 2010). The nanocomposites display good barrier properties and mechanical strength and developed heat resistance compared to pure polymers and conventional compounds (Sorrentino et al. 2007; Ray and Okamoto 2003; Ray et al. 2006; Okamoto 2003). When used in food packaging, nanocomposites must withstand the thermal storage, transport, and the stress of processing of food (Ray and Okamoto 2003; Thostenson et al. 2005). Besides, because of their developed mechanical properties, the nanocomposites can allow descending gauging, thus decreasing raw materials.

In recent years, biopolymer nanocomposite membranes based on chitosan have received substantial interest because of their greater mechanical and thermal properties, as compared with the polymer itself (Kumar et al. 2003; De Azeredo 2009; Rhim et al. 2013a). The greatest common methods used to explore nanocomposite structures of chitosan are both wide angle (WAXS) and small angle (SAXS), X-ray diffraction (XRD), transmission electron microscopy (TEM), SEM, IR, and atomic force microscopy (AFM) (Karger-Kocsis and Zhang 2005; Ray and Okamoto 2003). However, the TEM method is the desired technique to examine dispersions of nanoparticle in such formulations.

Chitosan-clay nanocomposite membranes are a class of mixed materials consisting of bio-matrices and fillers of nanoparticles of the nanoscale clays; montmorillonite (a hydrated aluminum silicate-layered clay) is of particular interest and has been broadly examined (Uyama et al. 2003; McGlashan and Halley 2003; Darder et al. 2003, 2005). Yamaguchi and others prepared a chitosan/hydroxyapatite nanocomposite membranes by coprecipitation method. According to the TEM interpretations, the chitosan/hydroxyapatite crystallites in the composites formed elliptic aggregations 230 nm in length and 50 nm in width (Yamaguchi et al. 2001). Xu and co-authors reported that incorporation of chitosan chains into interlayers of silicate, using solution intercalation, improved the mechanical properties of chitosan-layered silicate nanocomposite membranes (Xu et al. 2006). Nanocrystalline cellulose (NCC)-reinforced chitosan-based biodegradable films were prepared using casting technique, and the structure was examined by SEM (Khan et al. 2012). It was found that the NCC incorporation similarly seriously enhanced the barrier properties and the WVP was reduced by 27% for the optimal content of 5% NCC.

The nanoparticles of chitosan can be gained by ionic gelling, where positively charged chitosan forms electrostatic connections with polyanions used as cross-linking agents such as tripolyphosphate (TPP) (López-León et al. 2005; Caro et al. 2016; Zhan et al. 2017; Aguirre-Chagala et al. 2017). de Moura et al. prepared chitosan-TPP nanoparticles and combined them into hydroxypropyl methylcellulose (HPMC) films that significantly developed barrier and mechanical properties of films (de Moura et al. 2009).

Chitosan silver nanocomposite (chitosan/AgNP) membranes are always looked for bioactive biomaterials (Vimala et al. 2010; Murugadoss and Chattopadhyay 2007; Hosseini et al. 2016). Principally, two different types of technique have been used for the preparation of composite films of chitosan/AgNP, that is to say, the direct use of silver nanoparticles (Rhim et al. 2006, 2013b; Djerahov et al. 2016; Hosseini et al. 2016) and the use of reduction of silver ion (Ag^+) from $AgNO_3$ (Sanpui et al. 2008; Ahmad et al. 2009; Murugadoss and Chattopadhyay 2007). Most composites of chitosan/AgNP were prepared using the silver ion reduction method from $AgNO_3$. Silver nanoparticles are classically formed by reduction of silver salt precursors using chemical reducing agents (hydrazine, sodium borohydride, triethanolamine, and formamide); however, these chemicals pose hazards to mammals. However, physical techniques have been recommended for the reduction of silver salts by UV, γ-ray and microwave irradiation, heat treatment, photochemical process, and sonochemical process as green technologies (Yoksan and Chirachanchai 2010). Rhim and the co-authors prepared composite films of chitosan/AgNP by direct addition of Ag nanoparticles and chitosan/Ag (Ag-Ion) adsorbed zeolite in chitosan-forming solution and tested their antimicrobial activity (Rhim et al. 2006). The chitosan and silver nanocomposites inclosing 20% silver were produced by a simple chemical reduction technique and characterized by FTIR, X-ray diffraction, and SEM. The spherical morphology of the silver nanoparticles has been confirmed from the SEM image (Govindan et al. 2012). Regiel and co-authors prepared and characterized chitosan-silver nanocomposite films with evaluation of their antibacterial activity against *Staphylococcus aureus* (Regiel et al. 2012). The films were formed by solution casting using chitosan as a stabilizing and reducing agent for the in situ preparation of inserted silver nanoparticles. The authors described that the highest results were acquired with high-DDA chitosan, where fast decrease was preferred, leading to smaller nanoparticles, and a satisfactorily high viscosity of the polymer avoids nanoparticles resulting from unwanted agglomeration.

Youssef et al. effectively produced zinc oxide (ZnO) nanoparticle using hydrothermal technique in addition to silver (Ag) nanoparticle which was directly prepared through membranes of chitosan nanocomposite preparation (Youssef et al. 2015). The elasticity of the prepared films was enhanced through mixing hydroxyethyl cellulose with chitosan. The loading of the films with nanoparticles of silver and zinc oxide (Ag-NPs and ZnO) was carried out during the preparation of the films under optimal conditions. FT-IR, XRD, SEM, EDAX, TEM, SEM, XRD, and FT-IR examined the prepared nanocomposite film characterizations. Nanocomposite films of chitosan/ZnO and chitosan/Ag exhibited good antimicrobial activity against *Bacillus cereus*, *Escherichia coli*, *Listeria monocyte*, *Salmonella typhimurium*, and *S. aureus*, and the authors proposed that these films could be used in packaging applications. Olaniyan and colleagues synthesized and characterized chitosan-silver nanocomposite film for food packaging using air drying technique and co-mixing chemical technique (Olaniyan et al. 2016). The integrity of the nanohybrid morphology was established using X-ray diffraction, TEM, and UV-visible spectrophotometer. TEM analysis exposed spherical cap-shaped morphology without agglomeration, and nanoparticle identical size distribution was in the range of 20 nm.

The optimal preparation parameters of chitosan-silver nanoparticle composites through a simple method of green chemical reduction with different preparation factors and their antibacterial activity were studied (Sherif et al. 2017). Chitosan has been used to reduce silver nitrate and stabilize silver nanoparticles in the medium. For this reason, UV, FTIR, and transmission electron microscopy (TEM) techniques have been used in the study of the molecular and morphological properties of the resulting composites. In addition, a simple and effective method for producing highly fluorescent chitosan-silver films has recently been reported (Wang and Gao 2017). The reduction of silver was obtained by $NaBH_4$ and involves less time and particularly mild experimental conditions. The structure, surface, and fluorescence of nanoscale film properties were studied by SEM, FTIR, and differential scanning calorimetry.

3 Chitosan Membranes in Food Packaging and Preservation

Over the last few decades, it has become increasingly interesting to develop and use bioactive chitosan membranes and films in the food engineering owing to their biocompatibility, high biodegradability, and specific physicochemical properties. This is principally owing to the fact that chitosan displays high antimicrobial action toward pathogenic and deteriorating microorganisms and in order to develop food preservation and reduce the use of chemical preservatives (Sudarshan et al. 1992; No et al. 2002; Sagoo et al. 2002; Kong et al. 2010; Goy et al. 2009; Dutta et al. 2009; Gómez-Estaca et al. 2010). Furthermost foods are a combination of different composites, such as proteins, fats, carbohydrates, minerals, salts, vitamins, etc., and many of which can interact with the chitosan leading to loss or enhancement of antimicrobial activity.

In food applications, semipermeable chitosan films reduce absorption of water or the loss of moisture through the food matrix, loss of flavor, oxygen entry into food, and transport of solutes, thus prolonging the shelf life of products (Pereda et al. 2011; Zhang et al. 2011; Elsabee and Abdou 2013; El Ghaouth et al. 1991; No et al. 2007; Rojas-Graü et al. 2009; Dutta et al. 2009). Up to now, biodegradable and bio-edible membranes are not intended to completely substitute old-style packaging materials (Schou et al. 2005). Conversely, the usage of chitosan-based active membranes as a material of packaging is still one of the greatest hopeful ways to maintain effective food quality. Chitosan coatings are intended to slow the release of antimicrobial agents from the surface, which would require a smaller amount of antimicrobial agents (Badawy and Rabea 2009; Devlieghere et al. 2004; No et al. 2007; Ouattara et al. 2000; Coma et al. 2002; van den Broek et al. 2015). In addition, these films can also use as carriers releasing of essential oils that are affected by vapor contact with pathogens (Avila-Sosa et al. 2012; Zivanovic et al. 2005; Acevedo-Fani et al. 2015). One of the most observable active biofilms is the chitosan-based ones, which are linked to various substances such as animal and plant proteins, polysaccharides, and antimicrobial peptides, for example, divergicin (a new bacteriocin from *Carnobacterium divergens*) and nisin (Tahiri et al. 2004, 2009).

The antimicrobial action of chitosan diverges considerably with the nature of chitosan, target organism, and medium conditions (ionic strength, pH, and soluble substances), thus enhancing the reaction of active amine groups (Younes et al. 2014; Tan et al. 2013). However, although the information available in the literature on the antimicrobial potency of chitosan varies somewhat and sometimes has contradictory results, it is commonly agreed that molds and yeasts are the most sensitive to chitosan, followed by Gram-positive and Gram-negative bacteria. Because of its exceptional cationic nature, chitosan has the ability to hit to several food components like proteins, alginates, pectins, and polyelectrolytes such as polyphosphate and positively charged ionic species through donor/acceptor interaction. All of the above features might decrease the antimicrobial effect of chitosan (Kubota and Kikuchi 1998). Chitosan was compared with further biomolecule-based active films used as packaging materials, and the described results exhibited that chitosan has more chelation owing to its divalent minerals and antibacterial activity (Chen et al. 2002; Zhang et al. 2017). Membranes having an antimicrobial action which were prepared by combining numerous essential oils and organic acids in a chitosan matrix and preventing the capacity of these bio-sourced membranes to prevent the growth of native bacteria (enterobacteria and lactic acid bacteria) or inoculated bacteria such as *Serratia liquefaciens* and *Lactobacillus sakei* on surfaces of vacuum-packed and cured meat products were examined. Propionic or acetic acid release was primarily rapid, when the ionic concentration gradient among the inner of the polymer matrix and the exterior environment was great and then diminished as acid release developed. Simultaneously, it was indicated that the antimicrobial action of the biological membranes studied did not affect the activity and growth of lactic acid bacteria, though the growth of *Enterobacteriaceae* and *S. liquefaciens* was delayed or totally inhibited afterward 21 days of storage at 4 °C. Strongest inhibition was perceived on surfaces with minor water activity values (bologna) with slower acid release, and films containing cinnamaldehyde, owing to its superior antimicrobial action in these circumstances (Quintavalla and Vicini 2002). The chitosan/nylon-6 mixed membranes with Ag^+ had remarkable antibacterial action against Gram-negative and Gram-positive bacteria, and the antibacterial property of the mixed membranes was diminished with increasing nylon-6 content, which may be owing to the corresponding reduction in the content of chitosan and Ag^+ in the mixture (Ma et al. 2008). In another study, the effects of low molecular weight chitosan coatings (94% DDA) to control Gram-negative bacteria were very sensitive to the applied chitosan, whereas the sensitivity of Gram-positive bacteria is greatly variable (Devlieghere et al. 2004). Membranes based on yam starch (4%), chitosan (1%, 3%, and 5%), and glycerol (2%) have been prepared with good elasticity and exhibited a bactericidal effect against *S. enteritidis* with the maximum efficiency of the films of 5% chitosan (Durango et al. 2006). Dissimilar collagen/chitosan weight ratios were used to immobilize numerous acrylic acid grafts on nonwoven polypropylene fabric, to study swelling and antibacterial properties. The increase of chitosan in the blend with immobilization at the same pH, the diffusion coefficients, and water uptake were diminished. Furthermore, the antibacterial activity considerably increased with increasing quantity of chitosan in the blend (Wang and Chen 2005). The starch/chitosan blend film was also prepared using the compression molding method in a physical gel state, and the antibacterial action was encouraged even when the chitosan was 5%, owing to the degradation of chitosan by radiation (Zhai et al. 2004). There is another review on chitosan-based antimicrobial films used for food packaging applications, which also summarizes formation and antimicrobial activity of chitosan-based films used in food preservation. There are additional reviews on the antimicrobial potency of chitosan films in food packaging applications (Tripathi et al. 2008). Chitosan films inclosing 0%, 2.5%, 5%, 10%, and 20% propolis extract were established as active food packaging and considerably inhibited *P. aeruginosa*, *S. enteritidis*, *E. coli*, and *S. aureus* on contact surface below the film discs. The tensile strength of the film, the total phenol content, the elongation, and the antioxidant activity improved even though oxygen permeability and the WVP reduced with increasing propolis concentration (Siripatrawan and Vitchayakitti 2016).

Chitosan nanocomposite film prepared with organic clay (Cloisite 30B) has a strong antimicrobial activity against food poisoning bacteria (Gram-positive bacteria), suggesting that the antimicrobial activity was attributable to the quaternary ammonium groups in the reformed films (Rhim et al. 2006, 2013a). Later, Hong and Rhim demonstrated that the same clay of Cloisite 30B has a strong antibacterial activity to Gram-negative and Gram-positive bacteria, produced by the quaternary ammonium groups (Hong and Rhim 2008). These authors decided that the delayed biodegradability of this nanocomposite which is earlier detected by Lee and coworkers (2002) was accountable to the antimicrobial activity as well as the quaternary ammonium groups. Consequently, the biodegradability of bio-nanocomposites is able to use for the extension of biodegradable nanocomposite packaging materials with controlled biodegradation purpose (Souza and Fernando 2016). However, the good constancy and durability of bio-nanocomposite packaging materials must be preserved during their valuable shelf life to fulfill their packaging purposes. It is interesting to discriminate that the nanoparticles have two contrary effects on polymeric nanocomposites, i.e., stabilization and degradation as a purpose of treatment and eco-friendly conditions (Kumar et al. 2009). Hybrid bio-nanocomposites, for example, layered silicates incorporated into a polymeric medium, increase stability, and their biodegradability can be refined via variable quantity of solvents or via combining different types of organoclay-modified surfaces with diverse kinds of surfactants (Sozer and Kokini 2009; Sorrentino et al. 2007; Azeredo et al. 2017). Such advanced criteria of nanoparticles are very interesting and significant in the area of packaging industry based on the finale practice. The progress of antimicrobial activity of nanocomposite films based on biopolymers such as chitosan with AgNPs has extensively been tried to investigate the promising activity of silver ions. The exact mechanism of antimicrobial potency of AgNPs is not obviously recognized and is the topic of discussion. There are, however, many principles about the mechanism of action of AgNPs. For example, silver ions can be predicted to stick to the negatively charged wall of the bacterial cell loaded with cytotoxicity, alter the permeability of the cell wall, and induce protein clotting, cell lysis, and death. There is a "pit" pattern on the cell surface, and there is an accumulation of nanoparticles on the cell surface (Sondi and Salopek-Sondi 2004). The creation of free radicals by AgNPs can be considered additional mechanism by which cells die. The action of the silver ion is also associated to its capability to alter DNA duplication mechanisms and to induce irregularities in cytoplasmic content, membrane, size, and outer layers of the microbial cells (Kim et al. 2007; Prabhu and Poulose 2012; Rai et al. 2009). In addition, it has also been suggested that nanoparticles are able to release silver ions, which can interact with thiol groups of many vital enzymes and deactivate them (Feng et al. 2000; Matsumura et al. 2003).

Most data of antimicrobial action indicated that chitosan loaded with AgNPs and others displayed excellent activity, which can be used as packaging materials in food industry. For example, Ag-adsorbed chitosan and chitosan/zeolite composites showed strong action against bacteria of Gram-negative and Gram-positive (Rhim et al. 2006). Yoksan and Chirachanchai found that silver-chitosan-starch membranes improved the inhibitory action against *B. cereus*, *E. coli*, and *S. aureus*, suggesting that AgNP-loaded films can be used as antimicrobial food packaging resources (Yoksan and Chirachanchai 2010; Sanpui et al. 2008). Three-component films of chitosan/Ag/ZnO were prepared by a single-phase conversion technique, in which the AgNPs were prepared and formed in a single phase (Li et al. 2010). AgNPs and ZnO were homogeneously dispersed in chitosan, and the composite film showed excellent antimicrobial activities against a wide range of bacteria, fungi, and yeast. Vimala and others have explored the in situ manufacture of chitosan-poly (lactic acid) (PLA)-silver nanocomposite films as antimicrobial packaging materials (Vimala et al. 2011). The films proved noteworthy properties against *E. coli*, *Pseudomonas*, *Staphylococcus*, *Micrococcus*, *Candida albicans*, and *P. aeruginosa*. In addition, the authors also improved their effectiveness by curcumin encapsulation in these nanocomposite films, and the results presented high inhibition of *E. coli* growth related to curcumin and chitosan-PVA-silver nanocomposite film only. Nanocomposite films consisted of poly(lactic acid) (PLA) incorporated with thymol (6% and 8%) and Ag-NPs (1%) were prepared and considered as degradable and promising active packaging materials with low influence on the environment (Ramos et al. 2014, 2016, 2017). The addition of thymol and AgNPs improved the thermal, optical, and barrier properties of PLA, especially WVP, keeping the oxygen transmission rate values unchanged.

Several workers have reported effectiveness of chitosan films or membranes on storage stability of meat and its products. For example, Youn et al. distinguished that the shelf life of spicy beef treated with chitosan was strangely enhanced by decreasing the bacterial cell counts and preventing lipid oxidation throughout storage (Youn et al. 2004). Wu and co-authors in the study noted that wrapping with film of chitosan, wheat gluten, or soy protein was not effective in controlling the fat oxidation of precooked beef patties (Wu et al. 2000). The authors explained that a higher oxygen permeability might contribute to the higher thiobarbituric acid-reactive ingredients. In another study, plasticizing action of water molecules may have altered the barrier properties of edible membranes, resulting in higher oxygen permeability or even loss of film integrity (Caner et al. 1998).

Abdallah and colleagues applied chitosan on the best widespread outdated dried meat foodstuffs (pastirma) in demand to increase its sensory, physicochemical, and microbiological features (Abdallah et al. 2017). The results showed improved sensory properties, reduced moisture loss, and an antioxidant effect with lower shear strength values in chitosan-coated samples. Cui and co-authors reported that the chitosan film incorporated with a liposome-encapsulated phage displayed high antibacterial action against *E. coli* O157: H7 without affecting the sensory properties of beef (Cui et al. 2017). The chitosan monomethyl fumaric acid derivative considerably reduced the total number of viable bacteria, enterobacteria, lactic bacteria, psychrotrophic bacteria, and mold yeasts on beef during refrigeration storage, extending the shelf life of beef of about 8 days (Khan et al. 2017).

Muscle foodstuffs have little oxidation stability and are highly susceptible to ripening throughout manufacture and storage. Several studies have shown that lipid oxidation in meat and meat products can be minimized using antioxidants (Gray et al. 1996; Nissen et al. 2004; Barden and Decker 2016; Ladikos and Lougovois 1990). However, chitosan does not have an antioxidant activity. Taste is a significant reflection that can limit the use of some antioxidants in meat and meat products. Spices and herbs have been used in numerous kitchens to provide food, flavor, and tastiness. Various studies have shown the potential of antimicrobials and oxidation of spices and herbs such as basil, garlic, cloves, thyme, rosemary, coriander, ginger, mustard, and pepper (Sebranek et al. 2005; Shahidi 2012; Tipsrisukond et al. 1998). Atarés and Chiralt investigated the applications of essential oils in biodegradable films and coatings for the wrapping of active foods and concluded that these products can impart antioxidant and/or antimicrobial possessions to the films and that the components of the oil and the exact interactions with the polymer (Atarés and Chiralt 2016).

A mixture of mint extract (*Mentha spicata*) and chitosan was used as new preservatives of meat and meat products. The mint extract has a good antioxidant action, but a weak antimicrobial effect, while chitosan alone shows an antioxidant activity and high antimicrobial action. Earlier studies showed that the mint extract had a very good antioxidation possible equivalent to that of the synthetic antioxidant such as butylated hydroxytoluene (Kanatt et al. 2007, 2008; Zheng and Wang 2001). Kanatt and co-authors indicated that a 0.05% mixture of chitosan and mint was the minimum inhibitory concentration and was further active against Gram-positive bacteria of meat and meat products (Kanatt et al. 2008).

The combination of essential oils into chitosan films has the possibility to increase the microbiological safety of different foods including meat. Chitosan edible membranes with and without essential oils of basil or thyme have effectively protected meat from oxidation and been very effective in controlling microbial growth in minced meat (Bonilla et al. 2014). Ready-to-eat meat samples were directly treated with solutions including chitosan, lauric arginate ester (LAE), and nisin and were evaluated against *Listeria innocua* (Guo et al. 2014). The authors reported that antimicrobial coatings containing 1.94 mg/cm^2 of chitosan and 0.388 mg/cm^2 of arginate ester reduced *L. innocua* by approximately 4.5 log CFU/cm^2. Edible films prepared from three molecular weight of chitosan containing carvacrol exhibited antimicrobial efficacy against bacteria *Pseudomonas fragi*, *Shewanella putrefaciens*, and *Aeromonas hydrophila*, common spoilers of fish and seafood products (Fernández-Pan et al. 2015). Thus, in order to slow down the release rate of carvacrol, chitosan was proposed with low molecular weight and technical applications targeted for low-fat low-content products. An active film-forming formulation containing essential oil of *Origanum virens* was applied to the surface of two outmoded Portuguese sausages (*paínhos* and *alheiras*) throughout manufacturing with a positive overall acceptance of sensory analysis (Catarino et al. 2017). The coating caused slight variations in moisture content and texture profile with a strong inhibition of the overall microbial load, which extended the shelf life by about 20 and 15 days, respectively, for *paínhos* and *alheiras*.

4 Concluding Remarks

The use of chitosan in packaging and preservation of food is confirmed by the increasing request for natural additives and harmless with functional properties and increasing environmental problems. Current and future applications of chitosan in the food industry are likely to come from the medical and pharmaceutical fields, where it is widely used due to its antioxidant and antimicrobial properties. Chitosan molecule can be incorporated as an antimicrobial/antioxidant agent in recycled ingredients such as multilayer plastic packaging materials in which each polymer layer has a specific function. In addition, membranes or films of chitosan can be used to reduce moisture loss or water absorption by food matrix, oxygen permeability in food, loss of flavor, and transfer of solvents. The information presented in this chapter has shown the positive effects of pure chitosan or in combination with other polymers, essential oils, and other active products. Studies have shown that during storage, chitosan membranes provide protection against contamination and microbial damage to various foods and can extend the shelf life of foods.

Conflict of Interests The authors state that there is no conflict of interests concerning the publication of this chapter.

References

Abdallah MRS, Mohmaed MA, Mohamed HMH, Emara MMT (2017) Improving the sensory, physicochemical and microbiological quality of pastirma (a traditional dry cured meat product) using chitosan coating. LWT-Food Sci Technol 86:247–253

Acevedo-Fani A, Salvia-Trujillo L, Rojas-Graü MA, Martín-Belloso O (2015) Edible films from essential-oil-loaded nanoemulsions: physicochemical characterization and antimicrobial properties. Food Hydrocoll 47:168–177

Aguirre-Chagala YE, Pavón-Pérez LB, Altuzar V, Domínguez-Chávez JG, Muñoz-Aguirre S, Mendoza-Barrera C (2017) Comparative study of one-step cross-linked electrospun chitosan-based membranes. J Nanomater 2017: 14 pages

Ahmad MB, Shameli K, Darroudi M, Yunus WMZW, Ibrahim NA (2009) Synthesis and characterization of silver/clay/chitosan bionanocomposites by UV-irradiation method. Am J Appl Sci 6(12):2030

Ahmed S, Ahmad M, Ikram S (2016) Physicochemical characterization of gluteraldehyde crosslinked chitosan-gelatin films. Mater Focus 5(2):165–170

Ahmed J, Mulla M, Arfat YA (2017) Mechanical, thermal, structural and barrier properties of crab shell chitosan/graphene oxide composite films. Food Hydrocoll 71:141–148

Akhavan O, Ghaderi E (2010) Toxicity of graphene and graphene oxide nanowalls against bacteria. ACS Nano 4(10):5731–5736

Alexandre EMC, Lourenço RV, Bittante AMQB, Moraes ICF, do Amaral Sobral PJ (2016) Gelatin-based films reinforced with montmorillonite and activated with nanoemulsion of ginger essential oil for food packaging applications. Food Packag Shelf Life 10:87–96

Alishahi A, Aïder M (2012) Applications of chitosan in the seafood industry and aquaculture: a review. Food Bioprocess Technol 5(3):817–830

Appendini P, Hotchkiss JH (2002) Review of antimicrobial food packaging. Innov Food Sci Emerg Technol 3(2):113–126
Arora A, Padua G (2010) Nanocomposites in food packaging. J Food Sci 75(1):R43–R49
Arvanitoyannis I, Psomiadou E, Nakayama A, Aiba S, Yamamoto N (1997) Edible films made from gelatin, soluble starch and polyols, part 3. Food Chem 60(4):593–604
Arvanitoyannis IS, Nakayama A, Aiba S-i (1998) Chitosan and gelatin based edible films: state diagrams, mechanical and permeation properties. Carbohydr Polym 37(4):371–382
Arzate-Vázquez I, Chanona-Pérez JJ, Calderón-Domínguez G, Terres-Rojas E, Garibay-Febles V, Martínez-Rivas A, Gutiérrez-López GF (2012) Microstructural characterization of chitosan and alginate films by microscopy techniques and texture image analysis. Carbohydr Polym 87 (1):289–299
Atarés L, Chiralt A (2016) Essential oils as additives in biodegradable films and coatings for active food packaging. Trends Food Sci Technol 48:51–62
Auerbach B (1975) The structure of chitin and chitosan. In: OCEAN 75 Conference, IEEE, pp 454–456
Auerbach B (1978) Film-forming capability of chitosan. In: Proceedings of the First International Conference on Chitin and Chitosan. MIT, Cambridge, MA, pp 199–209
Austin PR (1982) Renatured chitosan and process of making same. Google Patents
Avila-Sosa R, Palou E, Munguía MTJ, Nevárez-Moorillón GV, Cruz ARN, López-Malo A (2012) Antifungal activity by vapor contact of essential oils added to amaranth, chitosan, or starch edible films. Int J Food Microbiol 153(1):66–72
Azeredo HM, Rosa MF, Mattoso LHC (2017) Nanocellulose in bio-based food packaging applications. Ind Crop Prod 97:664–671
Baba Y, Hirakawa H, Kawano Y (1994) Selective adsorption of precious metals on sulfur-containing chitosan derivatives. Chem Lett 1:117–120
Badawy MEI, Rabea EI (2009) Potential of the biopolymer chitosan with different molecular weights to control postharvest gray mold of tomato fruit. Postharvest Biol Technol 51 (1):110–117
Badawy MEI, Rabea EI, El-Nouby MA (2016) Preparation, physicochemical characterizations, and the antioxidant activity of the biopolymer films based on modified chitosan with starch, gelatin, and plasticizers. J Polym Mater 33(1):17
Badawy MEI, Rabea EI, AM El-Nouby M, Ismail RIA, Taktak NEM (2017) Strawberry shelf life, composition, and enzymes activity in response to edible chitosan coatings. Int J Fruit Sci 17 (2):117–136
Baldwin EA (1994) Edible coatings for fresh fruits and vegetables: past, present, and future. In: Krochta JM, Baldwin EA, Nisperos-Carrido M (eds) Edible coatings and films to improve food quality, vol 1. CRC Press, Washington, DC, p 25
Baldwin EA, Hagenmaier R, Bai J (2011) Edible coatings and films to improve food quality. CRC Press, Boca Raton
Barbosa-Cánovas G, Hendrickx M, Knorr D (2000) Innovative food science and emerging technologies. Innov Food Sci Emerg Technol 1(1):1
Barden L, Decker EA (2016) Lipid oxidation in low-moisture food: a review. Crit Rev Food Sci Nutr 56(15):2467–2482
Beppu M, Vieira R, Aimoli C, Santana C (2007) Crosslinking of chitosan membranes using glutaraldehyde: effect on ion permeability and water absorption. J Membr Sci 301(1):126–130
Blair HS, Guthrie J, Law TK, Turkington P (1987) Chitosan and modified chitosan membranes I. Preparation and characterisation. J Appl Polym Sci 33(2):641–656
Boesel LF (2015) Effect of plasticizers on the barrier and mechanical properties of biomimetic composites of chitosan and clay. Carbohydr Polym 115:356–363
Bonilla J, Atarés L, Vargas M, Chiralt A (2012) Effect of essential oils and homogenization conditions on properties of chitosan-based films. Food Hydrocoll 26(1):9–16
Bonilla J, Vargas M, Atarés L, Chiralt A (2014) Effect of chitosan essential oil films on the storage-keeping quality of pork meat products. Food Bioprocess Technol 7(8):2443–2450

Bourtoom T, Chinnan MS (2008) Preparation and properties of rice starch–chitosan blend biodegradable film. LWT-Food Sci Technol 41(9):1633–1641
Brockgreitens J, Abbas A (2016) Responsive food packaging: recent progress and technological prospects. Compr Rev Food Sci Food Saf 15(1):3–15
Bueno CZ, Moraes ÂM, de Sousa HC, Braga MEM (2016) Effects of supercritical carbon dioxide processing on the properties of chitosan–alginate membranes. J Supercrit Fluids 112:128–135
Butler B, Vergano P, Testin R, Bunn J, Wiles J (1996) Mechanical and barrier properties of edible chitosan films as affected by composition and storage. J Food Sci 61(5):953–956
Caner C, Vergano P, Wiles J (1998) Chitosan film mechanical and permeation properties as affected by acid, plasticizer, and storage. J Food Sci 63(6):1049–1053
Cao N, Fu Y, He J (2007) Mechanical properties of gelatin films cross-linked, respectively, by ferulic acid and tannin acid. Food Hydrocoll 21(4):575–584
Carneiro-da-Cunha MG, Cerqueira MA, Souza BW, Carvalho S, Quintas MA, Teixeira JA, Vicente AA (2010) Physical and thermal properties of a chitosan/alginate nanolayered PET film. Carbohydr Polym 82(1):153–159
Caro N, Medina E, Díaz-Dosque M, López L, Abugoch L, Tapia C (2016) Novel active packaging based on films of chitosan and chitosan/quinoa protein printed with chitosan-tripolyphosphate-thymol nanoparticles via thermal ink-jet printing. Food Hydrocoll 52:520–532
Catarino MD, Alves-Silva JM, Fernandes RP, Gonçalves MJ, Salgueiro LR, Henriques MF, Cardoso SM (2017) Development and performance of whey protein active coatings with Origanum virens essential oils in the quality and shelf life improvement of processed meat products. Food Control 80:273–280
Cha DS, Chinnan MS (2004) Biopolymer-based antimicrobial packaging: a review. Crit Rev Food Sci Nutr 44(4):223–237
Chandrasekar V, Coupland JN, Anantheswaran RC (2016) Release kinetics of nisin from chitosan-alginate complex films. J Food Sci 81(10):E2503–E2510
Chen X-G, Zheng L, Wang Z, Lee C-Y, Park H-J (2002) Molecular affinity and permeability of different molecular weight chitosan membranes. J Agric Food Chem 50(21):5915–5918
Clark DE, Green HC (1936) Alginic acid and process of making same. Google Patents
Clasen C, Wilhelms T, Kulicke W-M (2006) Formation and characterization of chitosan membranes. Biomacromolecules 7(11):3210–3222
Coma V, Martial-Gros A, Garreau S, Copinet A, Salin F, Deschamps A (2002) Edible antimicrobial films based on chitosan matrix. J Food Sci Chicago 67(3):1162–1169
Conte A, Cappelletti G, Nicoletti G, Russo C, Del Nobile M (2015) Environmental implications of food loss probability in packaging design. Food Res Int 78:11–17
Córdoba LJP, Sobral PJ (2017) Physical and antioxidant properties of films based on gelatin, gelatin-chitosan or gelatin-sodium caseinate blends loaded with nanoemulsified active compounds. J Food Eng 213:47–53
Cui H, Yuan L, Lin L (2017) Novel chitosan film embedded with liposome-encapsulated phage for biocontrol of *Escherichia coli* O157: H7 in beef. Carbohydr Polym 177:156–164
Darder M, Colilla M, Ruiz-Hitzky E (2003) Biopolymer–clay nanocomposites based on chitosan intercalated in montmorillonite. Chem Mater 15(20):3774–3780
Darder M, Colilla M, Ruiz-Hitzky E (2005) Chitosan–clay nanocomposites: application as electrochemical sensors. Appl Clay Sci 28(1):199–208
De Azeredo HM (2009) Nanocomposites for food packaging applications. Food Res Int 42 (9):1240–1253
de Moura MR, Aouada FA, Avena-Bustillos RJ, McHugh TH, Krochta JM, Mattoso LH (2009) Improved barrier and mechanical properties of novel hydroxypropyl methylcellulose edible films with chitosan/tripolyphosphate nanoparticles. J Food Eng 92(4):448–453
Debeaufort F, Quezada-Gallo J-A, Voilley A (1998) Edible films and coatings: tomorrow's packagings: a review. Crit Rev Food Sci 38(4):299–313

Devlieghere F, Vermeulen A, Debevere J (2004) Chitosan: antimicrobial activity, interactions with food components and applicability as a coating on fruit and vegetables. Food Microbiol 21 (6):703–714

Djerahov L, Vasileva P, Karadjova I, Kurakalva RM, Aradhi KK (2016) Chitosan film loaded with silver nanoparticles - sorbent for solid phase extraction of Al (III), Cd (II), Cu (II), Co (II), Fe (III), Ni (II), Pb (II) and Zn (II). Carbohydr Polym 147:45–52

Durango A, Soares N, Benevides S, Teixeira J, Carvalho M, Wobeto C, Andrade N (2006) Development and evaluation of an edible antimicrobial film based on yam starch and chitosan. Packag Technol Sci 19(1):55–59

Dutta P, Tripathi S, Mehrotra G, Dutta J (2009) Perspectives for chitosan based antimicrobial films in food applications. Food Chem 114(4):1173–1182

El Ghaouth A, Arul J, Ponnampalam R, Boulet M (1991) Chitosan coating effect on storability and quality of fresh strawberries. J Food Sci 56(6):1618–1620

Elsabee MZ, Abdou ES (2013) Chitosan based edible films and coatings: a review. Mater Sci Eng C 33(4):1819–1841

Feng Q, Wu J, Chen G, Cui F, Kim T, Kim J (2000) A mechanistic study of the antibacterial effect of silver ions on *Escherichia coli* and *Staphylococcus aureus*. J Biomed Mater Res 52 (4):662–668

Fernández-Pan I, Maté JI, Gardrat C, Coma V (2015) Effect of chitosan molecular weight on the antimicrobial activity and release rate of carvacrol-enriched films. Food Hydrocoll 51:60–68

Ferreira AR, Alves VD, Coelhoso IM (2016) Polysaccharide-based membranes in food packaging applications. Membranes 6(2):22

Gómez-Estaca J, de Lacey AL, López-Caballero M, Gómez-Guillén M, Montero P (2010) Biodegradable gelatin–chitosan films incorporated with essential oils as antimicrobial agents for fish preservation. Food Microbiol 27(7):889–896

Gontard N, Guilbert S, Cuq JL (1993) Water and glycerol as plasticizers affect mechanical and water vapor barrier properties of an edible wheat gluten film. J Food Sci 58(1):206–211

Goosen MF (1996) Applications of chitin and chitosan. CRC Press, Boca Raton

Govindan S, Nivethaa E, Saravanan R, Narayanan V, Stephen A (2012) Synthesis and characterization of chitosan–silver nanocomposite. Appl Nanosci 2(3):299–303

Goy RC, Britto DD, Assis OB (2009) A review of the antimicrobial activity of chitosan. Polímeros 19(3):241–247

Grande CD, Mangadlao J, Fan J, De Leon A, Delgado-Ospina J, Rojas JG, Rodrigues DF (2017) Advincula R chitosan cross-linked graphene oxide nanocomposite films with antimicrobial activity for application in food industry. In: Macromolecular Symposia, vol 1. Wiley Online Library

Gray J, Gomaa E, Buckley D (1996) Oxidative quality and shelf life of meats. Meat Sci 43:111–123

Guo M, Jin TZ, Wang L, Scullen OJ, Sommers CH (2014) Antimicrobial films and coatings for inactivation of Listeria innocua on ready-to-eat deli turkey meat. Food Control 40:64–70

Han D, Yan L, Chen W, Li W (2011) Preparation of chitosan/graphene oxide composite film with enhanced mechanical strength in the wet state. Carbohydr Polym 83(2):653–658

Hong S-I, Rhim J-W (2008) Antimicrobial activity of organically modified nano-clays. J Nanosci Nanotechnol 8(11):5818–5824

Hosseini SF, Rezaei M, Zandi M, Ghavi FF (2013) Preparation and functional properties of fish gelatin–chitosan blend edible films. Food Chem 136(3):1490–1495

Hosseini SF, Rezaei M, Zandi M, Farahmandghavi F (2016) Development of bioactive fish gelatin/chitosan nanoparticles composite films with antimicrobial properties. Food Chem 194:1266–1274

Huei CR, Hwa H-D (1996) Effect of molecular weight of chitosan with the same degree of deacetylation on the thermal, mechanical, and permeability properties of the prepared membrane. Carbohydr Polym 29(4):353–358

Ibarguren C, Céliz G, Díaz AS, Bertuzzi MA, Daz M, Audisio MC (2015) Gelatine based films added with bacteriocins and a flavonoid ester active against food-borne pathogens. Innov Food Sci Emerg Technol 28:66–72
Inoue K, Baba Y, Yoshizuka K (1993) Adsorption of metal ions on chitosan and crosslinked copper (II)-complexed chitosan. Bull Chem Soc Jpn 66(10):2915–2921
Jagannath J, Nanjappa C, Das Gupta D, Bawa A (2003) Mechanical and barrier properties of edible starch–protein-based films. J Appl Polym Sci 88(1):64–71
Jiménez A, Fabra MJ, Talens P, Chiralt A (2012) Edible and biodegradable starch films: a review. Food Bioprocess Technol 5(6):2058–2076
Jridi M, Hajji S, Ayed HB, Lassoued I, Mbarek A, Kammoun M, Souissi N, Nasri M (2014) Physical, structural, antioxidant and antimicrobial properties of gelatin–chitosan composite edible films. Int J Biol Macromol 67:373–379
Justin R, Chen B (2014) Characterisation and drug release performance of biodegradable chitosan–graphene oxide nanocomposites. Carbohydr Polym 103:70–80
Kamal MA, Bibi S, Bokhari SW, Siddique AH, Yasin T (2017) Synthesis and adsorptive characteristics of novel chitosan/graphene oxide nanocomposite for dye uptake. React Funct Polym 110:21–29
Kanatt SR, Chander R, Sharma A (2007) Antioxidant potential of mint (*Mentha spicata* L.) in radiation-processed lamb meat. Food Chem 100(2):451–458
Kanatt SR, Chander R, Sharma A (2008) Chitosan and mint mixture: a new preservative for meat and meat products. Food Chem 107(2):845–852
Karger-Kocsis J, Zhang Z (2005) Structure-property relationships in nanoparticle/semicrystalline thermoplastic composites. In: Mechanical properties of polymers based on nanostructure and morphology. CRC Press, Washington, DC, pp 547–596
Khan F, Ahmad SR (2013) Polysaccharides and their derivatives for versatile tissue engineering application. Macromol Biosci 13(4):395–421
Khan A, Khan RA, Salmieri S, Le Tien C, Riedl B, Bouchard J, Chauve G, Tan V, Kamal MR, Lacroix M (2012) Mechanical and barrier properties of nanocrystalline cellulose reinforced chitosan based nanocomposite films. Carbohydr Polym 90(4):1601–1608
Khan I, Tango CN, Oh DH (2017) Development and evaluation of chitosan and its derivative for the shelf life extension of beef meat under refrigeration storage. Int J Food Sci Technol 52 (5):1111–1121
Kim KM, Son JH, Kim SK, Weller CL, Hanna MA (2006) Properties of chitosan films as a function of pH and solvent type. J Food Sci 71(3):E119–E124
Kim JS, Kuk E, Yu KN, Kim J-H, Park SJ, Lee HJ, Kim SH, Park YK, Park YH, Hwang C-Y (2007) Antimicrobial effects of silver nanoparticles. Nanomedicine 3(1):95–101
Kołodziejska I, Piotrowska B (2007) The water vapour permeability, mechanical properties and solubility of fish gelatin–chitosan films modified with transglutaminase or 1-ethyl-3-(3-dimethylaminopropyl) carbodiimide (EDC) and plasticized with glycerol. Food Chem 103 (2):295–300
Kong M, Chen XG, Xing K, Park HJ (2010) Antimicrobial properties of chitosan and mode of action: a state of the art review. Int J Food Microbiol 144(1):51–63
Kowalczyk D, Kordowska-Wiater M, Nowak J, Baraniak B (2015) Characterization of films based on chitosan lactate and its blends with oxidized starch and gelatin. Int J Biol Macromol 77:350–359
Kubota N, Kikuchi Y (1998) Macromolecular complexes of chitosan. In: Dumitriu S (ed) Polysaccharides: structural diversity and functional versatility. Marcel Dekker Inc., New York, pp 595–628
Kumar S, Jog JP, Natarajan U (2003) Preparation and characterization of poly (methyl methacrylate)–clay nanocomposites via melt intercalation: the effect of organoclay on the structure and thermal properties. J Appl Polym Sci 89(5):1186–1194
Kumar AP, Depan D, Tomer NS, Singh RP (2009) Nanoscale particles for polymer degradation and stabilization—trends and future perspectives. Prog Polym Sci 34(6):479–515

Ladikos D, Lougovois V (1990) Lipid oxidation in muscle foods: a review. Food Chem 35 (4):295–314

Lazaridou A, Biliaderis CG (2002) Thermophysical properties of chitosan, chitosan–starch and chitosan–pullulan films near the glass transition. Carbohydr Polym 48(2):179–190

Lee KY, Mooney DJ (2012) Alginate: properties and biomedical applications. Prog Polym Sci 37 (1):106–126

Lee S-R, Park H-M, Lim H, Kang T, Li X, Cho W-J, Ha C-S (2002) Microstructure, tensile properties, and biodegradability of aliphatic polyester/clay nanocomposites. Polymer 43 (8):2495–2500

LeRoux MA, Guilak F, Setton LA (1999) Compressive and shear properties of alginate gel: effects of sodium ions and alginate concentration. J Biomed Mater Res 47(1):46–53

Li L-H, Deng J-C, Deng H-R, Liu Z-L, Li X-L (2010) Preparation, characterization and antimicrobial activities of chitosan/Ag/ZnO blend films. Chem Eng J 160(1):378–382

Lim H, Huang N, Loo C (2012) Facile preparation of graphene-based chitosan films: enhanced thermal, mechanical and antibacterial properties. J Non-Cryst Solids 358(3):525–530

Lin D, Zhao Y (2007) Innovations in the development and application of edible coatings for fresh and minimally processed fruits and vegetables. Compr Rev Food Sci Food Saf 6(3):60–75

Liu Y-L, Su Y-H, Lai J-Y (2004) In situ crosslinking of chitosan and formation of chitosan–silica hybrid membranes with using γ-glycidoxypropyltrimethoxysilane as a crosslinking agent. Polymer 45(20):6831–6837

López-León T, Carvalho E, Seijo B, Ortega-Vinuesa J, Bastos-González D (2005) Physicochemical characterization of chitosan nanoparticles: electrokinetic and stability behavior. J Colloid Interface Sci 283(2):344–351

Lozano-Navarro JI, Díaz-Zavala NP, Velasco-Santos C, Martínez-Hernández AL, Tijerina-Ramos BI, García-Hernández M, Rivera-Armenta JL, Páramo-García U, Reyes-de la Torre AI (2017) Antimicrobial, optical and mechanical properties of chitosan–starch films with natural extracts. Int J Mol Sci 18(5):997

Lucinda-Silva RM, Salgado HRN, Evangelista RC (2010) Alginate–chitosan systems: in vitro controlled release of triamcinolone and in vivo gastrointestinal transit. Carbohydr Polym 81 (2):260–268

Luo Y, Wang Q (2014) Recent development of chitosan-based polyelectrolyte complexes with natural polysaccharides for drug delivery. Int J Biol Macromol 64:353–367

Ma Y, Zhou T, Zhao C (2008) Preparation of chitosan–nylon-6 blended membranes containing silver ions as antibacterial materials. Carbohydr Res 343(2):230–237

Magnier L, Schoormans J (2015) Consumer reactions to sustainable packaging: the interplay of visual appearance, verbal claim and environmental concern. J Environ Psychol 44:53–62

Matsumura Y, Yoshikata K, Kunisaki S-i, Tsuchido T (2003) Mode of bactericidal action of silver zeolite and its comparison with that of silver nitrate. Appl Environ Microbiol 69(7):4278–4281

McGlashan SA, Halley PJ (2003) Preparation and characterisation of biodegradable starch-based nanocomposite materials. Polym Int 52(11):1767–1773

Meenakshi P, Noorjahan S, Rajini R, Venkateswarlu U, Rose C, Sastry T (2002) Mechanical and microstructure studies on the modification of CA film by blending with PS. Bull Mater Sci 25 (1):25–29

Mendes J, Paschoalin R, Carmona V, Neto ARS, Marques A, Marconcini J, Mattoso L, Medeiros E, Oliveira J (2016) Biodegradable polymer blends based on corn starch and thermoplastic chitosan processed by extrusion. Carbohydr Polym 137:452–458

Meng Q, Heuzey M-C, Carreau PJ (2014) Hierarchical structure and physicochemical properties of plasticized chitosan. Biomacromolecules 15(4):1216–1224

Mi F-L, Shyu S-S, Wu Y-B, Lee S-T, Shyong J-Y, Huang R-N (2001) Fabrication and characterization of a sponge-like asymmetric chitosan membrane as a wound dressing. Biomaterials 22 (2):165–173

Miller KS, Krochta J (1997) Oxygen and aroma barrier properties of edible films: a review. Trends Food Sci Technol 8(7):228–237

Mladenova EK, Dakova IG, Karadjova IB (2011) Chitosan membranes as sorbents for trace elements determination in surface waters. Environ Sci Pollut Res 18(9):1633–1643
Mukhopadhyay P, Chakraborty S, Bhattacharya S, Mishra R, Kundu P (2015) pH-sensitive chitosan/alginate core-shell nanoparticles for efficient and safe oral insulin delivery. Int J Biol Macromol 72:640–648
Murugadoss A, Chattopadhyay A (2007) A 'green' chitosan–silver nanoparticle composite as a heterogeneous as well as micro-heterogeneous catalyst. Nanotechnology 19(1):015603
Muzzarelli RA (1983) Chitin and its derivatives: new trends of applied research. Carbohydr Polym 3(1):53–75
Nagahama H, Maeda H, Kashiki T, Jayakumar R, Furuike T, Tamura H (2009) Preparation and characterization of novel chitosan/gelatin membranes using chitosan hydrogel. Carbohydr Polym 76(2):255–260
Nissen LR, Byrne DV, Bertelsen G, Skibsted LH (2004) The antioxidative activity of plant extracts in cooked pork patties as evaluated by descriptive sensory profiling and chemical analysis. Meat Sci 68(3):485–495
No HK, Park NY, Lee SH, Meyers SP (2002) Antibacterial activity of chitosans and chitosan oligomers with different molecular weights. Int J Food Microbiol 74(1):65–72
No H, Meyers S, Prinyawiwatkul W, Xu Z (2007) Applications of chitosan for improvement of quality and shelf life of foods: a review. J Food Sci 72(5):R87–R100
Okamoto M (2003) Polymer/layered silicate nanocomposites, vol 14. Smithers Rapra Publishing, Shrewsbury
Olaniyan OJ, Dare EO, Adetunji OR, Adedeji OO, Ogungbesan SO (2016) Synthesis and characterization of chitosan-silver nanocomposite film. Nano Hybrids 11:22–29
Ouattara B, Simard RE, Piette G, Bégin A, Holley RA (2000) Inhibition of surface spoilage bacteria in processed meats by application of antimicrobial films prepared with chitosan. Int J Food Microbiol 62(1):139–148
Pereda M, Ponce A, Marcovich N, Ruseckaite R, Martucci J (2011) Chitosan-gelatin composites and bi-layer films with potential antimicrobial activity. Food Hydrocoll 25(5):1372–1381
Prabhu S, Poulose EK (2012) Silver nanoparticles: mechanism of antimicrobial action, synthesis, medical applications, and toxicity effects. Int Nano Lett 2(1):32–41
Puttipipatkhachorn S, Nunthanid J, Yamamoto K, Peck G (2001) Drug physical state and drug–polymer interaction on drug release from chitosan matrix films. J Control Release 75 (1):143–153
Quintavalla S, Vicini L (2002) Antimicrobial food packaging in meat industry. Meat Sci 62 (3):373–380
Rabea EI, Badawy ME-T, Stevens CV, Smagghe G, Steurbaut W (2003) Chitosan as antimicrobial agent: applications and mode of action. Biomacromolecules 4(6):1457–1465
Rai M, Yadav A, Gade A (2009) Silver nanoparticles as a new generation of antimicrobials. Biotechnol Adv 27(1):76–83
Ramos M, Jiménez A, Peltzer M, Garrigós MC (2014) Development of novel nano-biocomposite antioxidant films based on poly (lactic acid) and thymol for active packaging. Food Chem 162:149–155
Ramos M, Fortunati E, Peltzer M, Jimenez A, Kenny JM, Garrigós MC (2016) Characterization and disintegrability under composting conditions of PLA-based nanocomposite films with thymol and silver nanoparticles. Polym Degrad Stab 132:2–10
Ramos M, Jiménez A, Garrigós MC (2017) Active nanocomposites in food contact materials. In: Nanoscience in food and agriculture. Springer, New York, pp 1–44
Rassu G, Salis A, Porcu EP, Giunchedi P, Roldo M, Gavini E (2016) Composite chitosan/alginate hydrogel for controlled release of deferoxamine: a system to potentially treat iron dysregulation diseases. Carbohydr Polym 136:1338–1347
Rathke TD, Hudson SM (1994) Review of chitin and chitosan as fiber and film formers. J Macromol Sci Polym Rev 34(3):375–437

Ray SS, Okamoto M (2003) Polymer/layered silicate nanocomposites: a review from preparation to processing. Prog Polym Sci 28(11):1539–1641

Ray S, Quek SY, Easteal A, Chen XD (2006) The potential use of polymer-clay nanocomposites in food packaging. Int J Food Eng 2(4):1556–3758

Regiel A, Irusta S, Kyzioł A, Arruebo M, Santamaria J (2012) Preparation and characterization of chitosan–silver nanocomposite films and their antibacterial activity against *Staphylococcus aureus*. Nanotechnology 24(1):015101

Rhim J-W, Hong S-I, Park H-M, Ng PK (2006) Preparation and characterization of chitosan-based nanocomposite films with antimicrobial activity. J Agric Food Chem 54(16):5814–5822

Rhim J-W, Park H-M, Ha C-S (2013a) Bio-nanocomposites for food packaging applications. Prog Polym Sci 38(10):1629–1652

Rhim J, Wang L, Hong S (2013b) Preparation and characterization of agar/silver nanoparticles composite films with antimicrobial activity. Food Hydrocoll 33(2):327–335

Rinaudo M (1992) On the abnormal exponents a ν and a D in Mark Houwink type equations for wormlike chain polysaccharides. Polym Bull 27(5):585–589

Rivero S, García M, Pinotti A (2009) Composite and bi-layer films based on gelatin and chitosan. J Food Eng 90(4):531–539

Robyt JF (2008) Starch: structure, properties, chemistry, and enzymology. In: Glycoscience. Springer, Berlin, pp 1437–1472

Rojas-Graü MA, Soliva-Fortuny R, Martín-Belloso O (2009) Edible coatings to incorporate active ingredients to fresh-cut fruits: a review. Trends Food Sci Technol 20(10):438–447

Sagoo S, Board R, Roller S (2002) Chitosan potentiates the antimicrobial action of sodium benzoate on spoilage yeasts. Lett Appl Microbiol 34(3):168–172

Salehi E, Daraei P, Shamsabadi AA (2016) A review on chitosan-based adsorptive membranes. Carbohydr Polym 152:419–432

Sanpui P, Murugadoss A, Prasad PD, Ghosh SS, Chattopadhyay A (2008) The antibacterial properties of a novel chitosan–Ag-nanoparticle composite. Int J Food Microbiol 124 (2):142–146

Santos CM, Mangadlao J, Ahmed F, Leon A, Advincula RC, Rodrigues DF (2012) Graphene nanocomposite for biomedical applications: fabrication, antimicrobial and cytotoxic investigations. Nanotechnology 23(39):395101

Schou M, Longares A, Montesinos-Herrero C, Monahan F, O'Riordan D, O'sullivan M (2005) Properties of edible sodium caseinate films and their application as food wrapping. LWT-Food Sci Technol 38(6):605–610

Schrieber R, Gareis H (2007) Gelatine handbook: theory and industrial practice. John Wiley & Sons, USA

Sebranek J, Sewalt V, Robbins K, Houser T (2005) Comparison of a natural rosemary extract and BHA/BHT for relative antioxidant effectiveness in pork sausage. Meat Sci 69(2):289–296

Shahidi F (2012) Flavor of meat and meat products. Springer Science & Business Media, New York

Shao L, Chang X, Zhang Y, Huang Y, Yao Y, Guo Z (2013) Graphene oxide cross-linked chitosan nanocomposite membrane. Appl Surf Sci 280:989–992

Shen Z, Kamdem DP (2015) Development and characterization of biodegradable chitosan films containing two essential oils. Int J Biol Macromol 74:289–296

Sherif H, Khalil S, Hegazi A, Khalil W, Moharram M (2017) Factors affecting the antibacterial activity of chitosan-silver nanocomposite. IET Nanobiotechnol 11:731–737

Shim S-K, Ryu J-J (1998) Synthesis of chelating adsorbent (2, 2′-iminodibenzoic acid-crosslinked chitosan) and adsorptivity of Pb (II), Cu (II), Cd (II). Analytical. Sci Technol 11:452–459

Shu X, Zhu K, Song W (2001) Novel pH-sensitive citrate cross-linked chitosan film for drug controlled release. Int J Pharm 212(1):19–28

Singh DK, Ray AR (1999) Controlled release of glucose through modified chitosan membranes. J Membr Sci 155(1):107–112

Siripatrawan U, Vitchayakitti W (2016) Improving functional properties of chitosan films as active food packaging by incorporating with propolis. Food Hydrocoll 61:695–702

Sondi I, Salopek-Sondi B (2004) Silver nanoparticles as antimicrobial agent: a case study on *E. coli* as a model for gram-negative bacteria. J Colloid Interface Sci 275(1):177–182
Sorrentino A, Gorrasi G, Vittoria V (2007) Potential perspectives of bio-nanocomposites for food packaging applications. Trends Food Sci Technol 18(2):84–95
Souza VGL, Fernando AL (2016) Nanoparticles in food packaging: biodegradability and potential migration to food – a review. Food Packag Shelf Life 8:63–70
Sozer N, Kokini JL (2009) Nanotechnology and its applications in the food sector. Trends Biotechnol 27(2):82–89
Srinivasa P, Ramesh M, Tharanathan R (2007) Effect of plasticizers and fatty acids on mechanical and permeability characteristics of chitosan films. Food Hydrocoll 21(7):1113–1122
Sudarshan N, Hoover D, Knorr D (1992) Antibacterial action of chitosan. Food Biotechnol 6 (3):257–272
Suyatma NE, Copinet A, Tighzert L, Coma V (2004) Mechanical and barrier properties of biodegradable films made from chitosan and poly (lactic acid) blends. J Polym Environ 12 (1):1–6
Tahiri I, Desbiens M, Benech R, Kheadr E, Lacroix C, Thibault S, Ouellet D, Fliss I (2004) Purification, characterization and amino acid sequencing of divergicin M35: a novel class IIa bacteriocin produced by Carnobacterium divergens M35. Int J Food Microbiol 97(2):123–136
Tahiri I, Desbiens M, Lacroix C, Kheadr E, Fliss I (2009) Growth of Carnobacterium divergens M35 and production of Divergicin M35 in snow crab by-product, a natural-grade medium. LWT-Food Sci Technol 42(2):624–632
Talón E, Trifkovic KT, Nedovic VA, Bugarski BM, Vargas M, Chiralt A, González-Martínez C (2017a) Antioxidant edible films based on chitosan and starch containing polyphenols from thyme extracts. Carbohydr Polym 157:1153–1161
Talón E, Trifkovic KT, Vargas M, Chiralt A, González-Martínez C (2017b) Release of polyphenols from starch-chitosan based films containing thyme extract. Carbohydr Polym 175:122–130
Tan H, Ma R, Lin C, Liu Z, Tang T (2013) Quaternized chitosan as an antimicrobial agent: antimicrobial activity, mechanism of action and biomedical applications in orthopedics. Int J Mol Sci 14(1):1854–1869
Thostenson ET, Li C, Chou T-W (2005) Nanocomposites in context. Compos Sci Technol 65 (3):491–516
Tikhonov VE, Radigina LA, Yamskov YA (1996) Metal-chelating chitin derivatives via reaction of chitosan with nitrilotriacetic acid. Carbohydr Res 290(1):33–41
Tipsrisukond N, Fernando L, Clarke A (1998) Antioxidant effects of essential oil and oleoresin of black pepper from supercritical carbon dioxide extractions in ground pork. J Agric Food Chem 46(10):4329–4333
Tripathi S, Mehrotra G, Dutta P (2008) Chitosan based antimicrobial films for food packaging applications. e-Polymers 8(1):1082–1088
Tsai H-A, Chen W-H, Kuo C-Y, Lee K-R, Lai J-Y (2008) Study on the pervaporation performance and long-term stability of aqueous iso-propanol solution through chitosan/polyacrylonitrile hollow fiber membrane. J Membr Sci 309(1):146–155
Tual C, Espuche E, Escoubes M, Domard A (2000) Transport properties of chitosan membranes: influence of crosslinking. J Polym Sci B Polym Phys 38(11):1521–1529
Uyama H, Kuwabara M, Tsujimoto T, Nakano M, Usuki A, Kobayashi S (2003) Green nanocomposites from renewable resources: plant oil– clay hybrid materials. Chem Mater 15 (13):2492–2494
Valenzuela C, Abugoch L, Tapia C (2013) Quinoa protein–chitosan–sunflower oil edible film: mechanical, barrier and structural properties. LWT-Food Sci Technol 50(2):531–537
van den Broek LA, Knoop RJ, Kappen FH, Boeriu CG (2015) Chitosan films and blends for packaging material. Carbohydr Polym 116:237–242
Vanin FM, Hirano MH, Carvalho RA, Moraes ICF, Bittante AMQB, do Amaral Sobral PJ (2014) Development of active gelatin-based nanocomposite films produced in an automatic spreader. Food Res Int 63:16–24

Varma A, Deshpande S, Kennedy J (2004) Metal complexation by chitosan and its derivatives: a review. Carbohydr Polym 55(1):77–93

Vieira RS, Beppu MM (2006) Dynamic and static adsorption and desorption of Hg (II) ions on chitosan membranes and spheres. Water Res 40(8):1726–1734

Vimala K, Mohan YM, Sivudu KS, Varaprasad K, Ravindra S, Reddy NN, Padma Y, Sreedhar B, MohanaRaju K (2010) Fabrication of porous chitosan films impregnated with silver nanoparticles: a facile approach for superior antibacterial application. Colloids Surf B: Biointerfaces 76(1):248–258

Vimala K, Yallapu MM, Varaprasad K, Reddy NN, Ravindra S, Naidu NS, Raju KM (2011) Fabrication of curcumin encapsulated chitosan-PVA silver nanocomposite films for improved antimicrobial activity. J Biomater Nanobiotechnol 2(01):55

Wang CC, Chen CC (2005) Anti-bacterial and swelling properties of acrylic acid grafted and collagen/chitosan immobilized polypropylene non-woven fabrics. J Appl Polym Sci 98 (1):391–400

Wang S, Gao Y (2017) Fabrication of chitosan/silver nanocomposite films and their fluorescence sensing of aluminum ions. Mater Technol 33:1–5

Wang L, Khor E, Wee A, Lim LY (2002) Chitosan-alginate PEC membrane as a wound dressing: assessment of incisional wound healing. J Biomed Mater Res A 63(5):610–618

Wu Y, Rhim J, Weller C, Hamouz F, Cuppett S, Schnepf M (2000) Moisture loss and lipid oxidation for precooked beef patties stored in edible coatings and films. J Food Sci 65 (2):300–304

Xu Y, Kim KM, Hanna MA, Nag D (2005) Chitosan–starch composite film: preparation and characterization. Ind Crop Prod 21(2):185–192

Xu Y, Ren X, Hanna MA (2006) Chitosan/clay nanocomposite film preparation and characterization. J Appl Polym Sci 99(4):1684–1691

Yadav SK, Jung YC, Kim JH, Ko YI, Ryu HJ, Yadav MK, Kim YA, Cho JW (2013) Mechanically robust, electrically conductive biocomposite films using antimicrobial chitosan-functionalized graphenes. Part Part Syst Charact 30(8):721–727

Yamaguchi I, Tokuchi K, Fukuzaki H, Koyama Y, Takakuda K, Monma H, Tanaka J (2001) Preparation and microstructure analysis of chitosan/hydroxyapatite nanocomposites. J Biomed Mater Res A 55(1):20–27

Yan XL, Khor E, Lim LY (2001) Chitosan-alginate films prepared with chitosans of different molecular weights. J Biomed Mater Res A 58(4):358–365

Yang JM, Su WY, Yang MC (2004) Evaluation of chitosan/PVA blended hydrogel membranes. J Membr Sci 236(1):39–51

Yang X, Tu Y, Li L, Shang S, X-m T (2010) Well-dispersed chitosan/graphene oxide nanocomposites. ACS Appl Mater Interfaces 2(6):1707–1713

Yin YJ, Yao KD, Cheng GX, Ma JB (1999) Properties of polyelectrolyte complex films of chitosan and gelatin. Polym Int 48(6):429–432

Yisong Y, Wenjun L, Tongyin Y (1990) Crosslinking chitosan membranes for pervaporation of alcohol-water mixtures. Polym Commun 31(8):319–321

Yoksan R, Chirachanchai S (2010) Silver nanoparticle-loaded chitosan–starch based films: fabrication and evaluation of tensile, barrier and antimicrobial properties. Mater Sci Eng C 30 (6):891–897

Youn S, Her J, Park S, Ahn D, Kim Y, Choi J (2004) Studies on the improvement of shelf-life in spicy beef meat using chitosan. J Korean Soc Food Scie Nutr 45:1233

Younes I, Sellimi S, Rinaudo M, Jellouli K, Nasri M (2014) Influence of acetylation degree and molecular weight of homogeneous chitosans on antibacterial and antifungal activities. Int J Food Microbiol 185:57–63

Youssef AM, Abou-Yousef H, El-Sayed SM, Kamel S (2015) Mechanical and antibacterial properties of novel high performance chitosan/nanocomposite films. Int J Biol Macromol 76:25–32

Yuan W, Dong H, Li CM, Cui X, Yu L, Lu Z, Zhou Q (2007) pH-controlled construction of chitosan/alginate multilayer film: characterization and application for antibody immobilization. Langmuir 23(26):13046–13052

Zeng X, Ruckenstein E (1998) Cross-linked macroporous chitosan anion-exchange membranes for protein separations. J Membr Sci 148(2):195–205

Zhai M, Zhao L, Yoshii F, Kume T (2004) Study on antibacterial starch/chitosan blend film formed under the action of irradiation. Carbohydr Polym 57(1):83–88

Zhan F, Sheng F, Yan X, Zhu Y, Jin W, Li J, Li B (2017) Enhancement of antioxidant and antibacterial properties for tannin acid/chitosan/tripolyphosphate nanoparticles filled electrospinning films: surface modification of sliver nanoparticles. Int J Biol Macromol 104:813–820

Zhang M, Li X, Gong Y, Zhao N, Zhang X (2002) Properties and biocompatibility of chitosan films modified by blending with PEG. Biomaterials 23(13):2641–2648

Zhang H, Wang M-l, Mao Y-t (2011) Advances in the application of chitosan coating in fresh-keeping of fruits and vegetables [J]. Guizhou Agric Sci 10:040

Zhang Y, Zhang H, Liu Y, Zhang Z, Ding C (2017) Chelating ability and microbial stability of an L-arginine-modified chitosan-based environmental remediation material. J Polym Environ 12:1–10

Zheng W, Wang SY (2001) Antioxidant activity and phenolic compounds in selected herbs. J Agric Food Chem 49(11):5165–5170

Zivanovic S, Chi S, Draughon AF (2005) Antimicrobial activity of chitosan films enriched with essential oils. J Food Sci 70(1):M45–M51

Chitosan-Based Nanocomposites in Food Packaging

Ritu Hooda, Bhawna Batra, Vijay Kalra, Joginder Singh Rana, and Minakshi Sharma

Abstract Since the last decade, there has been a strong drive of reducing packaging waste by replacing plastic food packaging materials with eco-friendly materials from renewable sources. Chitosan is the most abundant biopolymer made up of glucosamine (2-amino-2-deoxy-β-D-glucose) units linked by β-1,4-linkage. Because of its exceptional properties such as effective antimicrobial activity, high biodegradability, and low toxicity, its use has been elevated in food industries. Its shielding wall can delay ripening and prevent water loss as well as increase the shelf life of food products. Its role in foods could be viewed in broad categories based on its functions such as color stabilization, emulsification, antioxidant activity, and dietary fiber-like property aiding water holding and fat entrapment, thereby imparting health benefit. The innovation of nanotechnology in improving the useful properties of chitosan and the development of the chitosan-based materials have been done by merging of nanoparticles, metal oxides, and organic substances like acetic acid, lauric acid, cinnamaldehyde, and propionic acid into chitosan matrix so that it has remarkable applications in food packaging industries. This chapter deals with general information on chitosan and its structure, chitosan nanoparticle preparation by different methods, their characterization and application, chitosan-based bio-nanocomposites, and their role in food packaging.

Keywords Chitosan · Bionanocomposites · Food packaging

R. Hooda · M. Sharma (✉)
Department of Zoology, Maharshi Dayanand University, Rohtak, Haryana, India

B. Batra · J. S. Rana
Department of Biotechnology, DCRUST Murthal, Sonipat, Haryana, India

V. Kalra
Department of Biochemistry, PGIMS, Rohtak, Haryana, India

S. Ahmed (ed.), *Bio-based Materials for Food Packaging*,
https://doi.org/10.1007/978-981-13-1909-9_12

1 Introduction

Food packaging is necessary for goods, safeguarding, freedom from contamination, and shelter and for ease in providing information concerning the manufactured goods and the brand statement. Plastic packaging is used for above 50% of all European commodities for handing out, stocking up, shipping, defending, and preserving goods (Ferreira et al. 2016; Plastics Europe n.d.). The commercial hit of plastic is due to a blend of properties such as strength, flimsiness, solidity, flexibility, impermeability, and relief of sterilization. These properties construct plastic perfect materials for approximately all marketable and trade consumer (Plackett 2011). Common plastic used in food packaging are polyethylene, polypropylene and polyethylene terephthalate, polyvinyl chloride (PVC), and polystyrene (PS) due to their brilliant water barrier property (American Chemistry Concil 2015). The major trouble of using plastics is the post-user waste of packaging; packaging is undoubtedly the biggest donor (63%) of plastic waste (Tumwesigye et al. 2016). The recyclable plastic packaging materials are less than 14%; additionally various plastics are hard to use again (Lee et al. 2014). As the majority of plastic used is nonrecyclable, they stay in the surroundings for extended period of time, therefore harmful to human and environment well-being. As per knowledge and facts, there is a rapid growth in man-made polymers production and utilization, and it is estimated that utilization possibly will increase by four near the year 2100 (Tumwesigye et al. 2016). To defeat this environmental dilemma, eco-friendly polymers should be used from renewable sources. One of the encouraging substitute materials for food packaging is biopolymers. Biopolymers are chain-like molecules, formed by covalent bond in monomeric units. It has been considered like an alternate for plastic material prepared from petroleum due to their advance properties like renewability and biodegradability (Muratore et al. 2005; Tang et al. 2012). They left carbon dioxide (CO_2) and water (H_2O) and organic by-product behind degradation (Rhim et al. 2013; Tang et al. 2012). Currently, synthetic biopolymers are flourishing which includes polyvinyl alcohol (PVA), polylactic acid (PLA), polyglycolic acid (PGA), polybutylene succinate (PBS), and polycaprolactone (PCL) (Abdollahi et al. 2012). The improved properties such as tensile strength, durability, high luster, elasticity, and transparency of synthetic biopolymers make them sustainable for industry. Formerly, innate biopolymers, for example, chitosan, cellulose, agar, and starch, are the most frequent types of biopolymers which are used in food packaging applications. There are some drawbacks of biopolymer utilization including mechanical weakness in food packaging, poor barrier, and thermal properties as evaluated against the nonrecyclable plastic materials prepared as petroleum. Due to these limitations, a lot of research was done to develop mechanical and barricade properties of the biopolymers by using the nanocomposite concept (Arfat et al. 2014; Di Maio et al. 2014; Kanmani and Rhim 2014; Nafchi et al. 2013; Reddy and Rhim 2014; Sanuja et al. 2014; Trovatti et al. 2012; Sorrentino et al. 2007). Nanocomposites show amplified barrier and mechanical and heat resistance properties with reference to their neat polymers and usual composites (Ray and Okamoto

2003; Ray et al. 2006; Giannelis 1996; Thakur and Thakur 2016). Bio-nanocomposites are formed by a biopolymer continuous phase, reinforced with nanoparticles discontinuous phase within the range of nanometer (1–100 nm). For better food packaging and to lessen environmental crisis, the improvement of bio-nanocomposite materials is important.

2 Chitin and Chitosan

Chitin is a natural polymer of N-acetyl-D-glucosamine coupled with N-glucosamine residues via β-1,4 glycosidic bonds. After cellulose, it is the most abundant biomaterial found on the earth. Chitin can be extracted from the exoskeleton of arthropods, crustacean and cephalopod internal shells, fungi and yeast's cell walls, and the mollusk's radula. Due to its less solubility in water, it is comparatively hard to process and become less available to biological laboratories. Chitin can be formed either via enzymatic hydrolysis or fermentation method, but these routes are not so efficiently practicable on business scale yet (Babu et al. 2013; Dutta et al. 2008). Chitosan is a repeating unit of D-glucosamine linked by β (1–4) glycosidic bond. It is a de-*N*-acetylation form of chitin. Chitosan due to the occurrence of amino groups is soluble in aqueous acidic medium. The physicochemical and biological properties of chitosan are influenced by different factors such as degree of N-acetylating, temperature, concentration of alkali, and incubation time period. Chitosan has a broad range of applications in various areas (e.g., waste management, food processing, drug and medicine, agriculture, cosmetics, and textiles). Due to its solubility, recyclability, companionability, non-hazardousness, and non-immunogenic properties, chitosan is measured as a multipurpose copolymer which can be transformed into many ways, such as nanoparticles, films, and gels (Tangpasuthadol et al. 2003; Zhang et al. 2010; Kumirska et al. 2010).

2.1 *Structure of Chitin and Chitosan*

There are three dissimilar polymeric arrangements of chitin, namely, α-chitin, β-chitin, and γ-chitin, which are generally extracted from the exoskeletons of crustaceans (Tolaimate et al. 2003; Campana-Filho et al. 2007), squid pens, and fungi and yeast, respectively (Sannan et al. 1976). β-Chitin is very reactive, whereas α-chitin shows an advanced affinity for solvents (Kurita et al. 1974) and is the most familiar type (Noishiki et al. 2003). An alkaline treatment can simply convert β-chitin to α-chitin (Fukuda 1980).

Each C_6 structural unit of chitosan has two open hydroxyl groups (-OH) and one primary amino (-NH_2) (Fig. 1). Chitosan conveys a positive charge because of its easy accessibility of free amino groups, and therefore it interacts with many

Fig. 1 Structure of chitosan

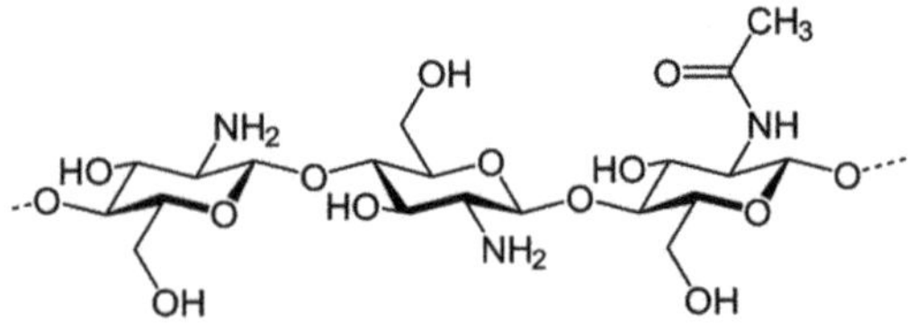

negatively charged polymers/surfaces and can show chelation with metal ions (Vargas and Gonzalez-Martinez 2010).

2.2 Chitosan in Foods

After the chitosan derivative has been acknowledged as generally recognized as safe (GRAS) by the US Food and Drug Administration (2011) for regular use in foods, there application is gaining attention. The function of chitosan derivative could be distinguishable in foods based on their role like (i) stabilizing property to facilitate color establishment, blending, and antioxidant action, (ii) nutritional fiber resembling assets helping water retention capacity and fat entrapment, and (iii) antimicrobial property, thus imparting health advantage. Recently numerals in patents have been passed on the dealing out of chitosan derivatives associated with application in food engineering (Agulló et al. 2003).

2.2.1 Stabilizing Property

Chitosan binds to the various categories of dyes (disperse, reactive, acid, and sulfur) with high affinity, because of their molecular configuration (Kumar 2000; Suman et al. 2010). The color stability of chitosan-made packaging of refrigerated ground beef has been considered earlier (Mokhtar et al. 2010). Chitosan when combined with rosemary extract in patties of fresh ground beef gives synergistic effect ensuing antioxidative defense and deep red color stabilization (Barreteau et al. 2006). Likewise, chitosan derived suitable for eating films and wrapping of fresh sliced vegetables and fruits have been concerned robustly in preservation and color stabilization (Mhurchu et al. 2004). In view of the broad diversity of functional characteristics of chitosan and its innate foundation, chitosan in different variety and permutation is one of the finest recognized food additives. Approximately three decades ago, Korea and Japan have permitted chitosan as a food additive (Kumar 2000). Generally in acidic food products, the low molecular weight chitosan and oligosaccharide chitosan have been used as food additive, enhancing their thickening and viscosity devoid of every allergic or toxic consequence (Mhurchu et al. 2004).

2.2.2 Nutritional Fiber and Health Advantage

One of the gifted applications of chitosan is its use as a dietary fiber in foods. Chitosan has dietary fiber-like properties like elevated water binding capacity and non-digestibility in upper intestinal tract. Oral administration efficacy of chitosan is determined by the chemical structure and the polymerization scale. Chitosan has been anticipated to have significant roles in fat entrapment and body weight management (Espinal-Ruiz et al. 2014), lowering lipid levels (Muzzarelli 1999) and glucose dialysis (Kumar 2000). Chitosan attaches to fatty acids and lipids in the acidic surroundings of the stomach, and such lipids associated with chitosan are tending to be excreted rather than digested (Yen et al. 2008). There are many formulations of chitosan-associated fibers that are patented and used as capsules and tablets like dietetic enhancement (Kerch 2015). The useful effect of chitosan and its derivatives in mature atherosclerosis, diabetes mellitus, cancer, hypercholesterolemia, cardiovascular diseases, and neurodegenerative diseases has been seen in advance (Liu et al. 2010). Liu et al. reveled to facilitate rats treated with streptozotocin and were also nurse by means of small and elevated molecular mass chitosan; as a result there was a considerable reduction in liver gluconeogenesis and the latent outcome of chitosan in reducing diabetic hyperglycemia signifying increased glucose uptake in skeletal muscle (Goy et al. 2009).

2.2.3 Antimicrobial Property

Chitosan exhibits antimicrobial activity against a broad range of bacteria (Gram negative, Gram positive), yeast, and mold, due to its capability to intermingle with cell membranes of these microorganisms (Rabea et al. 2003). The minimum inhibitory concentration (MIC) of chitosan against Gram-positive and Gram-negative bacteria has been found to be in the range of 100–2000 ppm. As acetylation is linked with antimicrobial outcomes, the difference in the results of MIC for the similar species was possibly due to difference in the acetylating of chitosan. Likewise, chitosan nanoparticles showed antifungal activities with MIC results in the range of 10–500 ppm against diversity of species. It was seen that higher efficacy of chitosan in inhibiting the escalation of microbes was depends on lesser the molecular mass and the higher the amount of acetylation. To elucidate the antimicrobial activity of chitosan, three possibilities have been known: (i) the chelating capability of chitosan, thus making an exterior obstacle resulting to suppression of necessary nutrients; (ii) piercing ability of chitosan into the nuclei resulting in suppression of mRNA and thereafter protein synthesis (Rabea et al. 2003; Qi et al. 2004); and (iii) increased membrane permeability due to the positively charged chitosan and the negatively charged microbial cell membrane (Jeon et al. 2014). Earlier information proposed to facilitate the antimicrobial capacity of chitosan is influenced by acidic environment decreased at pH 7; it possibly will be due to the hammering of positive charges on the amino group at neutral pH (Upadhyaya et al. 2014). In many juices counting apple, pomegranate, orange, and others, chitosan has been used like a clarifying agent.

3 Chitosan-Based Nanocomposites

Chitosan nanocomposites are a combination of chitosan with organic/inorganic nanoparticles in nano dimension, gaining significant consideration nowadays. Chitosan applications have contributed an immense role in agriculture, material science, and the medicinal area (Sorrentino et al. 2007). It is considered that it has improved the present tissue engineering tools especially in the improvement of fibers and composite and also developed novel methods to improve tissue repair. It is accepted that nano-reinforcement is extra appropriate for chitosan for escalating their functional properties when compared to generally used petroleum consequent polymers. Nano-reinforcement is more advanced in managing the price besides humanizing the properties of chitosan biopolymers (Lu et al. 2004). When isolated soy protein thermoplastics are added to the chitin whiskers, there tensile potency and water resistance quality were found to be improved (Sriupayo et al. 2005). Likewise, when chitosan films were added to chitosan whiskers, then tensile potency and water resistance ability of the chitosan films were improved by the whiskers (López-León et al. 2005a). Chitosan nanoparticles can be formed by ionic gelation (De Moura et al. 2008), and it was established that the mechanical and obstruction functions of the films can be increased by adding chitosan-tripolyphosphate nanoparticles in hydroxy-propionyl methyl cellulose membrane (Gecol et al. 2005). By assimilation of nano-reinforcement with the chitosan biopolymer, the poor functions of chitosan can be enhanced. Chitosan-based nanocomposites have a lot of applications in a range of field in human well-being care management like food industry, medicine, paper manufacturing, etc. Chitosan when combined with montmorillonite (MMT) forms nanocomposites which signify outstanding objects (Caner and Cansiz 2007). These nanocomposites, films and coating based on chitosan, offer capable improvement in food packaging manufacturing because of chitosan's capability to hold up the life of food products (Shahidi et al. 1999). The useful characteristics of nanocomposites derived from chitosan mostly pivot on the physical characteristics of nanofillers like shape, size, etc. Different types of nanofillers have been formed and discussed till now, and a nanofiller has to be of nanoscale in the range of r100 (Radhakrishnan et al. 2015). Therefore, physical and chemical properties along with the types, size, and surface areas of nanofillers could significantly improve the useful characteristics of the nanocomposites derived from chitosan. The improved characteristics of chitosan-based nanocomposites pivot on the proportion of particle measurement lengthwise to width between particles, platelets, and fibers. It was experimental that clay-based nanocomposite films enhanced their oxygen barricade capacity than neat chitosan film. Therefore the integration of clay into the chitosan matrix enhances the transport and barrier characteristics; hence, these clay-chitosan nanocomposite films appear appropriate in favor of food packaging and shielding coatings (Radhakrishnan et al. 2015). For the substitute of plastic resources, Mura et al. proposed the outstanding development in the mechanical characteristics of a nanostructured film formed of chitosan and methylcellulose nanoparticles that might substitute the plastic in packaging foods thus decreasing the ecological contamination done by plastics (Mura et al. 2011).

4 Various Methods for Preparation of Chitosan Nanoparticles Are Available Which Are as Follows

4.1 Emulsion Cross-Linking

Different group of researchers (Ohya et al. 1994; Yokohama et al. 1998; Kataoka et al. 2000) synthesized an emulsion of chitosan by emulsifying the aqueous solution of chitosan in oil. Using suitable surfactant, aqueous droplets are stabilized (Fig. 2). Cross-linker like glutaraldehyde is used to solidify the droplets produced after emulsification. Common cross-linkers used are glutaraldehyde, p-phthaldehyde, ascorbyl palmitate, and dehydroascorbyl palmitate.

For the first time, Ohya et al. (1994) described the synthesis of chitosan-gel nanospheres (CNSs) containing 5-fluorouracil (5-FU) or its derivatives (aminopentyl-carbamoyl- 5-FU or aminopentyl-ester-methylene-5FU) by emulsion method, and then cross-linkage was formed between glutaraldehyde and chitosan's amino groups. 5-FU derivatives were cross-linked with glutaraldehyde and then, using Schiff's base, concurrently immobilized to CNSs. The revolutionary studies reported the possibility of formation of even and reproducible chitosan nanoparticles so as to combine and deliver drugs. Regrettably, the harmful effects of chitosan nanoparticles, when cross-linked with glutaraldehyde, on viability of cell and

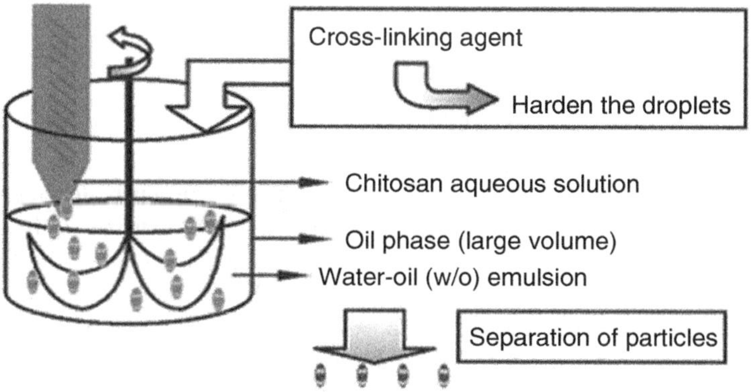

Fig. 2 Schematic presentation for synthesis of chitosan nanoparticles by emulsion cross-linking method. (Chenguang et al. 2007)

macromolecular drug integrity moved from the broad interest to less ruthless methods of forming nanospheres.

4.2 Ionotropic Gelation Method

This method is based on the principle of ionic bonding between positively charged amine groups of chitosan and negatively charged groups of polyanion-like tripolyphosphate (TPP) (Bodmeier et al. 1989; Pan et al. 2002). By varying proportion of chitosan and stabilizer, their size and surface charge can be manipulated (Calvo et al. 1997; Tang and Shi 2007). To produce at large scale, the chitosan/TPP weight ratio should be guarded and was established within the range of 3:1–6:1 (Prabaharan and Mano 2005; Gan et al. 2005). With increasing temperature, the average size of nanoparticle decreases (Tsai et al. 2008). Structural variations can also be introduced by changing the ionic strength such as by adding KCl, effecting ionic interactions between chitosan and TPP (López-León et al. 2005b).

4.3 Emulsion-Droplet Coalescence

Instead of using the cross-linkers to stabilize chitosan droplets, these are allowed to precipitate to coalesce with NaOH droplets. Gadolinium-bound chitosan nanoparticles (Gd-nano-CPs) were synthesized using this method (Tokumitsu et al. 1999; Van der Lubben et al. 2001) that combines both emulsion cross-linking and precipitation. A stable emulsion of chitosan was prepared using aqueous solution of chitosan in liquid paraffin oil and drug. Concurrently, the emulsion of chitosan containing aqueous solution of NaOH was prepared similarly. These two emulsions were mixed under high-speed stir; droplets of each emulsion would collide at random, coalesce and finally precipitate as small size particles.

4.4 Reverse Micellar Method

Rather than emulsion-based methods, the reverse micellization is better in producing ultrafine nanoparticles of about 100 nm/or even less, in comparison with the larger nanocarriers (> 200 nm) produced by emulsification techniques. For the first time, Mitra et al. (2001) produced chitosan nanoparticles by reverse micelles method. In this method, W/O droplets are formed in W/O system (Pileni 2006), in contrast to usual micelles that form in O/W environment. A W/O microemulsion is prepared using a lipophilic surfactant that is dissolved in an appropriate organic solvent, like n-hexane. For this, surfactant sodium bis (ethyl hexyl) sulfosuccinate (AOT) was dissolved in n-hexane. In AOT solution, chitosan (0.1% in acetic acid), doxorubicin-

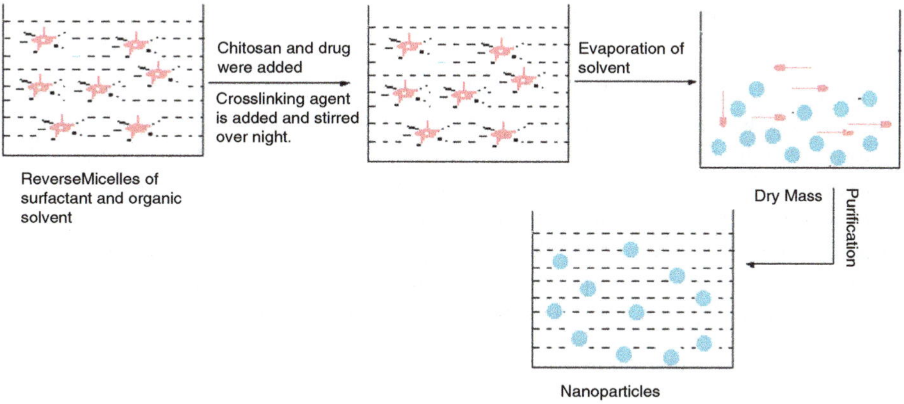

Fig. 3 Schematic representation of reverse micellar method. (Rajalakshmi et al. 2014)

dextran conjugate, liquid ammonia, and glutaraldehyde (0.01%) were added under continuous stirring. After that, nanoparticles were extracted following solvent evaporation (Fig. 3) (Mitra et al. 2001; Banerjee et al. 2002). An increase in the crosslinking rate leads to formation of larger particles (Banerjee et al. 2002).

Nevertheless, disadvantages, such as the difficult isolation of nanoparticles and the need for larger amounts of solvent, have been mentioned (Tang et al. 2007).

4.5 Polyelectrolyte Complex (PEC) Method

Polyelectrolyte complex method is based on ionic interactions between positively charged amine groups of chitosan and negatively charged carboxylic groups of Alg/sulfate groups of dextran sulfate leading to charge neutralization between cationic polymers and anionic group, which results in minimizing the hydrophilicity because of the self-assembly of the polyelectrolyte components (Tiyaboonchai 2003; Tiyaboonchai and Limpeanchob 2007; Erbacher et al. 1998; Sarmento et al. 2006). Their size may range from 50 to 700 nm.

4.6 Solvent Evaporation Method

In this method, in aqueous phase, polymer solution is emulsified followed by the evaporation of the polymer solvent which induces the precipitation of polymer as nanospheres are required. The formed nanospheres were filtered, washed using suitable solvent like petroleum ether, and finally dried. Drying is generally done by air or under vacuum. Metformin-loaded chitosan nanospheres were successfully produced by this method (Garud and Garud 2012).

4.7 Coprecipitation Method

By coprecipitation method, the broad degree of size evened chitosan-based nanoparticles was manufactured by fabrication of D,L-lactic acid on chitosan to exploit it as a drug transporter for extended drug release. For synthesizing the lactic acid-fabricated chitosan (LA-f-chitosan), the solvent was dehydrated and casted chitosan thin film possessing lactic acids. The LA-f-chitosan nanoparticles were formed by means of a coprecipitation route by LA-f-chitosan in ammonium hydroxide to make coacervate drops. Spherical and evenly distributed LA-f-chitosan nanoparticles having average diameter of 10 nm were prepared.

4.8 Complex Coacervation Method

The formation of chitosan–DNA nanoparticles was by means of coacervation among the amine groups (having positive charge) of chitosan and phosphate groups (having negative charge) of DNA (Bowman and Leong 2006; Chen et al. 2007; Leong et al. 1998). Various parameters such as surface charge, entrapment competence, particle size, and discharge features of the nanoparticles can be controlled by varying the weight percentage of the two polymers (Roy et al. 1999; Chen et al. 2003).

4.8.1 Method to Prepare Chemically Modified Chitosan Nanoparticles

Uchegbu et al. (1998) reported the production of chitosan nanovesicles of 300–600 nm by using palmitic acid-modified glycol chitosan with cholesterol. They exhibited better biocompatibility and hemocompatibility and stability in serum and bile salts. Furthermore, they were capable to stack bleomycin, a chemotherapeutic peptide. The loading procedure was performed via an ammonium sulfate gradient that drove the peptide into the carriers. Lee et al. (2001) prepared hydrophobically modified chitosan microspheres using 5.1 deoxycholic acid groups per 100 anhydroglucoses with 1-ethyl-3-(3-dimethylamino-propyl)-carbodiimide (EDC)-mediated coupling reaction.

5 Characterization of Chitosan Nanoparticles

The prepared chitosan nanoparticles are characterized in terms of zeta potential, particle size, surface analysis, and morphological study. The major challenge in characterization of chitosan nanoparticles is study of their particle size. Testing of particle size is critical to the implementation of strategy involved during industrialization.

5.1 Zeta Potential

Zeta potential testing is used to determine the magnitude of the electrostatic charges of the particles using ZetaMeter (Staunton, VA, USA). The principle of zeta potential is based on the laser Doppler electrophoresis. The colloidal particles will move toward the oppositely charged electrode on applying an electric field having motion proportional to the zeta potential. The particle motion is determined by laser Doppler anemometer (Rodrigues et al. 2012). Notably, the value of zeta potential displays the degree of electrostatic repulsion within charged nanoparticles. In colloidal system, higher value of zeta potential indicates the aggregation resistance of nanoparticles.

5.2 Dynamic Light Scattering (DLS)

Dynamic light scattering (DLS) gives the information about size distribution of the nanoparticles. It records the optical motions, because the floating particles use lucid light supply and are illuminated (Dubin et al. 1967). The light spread by floating nanoparticles has shifts in frequency separated to it from the time-dependent location or speed of the floating particles. The reason of the intensity of spread light to oscillate as a function of time was arbitrary to Brownian motion, and it has been found that nanoparticles are in constant dispersion rate. In photon correlation spectroscopy (PCS) instrument, the correlator used forms the correlation function G (τ) of the spread intensity. With an early correlation coefficient decay curve, the range of chitosan nanoparticles size should be ~300 nm and polydispersity index (PDI) 0.245.

5.3 Transmission Electron Microscopy (TEM)

TEM is taken to assess the shape, size, and uniformity of the nanoparticles. TEM image of chitosan nanoparticles depicts particles of spherical shape and having diameter of 300 nm, which can be compared to the size obtained by DLS measurements.

5.4 Surface Characterization

The surface characterization of nanoparticles is important to know the surface morphology of particles according to the desired application. The surface characteristics of nanoparticles are done by a scanning electron microscope (SEM). SEM images provide information about the nanoparticles surface. In SEM, nanoparticles are scanned with a focused beam of electrons, which collide with the nanoparticle atoms and generate signals that are detected with the help of highly sensitive detector. Prior to SEM observation, pretreatment of nanoparticles is done using thin film gold coating and vacuum treatment.

The more specific surface analysis of nanoparticles is done by the techniques X-ray photoelectron spectroscopy (XPS) and time-of-flight secondary ion mass spectrometry (ToF-SIMS). They reveal the origin identity of nanoparticles, their chemical bonding and high sensitivity. The XPS determines the uniform distribution of nanoparticles according to their size and also provides the chemical state of nanoparticles, while the ToF-SIMS shows the detailed examination of surface and near-surface composition (Rodrigues et al. 2012; Casettari et al. 2011).

5.5 Fourier Transform Infrared Spectroscopy (FTIR)

Fourier transform infrared spectroscopy is an important characterization technique for nanoparticles, which offers the chemical investigation of nanoparticles. The technique is considered as promising compared to other analytical techniques as it tells us about not only the structure of molecules but also about chemical bonding present in molecules. It is based upon the fact that nanoparticles exhibit a selective absorption within infrared region. They show close-packed absorption spectrum within this region, which is known as an IR absorption spectrum that can be comprehensive over a broad range of wavelength. The diverse absorption band deceit in an IR spectrum depicts the characteristic functional groups and chemical bonds occurring in nanoparticles. FTIR spectra of chitosan exhibit the well-built and broad peak in the 3500–3300 region due to hydrogen-bonded (O-H) stretching vibration. There were overlapping of peaks of primary amine N-H stretching and type II amide in the identical area. The peak at around 1150 cm^{-1} to C-O-C stretch asymmetric while the peak of C-N stretch type I amine was found at 1317 cm^{-1} (Casettari et al. 2011).

6 Conclusion

In food packaging the recyclable source-based biopolymers (cellulosic plastics, corn-derived plastics like polyhydroxyalkanoates (PHAs) and PLA, chitosan, and starch) are widely used to manufacture nanocomposites. The bio-nanocomposite barrier properties are improved in opposition to water vapor, CO_2, O_2, and flavor chemicals that have a foremost impact on increasing the shelf life of many processed and fresh foods. The degradability of the bio-nanocomposites is increased via appropriate selection of polymer matrix, which is as well a driver for the utilization of bio-nanocomposites. For particular application like in food packaging, the nanocomposites are used due to their much lighter weight than usual composite materials. The antimicrobial and antioxidative functions are added via formation of nanocomposites and a variety of types of nanoparticles like silver nanoparticles, nanoclays, silver zeolite, functional biopolymers like chitosan, and metal oxides. In spite of these advanced properties of bio-nanocomposites, there are a few security concerns regarding the utilization of nanocomposites like contact materials for food. A significant study is still necessary to assess the possible toxicity of nanocomposite materials, in addition to the environmental protection of their use. Bio-nanocomposite materials emerged to have an extremely brilliant future for a broad choice of applications in the food packaging and biomedical industries in addition to new active and smart food packaging by bio-functional properties.

References

Abdollahi M, Rezaei M, Farzi G (2012) A novel active bionanocomposite film incorporating rosemary essential oil and nanoclay into chitosan. J Food Eng 111(2):343–350

Agulló E, Rodríguez MS, Ramos V, Albertengo L (2003) Present and future role of chitin and chitosan in food. Macromol Biosci 3(10):521–530

American Chemistry Council (2015) Lifecycle of a plastic product. Available online: http://plastics.americanchemistry.com/Life-Cycle#uses. Accessed 2 Dec 2015

Arfat YA, Benjakul S, Prodpran T, Sumpavapol P, Songtipya P (2014) Properties and antimicrobial activity of fish protein isolate/fish skin gelatin film containing basil leaf essential oil and zinc oxide nanoparticles. Food Hydrocoll 41:265–273

Babu RP, O'connor K, Seeram R (2013) Current progress on bio-based polymers and their future trends. Prog Biomater 2(1):8

Banerjee T, Mitra S, Singh AK, Sharma RK, Maitra A (2002) Preparation, characterization and biodistribution of ultrafine chitosan nanoparticles. Int J Pharm 243:93–105

Barreteau H, Delattre C, Michaud P (2006) Production of oligosaccharides as promising new food additive generation. Food Technol Biotechnol 1:44(3)

Bodmeier R, Chen HG, Paeratakul O (1989) A novel approach to the oral delivery of micro- or nanoparticles. Pharm Res 6:413–417

Bowman K, Leong KW (2006) Chitosan nanoparticles for oral drug and gene delivery. Int J Nanomedicine 1:117–128

Calvo P, Remuñan-López C, Vila-Jato JL, Alonso MJ (1997) Chitosan and chitosan/ethylene oxide-propylene oxide block copolymer nanoparticles as novel carriers for proteins and vaccines. Pharm Res 14:1431–1436

Campana-Filho SP, Britto DD, Curti E, Abreu FR, Cardoso MB, Battisti MV, Sim PC, Goy RC, Signini R, Lavall RL (2007) Extraction, structures and properties of alpha-AND beta-chitin. Quím Nova 30(3):644 650

Caner C, Cansiz O (2007) Effectiveness of chitosan-based coating in improving shelf-life of eggs. J Sci Food Agric 87(2):227–232

Casettari L, Castagnino E, Stolnik S, Lewis A, Steven M, Illum HL (2011) Surface characterisation of bioadhesive PLGA/chitosan microparticles produced by supercritical fluid technology. Pharm Res 28:1668–1682

Chen Y, Mohanraj VJ, Parkin JE (2003) Chitosan-dextran sulfate nanoparticles for delivery of an anti-angiogenesis peptide. Lett Pept Sci 10:621–629

Chen Y, Mohanraj VJ, Wang F, Benson HA (2007) Designing chitosan-dextran sulfate nanoparticles using charge ratios. Am Assoc Pharmaceuticals Sci 8:E98

Chenguang L, Yulong T, Chengsheng L, Xiguang C, Lejun Yu (2007) Preparations, characterizations and applications of chitosan-based nanoparticles. J Ocean Univer China 6:237–243

De Moura MR, Avena-Bustillos RJ, McHugh TH, Krochta JM, Mattoso LH (2008) Properties of novel hydroxypropyl methylcellulose films containing chitosan nanoparticles. J Food Sci 73(7)

Di Maio L, Scarfato P, Milana MR, Feliciani R, Denaro M, Padula G, Incarnato L (2014) Bionanocomposite polylactic acid/organoclay films: functional properties and measurement of total and lactic acid specific migration. Packag Technol Sci 27(7):535–547

Dubin SB, Lunacek JH, Benedek GB (1967) Brownian motion using light scattering. Proc Natl Acad Sci U S A 57:1164

Dutta J, Dutta PK, Rinki K (2008) Current research on chitin and chitosan for tissue engineering applications and future demands on bioproducts. Current research and developments on chitin and chitosan in biomaterials science. Research Signpost, Trivandrum 167–86

Erbacher P, Zou S, Bettinger T, Steffan AM, Remy JS (1998) Chitosan-based vector/DNA complexes for gene delivery: biophysical characteristics and transfection ability. Pharm Res 15:1332–1339

Espinal-Ruiz M, Parada-Alfonso F, Restrepo-Sánchez LP, Narváez-Cuenca CE, McClements DJ (2014) Impact of dietary fibers [methyl cellulose, chitosan, and pectin] on digestion of lipids under simulated gastrointestinal conditions. Food Funct 5(12):3083–3095

Ferreira AR, Alves VD, Coelhoso IM (2016) Polysaccharide-based membranes in food packaging applications. Membranes 6(2):22

Fukuda H (1980) Polyelectrolyte complexes of chitosan with sodium carboxymethylcellulose. Bull Chem Soc Jpn 53(4):837–840

Gan Q, Wang T, Cochrane C, McCarron P (2005) Modulation of surface charge, particle size and morphological properties of chitosan-TPP nanoparticles intended for gene delivery. Colloids Surf B: Biointerfaces 44:65–73

Garud N, Garud (2012) A preparation and in-vitro evaluation of metformin microspheres using non-aqueous solvent evaporation technique. Trop J Pharm Res 11:577–583

Gecol H, Ergican E, Miakatsindila P (2005) Biosorbent for tungsten species removal from water: effects of co-occurring inorganic species. J Colloid Interface Sci 292(2):344–353

Giannelis EP (1996) Polymer layered silicate nanocomposites. Adv Mater 8:29–35

Goy RC, Britto DD, Assis OB (2009) A review of the antimicrobial activity of chitosan. Polímeros 19(3):241–247

Jeon SJ, Oh M, Yeo WS, Galvao KN, Jeong KC (2014) Underlying mechanism of antimicrobial activity of chitosan microparticles and implications for the treatment of infectious diseases. PLoS One 9(3):92723

Kanmani P, Rhim JW (2014) Physical, mechanical and antimicrobial properties of gelatin based active nanocomposite films containing AgNPs and nanoclay. Food Hydrocoll 35:644–652

Kataoka K, Matsumoto T, Tokohama M, Okano T, Sakurai Y (2000) Doxorubicin-loaded poly (ethylene glycol)- poly(b-benzyl-L-aspartate) copolymer micelles: their pharmaceutical characteristics and biological significance. J Control Release 64:143–153

Kerch G (2015) The potential of chitosan and its derivatives in prevention and treatment of age-related diseases. Mar Drugs 13(4):2158–2182

Kumar MN (2000) A review of chitin and chitosan applications. React Funct Polym 46(1):1–27

Kumirska J, Czerwicka M, Kaczyński Z, Bychowska A, Brzozowski K, Thöming J, Stepnowski P (2010) Application of spectroscopic methods for structural analysis of chitin and chitosan. Mar Drugs 8(5):1567–1636

Kurita K, Ishii S, Tomita K, Nishimura SI, Shimoda K (1974) Reactivity characteristics of squid β-chitin as compared with those of shrimp chitin: high potentials of squid chitin as a starting material for facile chemical modifications. J Polym Sci A Polym Chem 32(6):1027–1032

Lee KY, Kim JH, Kwon IC, Jeong SY (2001) Self-aggregates of deoxycholic acid-modified chitosan as a novel carrier of adriamycin. Colloid Polym Sci 278:1216–1219

Lee JH, Song NB, Jo WS, Song KB (2014) Effects of nano-clay type and content on the physical properties of sesame seed meal protein composite films. Int J Food Sci Technol 49:1869–1875

Leong KW, Mao HQ, Truong-Le VL, Roy K, Walsh SM, August JT (1998) DNA-polycation nanospheres as non-viral gene delivery vehicles. J Control Release 53:183–193

Liu SH, Chang YH, Chiang MT (2010) Chitosan reduces gluconeogenesis and increases glucose uptake in skeletal muscle in streptozotocin-induced diabetic rats. J Agric Food Chem 58 (9):5795–5800

López-León T, Carvalho EL, Seijo B, Ortega-Vinuesa JL, Bastos-González D (2005a) Physicochemical characterization of chitosan nanoparticles: electrokinetic and stability behavior. J Colloid Interface Sci 283(2):344–351

López-León T, Carvalho EL, Seijo B, Ortega-Vinuesa JL, Bastos-González D (2005b) Physicochemical characterization of chitosan 341 nanoparticles: electrokinetic and stability behavior. J Colloid Interface Sci 283:344–351

Lu Y, Weng L, Zhang L (2004) Morphology and properties of soy protein isolate thermoplastics reinforced with chitin whiskers. Biomacromolecules 5(3):1046–1051

Mhurchu CN, Poppitt SD, McGill AT, Leahy FE, Bennett DA, Lin RB, Ormrod D, Ward L, Strik C, Rodgers A (2004) The effect of the dietary supplement, chitosan, on body weight: a randomised controlled trial in 250 overweight and obese adults. Int J Obes 28(9):1149–1156

Mitra S, Gaur U, Ghosh PC, Maitra AN (2001) Tumor targeted delivery of encapsulated dextran-doxorubicin conjugate using chitosan nanoparticles as carrier. J Control Release 74:317–323

Mokhtar SM, Youssef KM, Morsy NE (2010) The effects of natural antioxidants on colour, lipid stability and sensory evaluation of fresh beef patties stored at 4 °C. J Agroaliment Process Technol 20(3):282–292

Mura S, Corrias F, Stara G, Piccinini M, Secchi N, Marongiu D, Innocenzi P, Irudayaraj J, Greppi GF (2011) Innovative composite films of chitosan, methylcellulose, and nanoparticles. J Food Sci 76(7)

Muratore G, Nobile D, Buonocore GG, Lanza CM, Asmundo N (2005) The influence of using biodegradable packaging films on the quality decay kinetic of plum tomato (PomodorinoDatterino®). J Food Eng 67(4):393–399

Muzzarelli RA (1999) Clinical and biochemical evaluation of chitosan for hypercholesterolemia and overweight control. EXS 87:293–304

Nafchi AM, Nassiri R, Sheibani S, Ariffin F, Karim AA (2013) Preparation and characterization of bionanocomposite films filled with nanorod-rich zinc oxide. Carbohydr Polym 96(1):233–239

Noishiki Y, Takami H, Nishiyama Y, Wada M, Okada S, Kuga S (2003) Alkali-induced conversion of β-chitin to α-chitin. Biomacromolecules 4(4):896–899

Ohya Y, Shiratani M, Kobayashi H, Ouchi T (1994) Release behavior of 5-fluorouracil from chitosan-gel nanospheres immobilizing 5-fluorouracil coated with polysaccharides and their cell specific cytotoxicity. Pure Appl Chem A 31:629–642

Pan Y, Li YJ, Zhao HY, Zheng JM, Xu H, Wei G, Hao JS, Cui FD (2002) Bioadhesive polysaccharide in protein delivery system: chitosan nanoparticles improve the intestinal absorption of insulin in vivo. Int J Pharm 249:139–147

Pileni MP (2006) Reverse micelles used as templates: a new understanding in nanocrystal growth. J Exp Nanosci 1:13–27

Plackett D (2011) Biopolymers: new materials for sustainable films and coatings. Wiley, Chichester

Plastics Europe (n.d.) Packaging. Available online: http://www.plasticseurope.org/use-of-plastics/packaging.aspx. Accessed 2 Dec 2015

Prabaharan M, Mano JF (2005) Chitosan-based particles as controlled drug delivery systems. Drug Deliv 12:41–57

Qi L, Xu Z, Jiang X, Hu C, Zou X (2004) Preparation and antibacterial activity of chitosan nanoparticles. Carbohydr Res 339(16):2693–2700

Rabea EI, Badawy ME, Stevens CV, Smagghe G, Steurbaut W (2003) Chitosan as antimicrobial agent: applications and mode of action. Biomacromolecules 4(6):1457–1465

Radhakrishnan Y, Gopal G, Lakshmanan CC, Nandakumar KS (2015) Molecular and genetic medicine

Rajalakshmi R, Aruna U, Vinesha V, Rupangada V (2014) Chitosan nanoparticles-An emerging trend in nanotechnology. Int J Drug Deliv 6:204–229

Ray SS, Okamoto M (2003) Polymer/layered silicate nanocomposites: a review from preparation to processing. Prog Polym Sci 28(11):1539–1641

Ray S, Quek SY, Easteal A, Chen XD (2006) The potential use of polymer-clay nanocomposites in food packaging. Int J Food Eng 18:2(4)

Reddy JP, Rhim JW (2014) Characterization of bionanocomposite films prepared with agar and paper-mulberry pulp nanocellulose. Carbohydr Polym 110:480–488

Rhim JW, Park HM, Ha CS (2013) Bio-nanocomposites for food packaging applications. Prog Polym Sci 38(10):1629–1652

Rodrigues S, da Costa AMR, Grenha A (2012) Chitosan/carrageenan nanoparticles: effect of cross-linking with tripolyphosphate and charge ratios. Carbohydr Polym 89:282–289

Roy K, Mao HQ, Huang SK, Leong KW (1999) Oral gene delivery with chitosan – DNA nanoparticles generates immunologic protection in a murine model of peanut allergy. Nat Med 5:387–391

Sannan T, Kurita K, Iwakura Y (1976) Studies on chitin, 2. Effect of deacetylation on solubility. Macromol Chem Phys 177(12):3589–3600

Sanuja S, Agalya A, Umapathy MJ (2014) Studies on magnesium oxide reinforced chitosan bionanocomposite incorporated with clove oil for active food packaging application. Int J Polym Mater Polym Biomater 63(14):733–740

Sarmento B, Martins S, Ribeiro A, Veiga F, Neufeld R, Ferreira D (2006) Development and comparison of different nanoparticulate polyelectrolyte complexes as insulin carriers. Int J Pept Res Ther 12:131–138

Shahidi F, Arachchi JK, Jeon YJ (1999) Food applications of chitin and chitosans. Trends Food Sci Technol 10(2):37–51

Sorrentino A, Gorrasi G, Vittoria V (2007) Potential perspectives of bio-nanocomposites for food packaging applications. Trends Food Sci Technol 18(2):84–95

Sriupayo J, Supaphol P, Blackwell J, Rujiravanit R (2005) Preparation and characterization of α-chitin whisker-reinforced chitosan nanocomposite films with or without heat treatment. Carbohydr Polym 62(2):130–136

Suman SP, Mancini RA, Joseph P, Ramanathan R, Konda MK, Dady G, Yin S (2010) Packaging-specific influence of chitosan on color stability and lipid oxidation in refrigerated ground beef. Meat Sci 86(4):994–998

Tang ZX, Shi LE (2007) Adsorption of neutral proteinase on chitosan nano-particles. Biotechnol Biotechnol Equip 21:223–228

Tang ZX, Qian JQ, Xi LE (2007) Preparation of chitosan nanoparticles as carrier for immobilized enzyme. Appl Biochem Biotechnol 136:77–96

Tang XZ, Kumar P, Alavi S, Sandeep KP (2012) Recent advances in biopolymers and biopolymer-based nanocomposites for food packaging materials. Crit Rev Food Sci Nutr 52(5):426–442

Tangpasuthadol V, Pongchaisirikul N, Hoven VP (2003) Surface modification of chitosan films: effects of hydrophobicity on protein adsorption. Carbohydr Res 338(9):937–942

Thakur VK, Thakur MK (2016) Handbook of sustainable polymers: processing and applications. CRC Press

Tiyaboonchai W (2003) Chitosan nanoparticles: a promising system for drug delivery. Naresuan Univ J 11:51–66

Tiyaboonchai W, Limpeanchob N (2007) Formulation and characterization of amphotericin B-chitosan-dextran sulfate nanoparticles. Int J Pharm 329:142–149

Tokumitsu H, Ichikawa H, Fukumori Y (1999) Chitosan gadopentetic acid complex nanoparticles for gadolinium neutron capture therapy of cancer: preparation by novel emulsion droplet coalescence technique and characterization. Pharm Res 16:1830–1835

Tolaimate A, Desbrieres J, Rhazi M, Alagui A (2003) Contribution to the preparation of chitins and chitosans with controlled physico-chemical properties. Polymer 44(26):7939–7952

Trovatti E, Fernandes SC, Rubatat L, Freire CS, Silvestre AJ, Neto CP (2012) Sustainable nanocomposite films based on bacterial cellulose and pullulan. Cellulose 19(3):729–737
Tsai ML, Bai SW, Chen RH (2008) Cavitation effects versus stretch effects resulted in different size and polydispersity of ionotropic gelation chitosan–sodium tripolyphosphate nanoparticle. Carbohydr Polym 71:448–457
Tumwesigye KS, Oliveira JC, Sousa-Gallagher MJ (2016) New sustainable approach to reduce cassava borne environmental waste and develop biodegradable materials for food packaging applications. Food Packag Shelf Life 7:8–19
Uchegbu IF, Schatzlein AG, Tetley L, Gray AI, Sludden J (1998) Polymeric chitosan-based vesicles for drug de livery. J Pharm Pharmacol 50:453–458
Upadhyaya L, Singh J, Agarwal V, Tewari RP (2014) The implications of recent advances in carboxymethyl chitosan based targeted drug delivery and tissue engineering applications. J Control Release 186:54–87
Van der Lubben IM, Verhoef JC, Borchard G, Junginger HE (2001) Chitosan and its derivatives in mucosal drug and vaccine delivery. Eur J Pharm Sci 14:201–207
Vargas M, Gonzalez-Martinez C (eds) (2010) Recent patents on food applications of chitosan. Recent Pat Food Nutr Agric 2(2):121–128
Yen MT, Yang JH, Mau JL (2008) Antioxidant properties of chitosan from crab shells. Carbohydr Polym 74(4):840–844
Yokohama M, Fukushima S, Uehara R, Okamoto K, Kataoka K (1998) Characterization of physical entrapment and chemical conjugation of adriamycin in polymeric micelle and their design for in vivo delivery to a solid tumor. J Control Release 50:79–92
Zhang J, Xia W, Liu P, Cheng Q, Tahi T, Gu W, Li B (2010) Chitosan modification and pharmaceutical/biomedical applications. Mar Drugs 8(7):1962–1987

Nanocomposites in Packaging: A Groundbreaking Review and a Vision for the Future

Sukanchan Palit and Chaudhery Mustansar Hussain

Abstract The world of nanotechnology and nanomaterials are witnessing dramatic challenges and moving from one visionary paradigm toward another. In the similar manner, nanocomposite applications are surpassing vast and versatile scientific boundaries. Material science and nanotechnology are two opposite sides of the coin today. Human civilization's immense scientific prowess, scientific prudence, and scientific validation will all lead a visionary way in the true emancipation of science of nanotechnology. The authors in this paper deeply elucidate on the success, the vast potential, and the deep scientific and technological ingenuity in the applications of nanocomposites and composites in the packaging domain. Technology, engineering, and science are today in the path of newer scientific regeneration and deep vision. Chemical engineering, environmental engineering, and many diverse areas of engineering science are connected with the science of nanotechnology. In this well-researched chapter, the authors focus on the needs of nanotechnology, nanomaterials, and composites to human society. Composite science and material science are the needs of civilization and scientific progress today. In this review paper, the authors comprehend the vision of the application of nanocomposites and polymer science in packaging. Packaging technology is highly advanced today but still not fully explored. This well-researched treatise explores the hidden scientific truth of application of polymer science, composite science, and material science in packaging and the vast vision behind it.

Keywords Vision · Technology · Composites · Science · Polymers · Packaging

S. Palit
Department of Chemical Engineering, University of Petroleum and Energy Studies, Dehradun, Uttarakhand, India

C. M. Hussain (✉)
Department of Chemistry and Environmental Sciences, New Jersey Institute of Technology, Newark, NJ, USA
e-mail: chaudhery.m.hussain@njit.edu

S. Ahmed (ed.), *Bio-based Materials for Food Packaging*,
https://doi.org/10.1007/978-981-13-1909-9_13

1 Introduction

Human civilization and progress today stands in the midst of deep scientific vision and vast scientific ingenuity. Technology and engineering science in similar manner stands in the juncture of scientific farsightedness and vast profundity. Material science, nanotechnology, and composite science are witnessing immense challenges and vast scientific understanding (Palit 2017b). Sustainability whether it is energy, environmental, social, or economic are facing immense challenges and deep scientific revelation. Global research and development initiatives should be targeted toward newer vision, newer innovations, and newer scientific instinct. The state of global scientific order is immensely dismal. Global climate change, the climate crisis, depletion of fossil resources, and the frequent environmental catastrophes are challenging the vast scientific firmament (Palit 2017b). In this paper, the authors target on the scientific advancements and the vast scientific ingenuity in the applications of nanocomposites in packaging and other diverse areas of engineering and science. Composite science and "smart materials" are the next-generation scientific endeavor today. Scientific endeavor in the field of composite science and material science needs to be envisioned and addressed as science and technology surges forward toward a newer visionary paradigm. Today, the scientific world stands in the juncture of deep scientific prudence and vast scientific stewardship. Global concerns for energy and environmental sustainability have urged the vast scientific community to gear forward toward newer scientific innovation and newer scientific instinct. In this paper the authors focus on the needs of composite science in packaging, the vast scientific vision, and the ingenuity behind nanocomposite applications in packaging. Packaging is a vast area of science and technology today. The need of polymer technology, composite science, and material science to human scientific progress is immense and groundbreaking in modern civilization and modern science. Energy, water, food, and shelter are the pivotal parameters toward the growth of civilization and human scientific pursuit today. In the similar manner, nanoscience, nanotechnology, and the vast area of nanomaterials are moving toward newer scientific regeneration. This chapter uncovers the scientific success, the purposeful and definite scientific vision, and the needs of composite science toward the furtherance of science and technology globally (Palit 2014b).

2 The Vision of This Study

The world of difficulties in the field of nanocomposites, composite science, and material science are immense and pathbreaking. Human scientific research forays today stand in the midst of deep revelation and introspection. Packaging science is itself a huge pillar with a purposeful vision of its own. The application of nanocomposites and composites to packaging science is a major pillar of this entire scientific endeavor. Mankind scientific wisdom, the truth, and the needs for energy and environmental sustainability will all lead a visionary way in the true emancipation of science and technology today. The vision, aim, and objective of this study are toward research forays in the field of nanocomposites, material science, and the vast world of nanotechnology (Palit 2017b). The challenges and the vision of science in the research pursuit in nanocomposites are vast and surpassing vast and versatile scientific frontiers. In this chapter, the authors deeply comprehend the vast necessity, the truth, and the scientific judgment in the application of nanocomposites and composite science in packaging and diverse areas of science and engineering (Palit 2017b).

3 What Do You Mean by Nanocomposites?

Nanotechnology is a visionary and promising area of scientific endeavor in modern civilization. Technology, engineering, and science of nanoscience and nanotechnology are surpassing vast scientific boundaries. "Nano-" can be defined as nanometer (10^{-9} m). The visionary and groundbreaking concept of nanotechnology was introduced by Richard Feynman, the noted physicist, in 1959 at a meeting of the American Physical Society (Palit 2017b). Since then the world witnessed immense scientific difficulties and vast scientific upheavals in the field of nanotechnology. Today the world of science and technology stands in the midst of revival and scientific vision. It has today become an interdisciplinary branch of applied science and technology. Technological challenges, foresight, and the deep scientific stewardship today are leading a visionary way in the true realization of nanotechnology science. Nanotechnology is the ability to work on a scale of about 1–100 nm. Human scientific vision and human scientific conscience are at its helm as nanotechnology surges forward. Because of their size, nanoparticles have proportionally larger surface area. The challenges and the vision of nanotechnology and nanocomposite applications are immense and path-reaching (Palit 2017b).

4 Scientific Doctrine and the Scientific Vision Behind Composite Science

Scientific vision in the field of composite science and nanocomposites is vast. Technology, engineering, and science of nanoscience and nanotechnology are surpassing vast scientific boundaries. Civilization's immense scientific truth, the scientific prowess of nanotechnology, and the vast scientific prudence will all lead an effective way in the true realization of nanocomposite science and composite science today. Today scientific doctrine in every branch of engineering science needs to be addressed and envisioned as science and technology surges forward toward a newer visionary era. Polymer technology, composite science, and the vast domain of "smart materials" are witnessing immense challenges and vast foresight. The world of science and engineering today stands in the juncture of deep restructuring and immense technological profundity (Palit 2017b). Climate change, frequent environmental calamities, and the grave concerns for sustainability are the pallbearers toward a newer scientific order globally today. Nanotechnology and nano-vision are the coinwords of today's research pursuit in engineering science and technology. In this paper, the authors focus on the scientific success, the scientific needs to human society, and the world of difficulties in the field of nanocomposite science and packaging today. Scientific progress is today globally in a state of immense catastrophe. Global warming, global climate change, and ozone layer depletion are transforming the face of mankind today. This paper veritably opens up newer scientific innovation, scientific instinct, and the deeper cause of environmental sustainability (Palit 2017b). Thus the world of scientific challenges will surely see a new day and a newer age in the field of science, technology, and engineering (Palit 2017b).

5 The World of Challenges, the Scientific Profundity, and the Vision Behind Polymer Science and Material Science

The vision behind polymer science and material science is groundbreaking in today's scientific scenario. Technology and engineering science today stands amidst vast vision and immense ingenuity. The scientific profundity, the scientific discernment, and the scientific wisdom are the needs of research endeavor today (Palit 2017b). Composite science and nanocomposite technology are surging ahead in today's scientific landscape in technology and engineering science. Packaging is one such example of nanocomposite applications. The application domain of nanocomposites is vast and versatile. Material science, polymer science, and nanocomposites are the newer visionary domains in the field of technology and engineering science today (Palit 2017b). Human civilization's immense scientific ingenuity, scientific farsightedness, and the vision of nanoscience and nanotechnology will surely lead a long way in the true emancipation of science and engineering today. Challenges, difficulties, and barriers of scientific endeavor in material science are immense as well as groundbreaking (Palit 2017b). This paper unfolds the scientific intricacies of nanotechnology applications in packaging in today's scientific age. The challenges and the vision of nanotechnology and composites are deeply elucidated in this paper. In this paper, the authors also reiterate the technological validation and scientific vision of nanotechnology applications in human society.

6 Significant Scientific Endeavor in the Field of Nanotechnology and Nanocomposites

Human research endeavor in composite science and polymer science are witnessing vast challenges as science and engineering moves forward. Technology and engineering science of polymer engineering and polymer technology are today challenging the vast scientific firmament (Palit 2017b). Today is the scientific world of composite science, polymer science, nanomaterials, and engineered nanomaterials. In this entire article, the authors reiterate on the success of nanotechnology applications in human society and the needs of science in scientific advancements of human civilization. Technological and scientific validation is the need of proliferation of science and engineering today. This global vision of material science is elucidated in details in this article (Palit 2014b, 2017b).

Salehi (2013, 2014) deeply discussed with deep scientific ingenuity and conscience current and future research trends in nanofiltration technology in food processing. Nanotechnology in food technology is veritably changing the scientific genre in the domain of science and technology. Membrane processing is the new coinword of scientific research pursuit today. Nanofiltration technology is still in the latent scientific stage, finding more and more applications in food processing/ technology, and is seen as an alternative to conventional techniques (Salehi 2014; Palit 2017b). The goal of this well-researched treatise is to present the recent endeavor and the future research trends of nanofiltration processes in the food industry (Salehi 2014). Recent research pursuit has deeply highlighted the potential of nanofiltration use in diverse areas, including water softening, wastewater treatment, vegetable oil processing, and beverage, dairy, and sugar industry (Salehi 2014). Science is today a huge pillar with a purposeful vision of its own. Human scientific research pursuit in membrane processing are visionary and engulfed with deep scientific acuity and farsightedness. The authors discussed pressure-driven membrane processes, water softening techniques, applications in wastewater treatment, and applications in beverage and dairy industries. A deep review on applications in sugar industry and vegetable oil processing industry is the other salient feature of this research endeavor (Salehi 2014; Palit 2017b).

United Nations Global Sustainable Development Report (2016) deeply elucidates with deep and cogent insight sustainable development goals and the visionary targets of sustainability. The status and the challenges of environmental and energy sustainability in today's world are immense and far-reaching. The authors discussed deeply the 2030 sustainability agenda, the infrastructure-inequality-resilience issues, perspectives of scientists on engineering and the sustainable development goals, and prioritization of emerging issues for sustainable development (United Nations Global Sustainable Development Report 2016). Scientific vision and the vast domain of scientific ardor are in the process of newer scientific rejuvenation. The targets of this treatise are the adoption of 2030 Agenda for Sustainable Development. The report was prepared specifically to present the discussions at the highest level political discussions on sustainable development in 2016 (United Nations Global

Sustainable Development Report 2016). Human scientific progress, scientific provenance, and vast revelation are the cornerstones of this well-researched report. The first chapter of this report explores the vast implications of leaving no one behind for the operationalization of the visionary concepts of sustainable development goals from a science-policy perspective (United Nations Global Sustainable Development Report 2016). The vision and the challenges of sustainable development whether it is energy or environment are deeply discussed in this entire treatise (Palit 2017b). The content of this report is based on the knowledge and expertise of scientists, engineers, and technologists from more than 20 United Nations bodies (United Nations Global Sustainable Development Report 2016). Technological and scientific validation, the needs of science and engineering, and the future of sustainability science will surely be the forerunners toward a newer visionary eon of science. The ambition to endeavor to reach the extremes of research pursuit is an important aspect of the 2030 agenda (Palit 2017b). This report also exemplifies interlinkages between infrastructure, inequality, and resilience. Scientific vision, scientific acuity, and articulation are the cornerstones of this report. The vast effect of infrastructure on resilience is an area of grave concern that has received much attention by the scientific community (United Nations Global Sustainable Development Report 2016). The entire report deeply enumerates the scientific success, the scientific ingenuity, and the vast needs of human scientific endeavor in modern science and modern human civilization (United Nations Global Sustainable Development Report 2016).

Ali and Sinha (2014) discussed with cogent and lucid insight the challenges in nanotechnology innovation in India. Nanotechnology is one of the promising areas of scientific endeavor in the twenty-first century, and today it is a groundbreaking and far-reaching area of science. It is an interdisciplinary science domain with vast applications in biotechnology, applied mathematics, computer science, electronics, communication, medical and food, energy production, and new materials (Ali and Sinha 2014; Palit 2017b). Nanomaterials and engineered nanomaterials are the frontiers of scientific research endeavor today. Innovation has immense impact on the economic advancement of a nation. Nanotechnology has received much global interest, and many national governments are making large investments in nanotechnology research and development initiatives (Ali and Sinha 2014). Science and technology are the visionary domains of scientific endeavor today. This well-researched paper is an attempt to explore the nanotechnology research and development initiatives in the area of energy, water, food, shelter, education, and agriculture sector in the Indian context (Ali and Sinha 2014). It also highlights the outcome of nanotechnology in the publications and patent numbers in the environmental, health, and safety issues in India. Nanotechnology is today a latent domain and encompasses disciplines such as applied physics, material science, physical chemistry, physics of condensed matter, biochemistry, molecular biology, composite science, biotechnology, and polymer engineering. Scientific revelation and scientific provenance are the pillars of research pursuit in this paper. India's research and development initiatives need to be envisioned with the passage of time (Ali and Sinha 2014). This entire report deeply enumerates on the opportunities and challenges in research and development forays in the field of nanotechnology and other branches of nanoscience in present-day India (Ali and Sinha 2014; Palit 2017b).

Yadav (2017) elucidated with lucid insight the potential of nanotechnology for agriculture and food engineering. In the current revolution in food sector, nanotechnology is one of the promising tools. Food technology is one of the ever-growing and revolutionary areas of scientific innovation today. It is also a promising tool for the agri-food and agriculture sector. The authors in this paper deeply discussed nanotechnology in agriculture, nanotechnology in the food sector, and safety concerns and regulatory laws. Human scientific research forays today are in the path of newer scientific regeneration. Nanotechnology, a recently developed field of science, is a branch of science and engineering which encompasses physics, chemistry, biology, and engineering sciences and now rapidly growing into electronics, automobiles, agriculture, food, and various other industrial and engineering systems. Scientific profundity and deep scientific vision stands as a major pillar of this entire research pursuit (Yadav 2017).

7 Significant Scientific Endeavor in the Field of Application of Nanocomposites in Packaging

Nanocomposites are the future generation smart materials and eco-materials. Nanotechnology is today integrated with diverse areas of science and technology. The world of difficulties and barriers, the vision of nanotechnology, and the needs of human society are all the torchbearers toward a new era in science and technology. Nanocomposites and its vast and versatile applications are today surpassing scientific boundaries. Material science and polymer science today are in the critical juncture of deep scientific regeneration and vast scientific rejuvenation. In this paper, the authors focus on the scientific potential, the scientific ingenuity, and the vision of the applications of nanocomposites in diverse areas of science and engineering (Palit 2017b).

de Azeredo (2009) discussed with deep insight and scientific conscience nanocomposites for food packaging uses and applications. Most materials used in packaging today are non-biodegradable resulting in immense environmental engineering problems (de Azeredo 2009). Environmental engineering science today stands in the crucial juncture of scientific comprehension and vast vision. Several biopolymers have been used to develop materials for eco-friendly food packaging and to do research in the area of eco-materials with a clear vision toward vast emancipation of environmental engineering science. Technological and scientific advancements in the field of nanocomposites are thus witnessing immense scientific vision and scientific fortitude. Most reinforced materials have poor matrix-filler interactions, which tend to enhance with decreasing filler dimensions. Thus the ingenuity and the scientific vision of nanocomposites and composite science. The use of fillers with at least one nanoscale dimension leads to the formation of nanocomposites or composites (de Azeredo 2009). Nanoparticles have proportionally larger surface area than the microscale counterparts (de Azeredo 2009). These nanocomposites when added to polymers have other functions such as antimicrobial activity, enzyme immobilization, sensing, etc. The path toward scientific regeneration in the field of biopolymers and nanocomposites is immense and certainly groundbreaking. This paper vastly unfolds the scientific ingenuity in the field of nanocomposites and the wide world of polymer science. Technological profundity and ingenuity, the scientific needs of human society, and the futuristic vision of nanotechnology will definitely open a chapter in the field of science and technology today. By using nanotechnology techniques, it is definitely possible to reassemble molecules into objects, along several length scales as nature always does (Palit 2017b). Human scientific regeneration in the field of nanotechnology and nanocomposites is witnessing immense vision and forbearance as science and engineering surges forward. In present-day human civilization and human scientific progress, most materials used for packaging are non-biodegradable resulting in a serious environmental issue (Palit 2017b). New bio-based materials have been envisioned to develop edible and biodegradable films with a clear vision to reducing environmental and industrial waste. The challenges, vision, and the goals of science and technology are immense in modern science and modern civilization today (Palit 2017b). Scientific research pursuit in biopolymers and nanocomposites needs to be addressed and re-envisioned with the passage of time. This challenge is deeply enumerated in this paper. The authors in this paper discussed nanoreinforcements, structure, properties, and types of composites and clays and silicates (de Azeredo 2009; Palit 2017b). The other salient features of this treatise are the techniques to improve the compatibility of clays with polymers and the applications of clay nanocomposites (de Azeredo 2009). The other pillars are cellulose-based nanoreinforcements, their structure and obtainment and applications and effects on polymer matrices (Palit 2017b). Scientific vision, scientific ardor, and deep scientific foresight are the pillars of this well-researched paper (Palit 2014b, 2017b).

Bratovcic et al. (2015) discussed with immense lucidity and cogent insight application of nanocomposite materials in food packaging. The world of challenges in polymer technology, the futuristic vision of nanocomposite applications, and the wide world of composite science will all lead a long way in the true realization of modern science. The term “nano” can be defined as nanoparticle size from 1 to 100 nanometers. The terminology “nanotechnology” was first propounded by Norio Taniguchi in 1974. Human scientific vision and scientific forbearance were immensely important as science and engineering surged forward toward a newer eon. The primary function of packaging is to maintain quality and safety of products during transport and storage as well as to extend its viability by preventing unwanted agents such as microorganisms and chemical contaminants. The entire paper touched upon the application of polymers and biopolymers in the vast world of packaging. The barrier applications of polymer nanocomposites, nanocomposite formation, the domain of biopolymers, nanocoating, antimicrobial systems, and the innovative world of intelligent packaging are enumerated in lucid details in this paper (Bratovcic et al. 2015). Technological and scientific validation, the needs for polymer science in human society, and the deep scientific evolution of biopolymers are the salient features of this paper. Nano-polymers and nanosensors are the next-generation smart materials and have immense applications in packaging as biodegradability stands as a primordial scientific issue. The entire treatise unfolds the scientific intricacies, the scientific vision, and the deep scientific insight in biopolymers and nanocomposites (Palit 2017b).

Ray et al. (2006) discussed with vast scientific insight the vast and emerging use of polymer-clay nanocomposites in food packaging. Scientific research pursuit, the vast scientific vision, and scientific introspection will all lead a visionary way in the true realization of nanotechnology in human society. With today's advancement and research forays in nanoscience, polymer-clay nanocomposites have emerged as a promising and novel food packaging material due to its several benefits such as enhanced mechanical, thermal, and barrier properties (Ray et al. 2006). This paper discusses with immense lucidity and vision the potential use of these polymer composites as novel food packaging materials with emphasis on preparation, characterization, and future visionary prospects. Nanocomposite science is today in the avenue of newer scientific regeneration and is surpassing vast scientific boundaries (Ray et al. 2006). To meet the vast needs of customers, food must be safe, of consistently good quality and immense sensory issues, and inexpensive and should have good shelf life. Here comes the importance of polymer technology and composite science. These vast issues have led to extensive investigations and research and development forays in suitable packaging for food items (Ray et al. 2006; Palit 2017b). Technology and engineering science of polymers and smart materials are in the process of immense scientific vision today. In the recent age, a new and emerging class of polymer-clay composites has been developed. The march of composite science and nanocomposites are challenging diverse areas of engineering science today (Ray et al. 2006). In this paper, the authors touched upon preparation and characterization of nanocomposites, properties of nanocomposites, thermal stability, and flammability reduction (Ray et al. 2006). The authors also deeply delineated biodegradable polymer-clay nanocomposites and the vast future prospects attached to these nanocomposites (Ray et al. 2006).

Muller et al. (2017) with deep insight and scientific conscience reviewed the processing and properties of polymer nanocomposites and their vast and versatile applications in the packaging, automotive, solar energy, and renewable energy fields (Palit 2017b). Human scientific prowess, the vast scientific ingenuity, and the futuristic vision of technology are all the pallbearers toward a newer era in the field of nanotechnology and composite science. For the last few decades, nanocomposite materials have been extensively studied in the scientific papers as they provide immense property enhancements, even at low nanoparticle content (Muller et al. 2017; Palit 2017b). Their performance depends on a number of properties, but the nanoparticle dispersion and distribution state stands as a major challenge in order to obtain the full nanocomposites' potential in terms of flame retardance, mechanical and thermal properties, etc. (Muller et al. 2017). This review deeply reviewed more in-depth literature on the properties and materials of immense importance in three target sectors: packaging, solar energy, renewable energy, and automotive (Muller et al. 2017). Technological profundity, scientific ardor, and vast scientific validation of nanocomposite applications are the pillars of this well-researched treatise. The authors also lucidly discussed processing techniques and application domain of nanocomposites (Palit 2017b). Today composite science and technological ingenuity are the two opposite sides of the coin. The challenge and the vision of scientific research pursuit in the field of polymer science and composite science need to be revamped and reorganized as science and engineering surges forward (Muller et al. 2017; Palit 2017b).

de Azeredo et al. (2011) discussed deeply in a comprehensive review nanocomposites in food packaging. A nanocomposite is a multiphase substance from the combination of two or more components including a matrix and a discontinuous nano-dimensional phase with at least one nano-sized dimension (i.e., with less than 100 nm) (de Azeredo et al. 2011). This entire treatise is a comprehensive reflection of the applications of nanocomposites. The authors in this paper discussed with deep scientific conscience nanoreinforcements in food packaging materials, nanoclays, cellulose nanoreinforcements, and nanocomposite active packaging. The salient features of this paper are a brief discussion on nanocomposite smart food packaging. Future trends in nanocomposite research, the vast technological vision, and the deep scientific intricacies are the other scientific pivots of this paper (de Azeredo et al. 2011).

Human scientific understanding and scientific acuity in the field of nanocomposites are today advancing as science and engineering gears forward toward newer and promising challenges. Challenges in research pursuit in composite science are immense and far-reaching. In this chapter, the authors vastly elucidate on the scientific needs, the scientific fortitude, and the vast vision in the field of packaging and polymer science (Palit 2017b, 2018).

8 Significant Research Endeavor in the Field of Polymer Science and Composite Science

Polymer science and composite science are today in the juncture of immense scientific rejuvenation. Vast scientific conscience, scientific insight, and scientific farsightedness are the needs of scientific research pursuit in modern civilization today. Human scientific advancement's immense prowess, the technological validation, and the vision of engineering science and technology will all today lead an effective way in the true emancipation of polymer science and composite science (Palit 2018).

Geise et al. (2010) deeply discussed with vast scientific conscience the role of polymer science in water treatment and industrial wastewater treatment by membranes. Two of the greatest challenges facing the twenty-first century involve providing clean water and energy, two highly interrelated resources, at cheaper costs. Science and engineering of polymer technology are huge pillars with a definite vision of its own (Giese et al. 2010). The challenges of polymer science applications are immense and far-reaching. Membrane technology is expected to dominate the water purification and industrial wastewater treatment scenario owing its energy efficiency and cheaper costs (Palit 2017b). Mankind immense scientific girth and determination, the profundity of science and engineering, and the vision of membrane science will all lead a visionary way in the true realization of engineering and technology in the present century (Giese et al. 2010). Membrane science is veritably aligned with water purification and water treatment by an unsevered umbilical cord. There is a need for improved and effective membranes that have higher flux, are more selective, and are less prone to various types of fouling (Giese et al. 2010). Membrane separation processes is the need of chemical engineering science today. This article envisions and envisages the nature of the global water issue and reviews the state-of-the-art membrane science and technology. The vision and the goals of science, engineering, and technology will surely lead an effective way in the true emancipation of membrane separation processes (Palit 2017b). In this paper, extensive background research and development techniques are provided to help the scientists and engineers understand the fundamental problems and technologies involved in membrane separation processes (Giese et al. 2010; Palit 2017b).

Klemm et al. (2006) discussed with deep and cogent insight nanocelluloses as new polymers and smart materials in research and application. Nanocellulose is a fascinating, invigorating, and sustainable polymeric raw material characterized by interesting properties such as hydrophilicity, chirality, and biodegradability (Klemm et al. 2006). Technological and scientific profundity and vision are at its helm as regards advances in polymer science in today's human civilization. Scientific research pursuit today needs to be envisioned and reframed as science and technology surges forward. Technological advancements have transformed the scientific scenario in polymer science today. The authors pointedly focus on types of nanocelluloses, nanocellulose membranes and composites in technical applications, development of medical devices, and bacterial nanocellulose in veterinary medicine and in synthetic chemistry (Klemm et al. 2006; Palit 2017b). Human scientific and technological research forays, the needs of human science and society, and the world of challenges will all lead a successful way in the true emancipation of nanocellulose science today. This article pointedly focuses on the deep scientific success, the scientific discernment, and the vision in the field of polymer science and nanotechnology in decades to come (Klemm et al. 2006; Palit 2017b).

Prashanth et al. (2017) deeply discussed with immense lucidity and scientific conscience fiber-reinforced composites. Fiber-reinforced composites are basically axial particulates embedded on fitting matrices in polymers. Technological and engineering ingenuity, the vast world of scientific and technological validation, and the vision of polymer science will all lead a visionary way in the true emancipation of fiber-reinforced composites and its properties (Prashanth et al. 2017). The primary goal and the objective of fiber-reinforced composites is to obtain materials with high strength along with higher elastic modulus. The vast world of polymer science is highly advanced today (Prashanth et al. 2017). In this article, the authors deeply present a comparative account on various kinds of synthetic fibers with special emphasis on carbon fibers. In this well-researched article, the authors focus on important types of fiber reinforcements, glass fibers, carbon fibers, Kevlar fibers, and their various synthetic properties. Human scientific ardor, the technological advancements in polymer science, and the vision of composite science are the veritable pillars toward a knowledge dimension in the field of advanced polymer science today (Prashanth et al. 2017).

Science and engineering are today in the crucial juncture of vision and scientific insight. Science of polymers and nanocomposites are overpowering vast and versatile scientific and engineering frontiers. The need of the human society of polymer science is immense and groundbreaking. This chapter opens up new vision and newer innovations in the field of composite science, polymer science, and polymer engineering in decades to come (Palit 2017b).

9 The Challenge and the Vision of Energy and Environmental Sustainability

Energy engineering and environmental engineering are the areas of scientific research which need to be envisioned with the passage of time. Global water shortage, climate change, and frequent environmental catastrophes are challenging the vast scientific firmament in today's modern human civilization. In such a crucial juncture of history and time, energy and environmental sustainability assumes vast importance (Palit 2014b, 2017b). Holistic sustainability is the immediate need of the hour for human civilization today. Global water shortage is an environmental crisis of immense proportions. Heavy metal and arsenic groundwater and drinking water contamination is a veritable curse to mankind and human progress (Palit 2014b, 2017b). In developing and developed countries around the world, the need for a comprehensive water research and development initiative is immense and challenging. Engineering science and technology has few answers to the growing concerns for global water crisis and global environmental sustainability. In this well-researched chapter, the authors rigorously pronounce the success of composite science and materials technology toward a truer vision of science and technology. The march of science and engineering and the true vision of sustainable development are the two opposite sides of the coin today (Palit 2017b). Today nanocomposite science and composite technology are the scientific necessities of the future of human civilization and human progress. Environmental and energy sustainability are the visionary coinwords of the entire domain of industrial pollution control. Energy security, water science, and industrial wastewater pollution control are the veritable needs of human scientific progress today. This vision is enumerated with deep insight in this entire treatise. Human scientific challenges, the scientific prowess, and the vision of material and polymer science will all lead a long way in the true emancipation of energy engineering and electrical engineering today. Energy and environmental sustainability are the needs and vision of human society today. This vision is deeply enumerated in minute details and deep vision in this treatise. Technological validation of nanocomposite applications and the needs of scientific progress are the other pivots of this entire paper (Palit 2017b).

10 The Need for Polymer Science and Material Science to Human Society

Polymer science and material science are today in the process of newer rejuvenation and deep farsightedness. Nanomaterials and engineered nanomaterials are the smart materials of today's world of scientific research forays. Human civilization today is in the midst of deep scientific revival and vision. Material science and nanotechnology are the veritable needs of the society and human scientific regeneration. Material science and composite science need to be vehemently addressed and envisioned with the passage of time (Palit, 2014b, 2017b). Polymer science, polymer engineering, and material science also need to be envisaged and revamped with the march of modern science. The need for polymer science is immense and far-reaching today. Polymer science, composite science, and material science are today integrated with diverse branches of engineering and science. Nanotechnology and material science have today immense impact on human scientific progress (Palit 2017b). Environmental engineering, chemical process technology, and petroleum engineering today stand in the midst of vast rejuvenation and immense scientific vision with the grave concerns for global climate change. Mankind and its research endeavor need to be envisioned and reframed as science and engineering dives deep into the scientific realms of nanotechnology. Global research and development forays in nanotechnology are witnessing immense revamping in the last three decades. This research paper opens up new scientific innovations and instincts in the field of material science, composite materials, and nanotechnology. The authors repeatedly pronounce and proclaim the needs of scientific research endeavor in the field of nanotechnology and material science in decades to come. The world of science, technology, and engineering will surely witness a new dawn and a new beginning if global environmental engineering and petroleum engineering needs are met with vision and scientific forbearance (Cheryan 1998; Hashim et al. 2011; Shannon et al. 2008; Palit 2014b, 2017b).

11 Modern Science, Polymer Science, and the Vast Vision for the Future

Nanotechnology, nano-engineering, and material science are the visionary areas of modern scientific civilization today. Modern science is in the process of a new dawn in applications, vision, and scientific ingenuity. Polymer technology and nanotechnology are two opposite sides of the coin. Modern science in this respect is in the midst of scientific introspection and deep adjudication. Today nanomaterials and engineered nanomaterials are the coinwords of scientific endeavor in modern science. Smart materials and eco-materials are the necessities of human society as human progress surges forward. The vision for modern science is vast and versatile (Shannon et al. 2008; Palit 2016a, b). Nanotechnology, nanomaterials, and composite science are the next-generation science and are challenging the vast scientific firmament of human scientific research forays. Environmental engineering and petroleum engineering are facing immense challenges as mankind gears forward toward a newer future of scientific genre and vision. Every branch of engineering science is challenged today as grave global concerns for energy, electricity, and environment assume immense importance. Modern science in such a crucial juncture of history and time needs to be envisioned and reframed as mankind witnesses scientific difficulties and scientific challenges. Human scientific ingenuity and deep scientific profundity in engineering science and technology are in a state of immense difficulties and in a dismal state of affairs. Modern science thus is in a state of immense conundrum. This chapter repeatedly points out toward the need of judgment and scientific truth toward the furtherance of polymer science, composite science, and nanotechnology. Nanocomposite applications are transforming the face of scientific research pursuit. Civilization's immense girth and determination, the futuristic vision of nanoscience, and the human needs of applied science and applied chemistry will all go an effective way in the true emancipation of engineering and science today. This chapter uncovers the scientific intricacies and the scientific needs of the science of nanotechnology with a deep vision toward regeneration and rejuvenation of composite science (Palit 2014a, b, 2015, 2017a, 2018).

12 Future Recommendations of this Study and the Future Flow of Scientific Thoughts

The future of material science, materials engineering, and nanotechnology are bright, inspiring, and groundbreaking. Human civilization's immense scientific girth and determination, the challenges in engineering pursuit, and the vision of nanotechnology will all lead an effective way in the realization of composite science, polymer science, and material science. Scientific endeavor also today is in the midst of deep scientific revelation and profundity. The world of science and technology today are in the midst of scientific truth, scientific wisdom, and deep scientific conscience. The future flow of scientific thoughts should be directed toward more applications of science and engineering of nanotechnology. Nanotechnology is today integrated toward diverse branches of science and engineering such as petroleum engineering, environmental engineering, and chemical engineering. Novel separation processes such as membrane science and nanofiltration are challenging the fabric of environmental engineering (Palit 2014a, b, 2017b, 2018). Global water crisis, climate change, frequent environmental catastrophes, and depletion of fossil fuel resources are transforming the face of human scientific endeavor today (Palit 2017b). The area of environmental engineering integrated with nanotechnology needs to be thoroughly addressed and envisioned with the passage of history and the visionary timeframe. The future of global science, engineering, and technology are immensely bright and pathbreaking. Today is the age of nuclear science and space technology (Palit 2014a, b, 2017b, 2018). Every branch of science and technology needs to be integrated with nanotechnology. Mankind thus will witness a newer scientific revival and vision (Palit 2014a, b, 2017b, 2018).

13 Conclusion and Future Scientific Perspectives

Human civilization and scientific research initiatives today are in the crossroads of introspection, truth, and vision (Palit 2017b). Material science and nanotechnology are the two opposite sides of the coin. In this chapter, the authors reiterate the success of science, the vast scientific stewardship, and the scientific necessity of nanotechnology to the progress of scientific research today (Palit 2017b). Deep deliberation of energy and environmental sustainability are the other pillars of this well-researched treatise. The authors, with deep scientific conscience, tread a visionary path toward the scientific evolution of nanoscience, nano-engineering, and nanotechnology. Scientific perspectives in the field of nanotechnology are vast and versatile. Future scientific perspectives should be targeted toward newer innovations and successful realization of sustainable development and holistic sustainability. Holistic realization of sustainability is the necessity of the hour. Mankind vast scientific prowess, the world of scientific challenges in the domain of nanotechnology, and the vast vision of composite science will all lead a visionary way in the true emancipation of nanocomposite technology and the vast world of nanoscience today. Technological advancements are the necessities of scientific vision today. In this chapter, the authors deeply elucidate on the visionary world of composite science and material science with a clear mission toward furtherance of science and engineering.

References

Ali A, Sinha K (2014) Exploring the opportunities and challenges in nanotechnology innovation in India. J Soc Sci Policy Implications 2(2):227–251

de Azeredo HMC (2009) Nanocomposites for food packaging applications. Food Res Int 42:1240–1253

de Azeredo HMC, Mattoso LHC, Mc Hugh TH (2011) Nanocomposites in food packaging-a review. In: Reddy B (ed) Advances in diverse industrial applications of nanocomposites. InTech Press, Croatia

Bratovcic A, Odobasic A, Catic S, Sestan I (2015) Application of polymer nanocomposite materials in food packaging. Croat J Food Sci Technol 7(2):86–94

Cheryan M (1998) Ultrafiltration and microfiltration handbook. Technomic Publishing Company Inc, USA
Giese GM, Lee H-S, Miller DJ, Freeman BD, Mcgrath JE, Paul DR (2010) Water purification by membranes: the role of polymer science. J Polym Sci B Polym Phys 48:1685–1718
Hashim MA, Mukhopadhayay S, Sahu JN, Sengupta B (2011) Remediation technologies for heavy metal contaminated groundwater. J Environ Manag 92:2355–2388
Klemm D, Schumann D, Kramer F, Hebler N, Hornung M, Schmauder H-P, Marsch S (2006) Nanocelluloses as innovative polymers in research and application, advances in polymer science, 205, 49–96. Springer, Berlin
Muller K, Bugnicourt E, Latorre M, Jorda M, Sanz YE, Lagaron JM, Miesbauer O, Bianchin A, Hankin S, Bolz U, Perez G, Jesdinszki M, Lindner M, Scheuerer Z, Castello S, Schmid M (2017) Review on the processing and properties of polymer nanocomposites and nanocoatings and their applications in the packaging, automotive and solar energy fields. Nanomaterials 7 (74):1–47
Palit S (2014a) Frontiers of nanofiltration, ultrafiltration and the future of global water shortage- a deep and visionary comprehension. Int Lett Phys Chem Astron 38:120–131
Palit S (2014b) Future vision of advanced oxidation process and its immediate efficacy- a deep, insightful comprehension and a far-reaching review, international letters of physics. Chem Astron 33:136–145
Palit S (2015) Advanced oxidation processes, nanofiltration, and application of bubble column reactor, Nanomaterials for Environmental Protection, Boris I. Kharisov, Oxana. V Kharissova, Rasika Dias H.V., Wiley, Hoboken. 207–215
Palit S (2016a) Nanofiltration and ultrafiltration- the next generation environmental engineering tool and a vision for the future. Int J Chem Tech Res 9(5):848–856
Palit S (2016b) Filtration: frontiers of the engineering and science of nanofiltration-a far-reaching review. In: Ortiz-Mendez U, Kharissova OV, Kharisov BI (eds) CRC Concise Encyclopedia of Nanotechnology. Taylor and Francis, Boca Raton, pp 205–214
Palit S (2017a) Advanced environmental engineering separation processes, environmental analysis and application of nanotechnology: a far-reaching review, (Chapter-14). Advanced Environmental Analysis: Application of Nanomaterials, Royal Society of Chemistry Detection Science. In Chaudhery Mustansar Hussain, Boris Kharisov (eds)
Palit S (2017b) Application of nanotechnology, nanofiltration and drinking and wastewater treatment- a vision for the future. In: Grumezescu AM (ed) Chapter 17, Book-water purification. Academic Press (Elsevier), USA, pp 587–320
Palit S (2018) Recent advances in corrosion science: a critical overview and a deep comprehension. In: Kharissov BI (ed) Chapter-11, Book-direct synthesis of metal cComplexes. Academic Press (Elsevier), Netherlands, pp 379–411
Prashanth S, Subbaya KM, Nithin K, Sacchidananda S (2017) Fiber reinforced composites-a review. J Mater Sci Eng 6(3):1–6
Ray S, Quek SY, Easteal A, Chen XD (2006) The potential use of polymer- clay nanocomposites in food packaging. Int J Food Eng 2(4):5
Salehi F (2014) Current and future applications for nanofiltration technology in the food processing. Food Bioprod Process 92(2):161–177
Shannon MA, Bohn PW, Elimelech M, Georgiadis JA, Marinas BJ (2008) Science and technology for water purification in the coming decades. Nat Publ Group:301–310
United Nations Global Sustainable Development Report (2016) Department of economic and social affairs. New York
Yadav SK (2017) Realizing the potential of nanotechnology for agriculture and food technology. J Tissue Sci Eng 8(1):195

Website References

https://en.wikipedia.org/wiki/Nanocomposite
https://en.wikipedia.org/wiki/Packaging_and_labeling
https://fog.ccsf.edu/.../14_CompositeMaterials/03_Fiber-reinforcedComposites.html
https://www.azonano.com/article.aspx?ArticleID=1832
https://www.azonano.com/article.aspx?ArticleID=3035
https://www.intechopen.com/.../fiber-reinforced.../introduction-of-fibre-reinforced-po...
https://www.nature.com › *subjects*
https://www.ncbi.nlm.nih.gov/pmc/articles/PMC4283176/
https://www.omicsonline.org/.../nanotechnology-a-new-approach-in-food-packaging-
https://www.slideshare.net/ShivaniPandya/nanocomposite
www.safenano.org/news/intheknow/in-the-knowon-food-packaging
www.tifac.org.in/index.php?option=com_content&id=523:nanocompos
www.understandingnano.com/nanocomposites-applications.html

CPSIA information can be obtained
at www.ICGtesting.com
Printed in the USA
LVHW050045131118
596929LV00005B/305/P

9 789811 319082